Innovation and Technology

Innovation has revolutionized the world economy, yet it remains often misunderstood. This textbook seeks to elucidate the nature and impact of innovation for both undergraduate and graduate students.

Innovation and Technology examines the impact of innovation on both economic theory and the real world. It addresses the topic at the level of policy and also drills down to provide analysis of firms. This book moves beyond the plethora of specialized studies on the subject and formulates a unified and comprehensive approach, encompassing the topic's huge breadth and scope. Issues such as innovation, information, knowledge, finance, intellectual property rights, incentives and regulation are featured.

Designed for MBA, Economics and Business students, this textbook will be useful to those interested in innovation, entrepreneurship and the economics of technology.

Nikos Vernardakis is Professor Emeritus, University of Patras, Greece.

Routledge Advanced Texts in Economics and Finance

1. **Financial Econometrics**
   Peijie Wang
2. **Macroeconomics for Developing Countries (Second Edition)**
   Raghbendra Jha
3. **Advanced Mathematical Economics**
   Rakesh Vohra
4. **Advanced Econometric Theory**
   John S. Chipman
5. **Understanding Macroeconomic Theory**
   John M. Barron, Bradley T. Ewing and Gerald J. Lynch
6. **Regional Economics**
   Roberta Capello
7. **Mathematical Finance**
   Core Theory, Problems and Statistical Algorithms
   Nikolai Dokuchaev
8. **Applied Health Economics**
   Andrew M. Jones, Nigel Rice, Teresa Bago d'Uva and Silvia Balia
9. **Information Economics**
   Urs Birchler and Monika Bütler
10. **Financial Econometrics (Second Edition)**
    Peijie Wang
11. **Development Finance**
    Debates, Dogmas and New Directions
    Stephen Spratt
12. **Culture and Economics**
    On Values, Economics and International Business
    Eelke de Jong
13. **Modern Public Economics (Second Edition)**
    Raghbendra Jha
14. **Introduction to Estimating Economic Models**
    Atsushi Maki
15. **Advanced Econometric Theory**
    John Chipman
16. **Behavioral Economics**
    Edward Cartwright
17. **Essentials of Advanced Macroeconomic Theory**
    Ola Olsson
18. **Behavioral Economics and Finance**
    Michelle Baddeley
19. **Applied Health Economics (Second Edition)**
    Andrew M. Jones, Nigel Rice, Teresa Bago d'Uva and Silvia Balia

20. **Real Estate Economics**
A Point to Point Handbook
Nicholas G. Pirounakis

21. **Finance in Asia**
Institutions, Regulation and Policy
Qiao Liu, Paul Lejot and Douglas Arner

22. **Behavioral Economics (Second Edition)**
Edward Cartwright

23. **Understanding Financial Risk Management**
Angelo Corelli

24. **Empirical Development Economics**
Måns Söderbom and Francis Teal with Markus Eberhardt, Simon Quinn and Andrew Zeitlin

25. **Strategic Entrepreneurial Finance**
From Value Creation to Realization
Darek Klonowski

26. **Computational Economics**
A concise introduction
Oscar Afonso and Paulo Vasconcelos

27. **Regional Economics (Second Edition)**
Roberta Capello

28. **Game Theory and Exercises**
Gisèle Umbhauer

29. **Innovation and Technology**
Business and Economics Approaches
Nikos Vernardakis

Innovation and Technology

Business and economics approaches

Nikos Vernardakis

Routledge
Taylor & Francis Group
LONDON AND NEW YORK

First published 2016
by Routledge
2 Park Square, Milton Park, Abingdon, Oxon OX14 4RN

and by Routledge
711 Third Avenue, New York, NY 10017

Routledge is an imprint of the Taylor & Francis Group, an informa business

British Library Cataloguing in Publication Data
A catalogue record for this book is available from the British Library

Library of Congress Cataloging in Publication Data
A catalog record for this book has been requested

ISBN: 978-0-415-67679-3 (hbk)
ISBN: 978-0-415-67680-9 (pbk)
ISBN: 978-1-315-72395-2 (ebk)

Typeset in Bembo
by Saxon Graphics Ltd, Derby

To my wife Melita for all her love and support.

Contents

Detailed contents

Tables

Figures

Boxes

Preface

> We are all fluent in innovation …
> (says a recent advertisement)

Nowadays, many advertisements, whether they refer to financial "products", tourist packages or car rentals, include the word "innovation" – or "innovative" – preferably accompanied by the equally mystical term "disruptive". We have come rather late in taking an interest in innovation, even more so in attempting to decipher what it is all about; the truth is that it is still not well understood, as it keeps evolving. We keep understanding more and more types and dimensions of innovation (ways of conducting research, actors, sources, agents, search methods, incentives, financing, means of appropriation, etc.), but by now we know that there will always be more to discover.

There is nothing automatic about the process of innovation – all along the links in the chain from conception of an idea to its final fruition. Some of the links along this chain have been gradually simplified, broken down into smaller ones and assisted in their execution, thus rendering the whole process more accessible. One does not have to repeat the heroic deeds of the romanticized intrepid innovator-turned-entrepreneur of the past. Instead, one can hire a 3D modeler to design one's own idea and have it printed on a high-end 3-D printer at Shapeways. Similarly, in digital services and products, the basic building blocks, the "technologies of startup production", have become cheap, easily accessible and can be combined and recombined. Further down the line, there are incubators like Seedcamp and accelerators, that is, schools for startups, like Y Combinator and TechStars. For the more ambitious startups, there is always SU Labs, an outgrowth of Silicon Valley's Singularity University. A further simplification along the innovation path comes in financing with "crowd sourcing".

Yet innovation, the new mantra, remains misunderstood to a considerable extent. It is often confused with the hype and its fragility is not always recognized, nor is it seen as a part and result of a whole ecosystem.

This work is an attempt to show what innovation is all about, how it is evolving, how and why it has affected and is still affecting the world we live in, its economy down to the firm and what the state can do about it.

Some of its major points include the following:

It states and substantiates (1) the fact that innovation and the innovation process itself, which are constantly evolving – in terms of notion, breadth, patronage, actors, environment, supporting links and institutions – have been changing our lives, the world and its economy at every level.

In doing so, innovation has been instrumental in bringing about (2) a new international division of production and employment based on a hierarchy of products in terms of technology intensity corresponding to a hierarchy of technological frontrunners followed by technological laggards at an increasing distance. High-technology-intensity products produced by the frontrunners command a high value added and compete through innovation, in contrast with low-technology-intensity products, which are produced by the laggards, command a low value added and compete on price.

Innovation has changed (3) the substance of comparative advantage, from being based on natural resources to knowledge, and has enhanced globalization. Through a process of "creative destruction" it is (4) rejuvenating the capitalist system by replacing the old with the new – bricks and mortar establishments with e-businesses – creating, destroying, merging and separating sectors and firms.

Innovation (5) drives all activities towards higher technology intensity levels with heightened competition and oligopolistic practices and behaviors, and as they do, they increasingly obey non-traditional economic principles. The firm (6) is constantly transformed through new challenges and opportunities and innovation imposes upon it a number of imperatives, which become more demanding the more the firm moves into high-technology territory. High-tech firms are (7) a breed apart, and deserve to be treated as such in business and management textbooks, as they obey increasing returns, face narrow windows of opportunity, oligopolistic situations and stiff competition, "winner takes all" outcomes and increased chances of falling prey to disruptive innovation.

Gradually, the firm (8) has moved from a producing entity to a learning one, and faces more ephemeral opportunities in fast-growing markets. Internally, the firm learned to value its human capital more. As to external environment, firms tend to congregate in ephemeral groupings and in "galaxies" of cooperation. Competition is accompanied by cooperation; we have a state of "co-opetition".

A dimension, which imposes its presence almost from the start, consists of the theoretical differences between Economic Theory (or Mainstream Economic Theory) and heterodox economics (or evolutionary or Schumpeterian economics) favored in this book (9). Besides changing the world, innovation has impacted Economic Theory which is supposed to explain developments, ex post, and which has responded half-heartedly and to a much lesser degree than it should, and which is still striving to catch up. With its rigid assumptions Microeconomic Theory has not developed the tools to go beyond process innovation, Growth Theory is unable to explain growth achieved, International Trade Theory cannot explain volume and direction of trade, and convergence as predicted by theory runs against experience.

(10) In contrast with Mainstream Economic Theory, evolutionary economics asserts that innovation is not exogenous to the system but exists within it, at its heart, and has two-way causation. It asserts the differentiation between knowledge and information and puts emphasis on learning; it introduces the notion of time in theory; it is against the notion of equilibrium; it accepts the existence and important role of increasing returns; and the role of prior experience. Evolutionary economics is also known as the economics of an imperfect world: perfect competition is a utopia; firms do not behave as monoliths; taking a decision takes time. Most importantly, "it is often market imperfections – knowledge asymmetries, market imperfections, monopolies and conditions of disequilibrium – which fuel innovation and economic growth" (Bryant and Wells 2000: 1).

(11) An overall verdict on theory in general: we are still short of a unifying theory on the origins of innovation and a convincing theory of diffusion across sectors. Parts of

Economic Theory need rewriting if it aspires to continue playing the role of analytical tool in relation to innovation in the future.

(12) For innovation to keep alive, sufficient, efficient and meaningful to society, we need an innovation policy whose aims of generating growth, raising income and creating employment are best served through the provision of a National System of Innovation (NSI), a well as a functioning ecosystem which extends beyond meeting individual "market failures" (failures to provide adequate innovation effort). It needs to be a system that is hospitable to innovators through intellectual property rights (IPRs) and skilled immigration laws; that has a productive structure biased towards innovation; that is able to withstand external threats like globalization; that is balanced between variety and selection; and that is supplemented by standards, and updated regulation and antitrust laws.

(13) And on this last issue, we should remember that competition is not uniformly a stimulus to innovation and hence each point of policy should be considered on its merits. The regulatory system, which is already under strain with present problems, especially in cases involving high technology, should be alerted and made fit to deal with anticipated future developments. We should also remember that standards can play an important role in Innovation Policy; that we need to design new laws on antitrust in the future; and that we need to understand better a different kind of competition than the one we have been used to, namely "dynamic competition", in Schumpeter's words " the gales of creative destruction".

(14) However, with some notable exceptions the world over, firms have been mostly left to themselves to decide on the way, the direction and the volume of innovation. But firms do not seem particularly engaged, their R&D is a pale reminder of their past and they are less and less preoccupied with basic research. (15) More recently, innovations have come to be treated as synonymous with "applications" (apps), and startups as synonymous with Internet-related startups. But startups, whether Internet-related or otherwise, are proving somewhat less than miraculous as they do not deliver as expected in terms of jobs. Even venture capital firms (VCs) seem to be losing their original shine as they are turning into copycats and bring old ideas to new markets.

(16) Even worse, lately, innovation has become more marginal, and more laborsaving. Software of the digital age at the heart of present-day innovations is eating away at jobs created in the analogue age. Together with globalization, which it enhances, innovation diminishes employment and labor income (hence income share too) in the West and contributes to the rise of inequality. As of now, the most pressing problem is the waning of the benefits of innovation in terms of jobs and income.

(17) Such developments render the presence of the state crucially important. The presence of the state is required to enforce the hospitable and efficient ecosystem described above, to orchestrate the whole system, coordinate, organize, direct, and above all support it constantly and on a long-term basis, and to make sure that these changes do happen. For the technological frontrunner(s) the state is expected also to play the role of midwife and initiator by extending old markets and creating new ones. Countries with a vibrant entrepreneurial community, especially those that are small, already developed and specialized, with R&D covered by large multinationals, may need less state intervention. On the other hand, technological laggards lacking a vibrant entrepreneurial community and latecomers to development can follow the highly successful Israeli model relying heavily on state intervention. China is expected soon to become the biggest spender on R&D, but the prospects of innovation at world level will depend on whether the US continues to play its role of explorer and pathfinder.

Innovation and Technology: Business and Economics Approaches is neither an encyclopedia on innovation nor a collection of tech gadget prospectuses. It is a textbook on technology and innovation and on their implications on both Economic Theory and on the real world down to the level of firms, thus straddling the disciplines of both economics and business. Policy is an additional dimension considered both in relation to innovation per se and vis-à-vis the implications of innovation on firms and on the economy at large.

The purpose of the present work is not to add to the list of specialized studies on this subject matter but to suggest a unified and comprehensive approach for it. It is a textbook primarily meant for MBA students but also for mature students of economics or business in core courses as well as students in elective courses from the engineering schools or the sciences. It requires no prior knowledge of economics or business. For the reader at large, the non-academic, s/he may be interested in what innovation is all about: where does it come from; does it have any rules; where is it heading; how does it affect our lives, the world, the firm; can we do something about it; should we?

It is a textbook, which means that it lends itself to teaching. The links and passage from one chapter (or part) to another are clearly underlined and understood and so is the book's direction. Personal experience has taught this writer the power of illustrations and examples. Beside the seven tables supporting the arguments in chapter 1, there are 21 figures. In addition, there are 16 framed pieces of text covering items which correspond to examples but which could reach the level of case studies: "Better Places", the shortest one refers to an Israeli innovation that is difficult to classify. The most extensive is the one on the evolution of the US motion picture industry. Some of these cases derive from this writer's previous research, including the "Scandinavian alarm", the "bar code", "urea", "competing technologies and market power", and the "US motion picture industry". There are two framed pieces of text providing theoretical explanations for topics that a reader with no background in economics would not normally have previously encountered: increasing/decreasing returns and microeconomics and technological change. There are also two multiple-page frames, each delineating an item that falls within a whole yet is an entity in itself: on finance and IPRs, both within the NSI.

As a textbook it is modular and it can be taught either in series or by focusing on a single chapter or group of chapters, possibly supplemented by articles. An attempt in this direction was to make each chapter as self-contained as possible at the risk of some repetition.

It has a very wide scope. For instance, Innovation Policy is usually treated on its own. It presents the state of the art in innovation. It explains why some issues have fallen out of fashion and discusses research results and, where appropriate, tendencies. On issues that may be considered controversial it takes a very clear stand after providing ample explanation.

Part 1 consists of the introductory chapter 1, which sets the stage, brings the reader quickly to the present world situation and provides a bird's eye view of innovation today and some remarks as to future trends.

Part 2 consists of chapters 2 and 3, the theoretical background for innovation and diffusion. Chapter 2, devoted to innovation, gathers the essentials needed for its understanding: relevant facts, characteristics, relationships, origins, evolution, drivers and theory, where appropriate. As innovation, technology and science are linked with knowledge and learning, it is necessary to discuss these concepts and the relations between them before focusing on innovation itself. Thus, we first discuss the relation between technology and science. Next, we present the relation between invention and innovation, and finally there is a discussion of the relation between knowledge, learning and innovation. This analysis helps the understanding of theory and provides some essential definitions. There follows a discussion of the nature and special characteristics of the innovation process, of the origins and evolution of innovation, and of the creation of technological paradigms and trajectories. Finally, we address technological opportunities, incentives and differences among sectors. The chapter ends with a discussion on the state of the art on the relationship linking innovation, size of firm and market structure.

Chapter 3, devoted to diffusion, first discusses intra-sectoral diffusion (within the same sector), starting with the five stages of adoption, a process wrongly assumed to be instantaneous by Economic Theory; then it presents a few stylized facts and the determinants of the rate of adoption. Next comes the diffusion curve and a critique of existing diffusion models – epidemic and probit. This is followed by direct and indirect network effects, feedback mechanisms, path dependence, "success brings success" and standards. The following topic is inter-sectoral (across sectors) diffusion, a subject usually absent from textbooks. DeBresson's theory based on circulation and dominant technological requirements covers some distance, but it has flaws. Technology transfer, and conditions for successful transfer are discussed next. Technological espionage, an unwarranted type of transfer of technology, closes the chapter, together with a few words on the modes of technology transfer and diffusion of embodied technology.

Part 3 refers to the impact of innovation on Economic Theory and on the real world. Chapter 4 concerns itself with Economic Theory, chapter 5 deals with the macro and meso levels of the economy (i.e. national and international levels and the sectoral level), and chapter 6 covers the impact on firms.

In chapter 4, we argue that Economic Theory, with a mission to explain (ex post) how the real economy works, has been forced to change as well (to some extent but not enough) to keep abreast with the induced transformations. At first, Economic Theory hardly took notice of innovation and by the time it finally did decide to deal with it, it was done less than whole-heartedly; by then hypotheses had already been adopted that are both rigid and hard to replace without destroying a major part of the entire logical edifice.

The Classical economists were interested in technological change but in a different more spherical way than at present with the emphasis put on the impact rather than the generation of the process, while the Neoclassical economists had no real interest in it. Schumpeter left a legacy on the topic of innovation that can hardly be met by the rest of economists. He is responsible for placing innovation not only within the economic system but at its heart: for the "Shumpetarian entrepreneur"; for the "gales of creative destruction"; and for dynamic competition. He wanted to "answer the question how the economic system generates the force which incessantly transforms it".

The belated advent of the economics of technology covered some but still not all of the ground. Yet it already produces a much better understanding in the less than perfect world we live in than Mainstream Economic Theory with its rigid and simplistic hypotheses. Developments in convergence do not support this theory and the same is true for comparative advantage, growth and trade. The only type of innovation that Microeconomic Theory can address is methods of production. Growth cannot be accounted for and neither can trade be explained. The way the firm is portrayed ignores the concept of the firm as the main generator of innovation. Time is ignored while information and knowledge are one and the same. We should realize that it is getting more and more difficult to live with the concept of equilibrium in a world heading for increasing returns and "winner takes all" or most. Parts of Economic Theory will have to be rewritten.

In chapter 5, we first address the repercussions at the macro level, that is, at the national and international levels. Then we move one level down and examine the effects of innovation at the middle level, i.e. the level of economic sectors. Using a historical narrative the text shows the impact of innovation on the world economy and national economies, explains the big cycles, periods of growth and recession, frontrunners and laggards up to de-industrialization, as well as presenting some of the most prominent theories on cycles.

We then descend one level and reach the level of the sectors of the economy, which grow, rise, fall and move under the overall economy and the cycle. The analysis discusses the way in which innovation impacts sectoral composition; technological opportunities, market structure and performance; appropriation conditions, performance; performance and evolution; and evolution of market structure. The product life cycle is also discussed. The text analyzes the mechanisms of innovation and their impact on sectors, their borders, their mergers, separations and internationalization. Finally, we deal with the impact of innovation on the evolution of sectors: emergence of new sectors, reinvigoration of existing sectors, substitution, death of sectors, competition.

Chapter 6 deals with innovation's impact at the level of firms. Innovation has generated new opportunities (as well as some threats), and in order to adapt and take advantage of them the firm has had to change dramatically in all its aspects: size, functions, organization, location, use of resources, aims, values and the way it sees itself and the environment in which it has to survive and with which it must cooperate.

Overall, technological change has imposed on the firm five broad direct imperatives plus an indirect one, as rules to comply with just for survival. In connection with innovation, still more directives are to be observed by all firms. For high-tech firms the demands are even more stringent as the threats multiply, and opportunities get harder and harder to discern, and require added boldness, deep pockets and knowledge. They operate according to different economic principles than the rest of firms, have narrower windows of opportunities, face higher competition and faster technological change and are more liable to fall victim to disruptive technologies.

The concept of the firm has evolved from a family-owned small enterprise, producing a single product in one location to the decentralized multi-departmental, multi-product, multinational firm, carrying out its own research but also sharing knowledge with other firms and striving to turn into a learning organization.

The firm used to attempt to optimize mass production of specific goods, and the challenge for its management was to forecast demand in slowly rising markets. Internally, respect for hierarchy and loyalty were valued most. Externally, the firm's relations with its environment could be described as "business is war". Under the pressure for competitiveness the firm "rationalized" by keeping its core activities and abandoning activities considered more distant from its heart. These were either sold to other firms through mergers and acquisitions or were "outsourced", that is, performed by outsiders to the firm, more efficiently. Rationalization and outsourcing resulted in trimming the size of the firm and in flattening its structure.

Gradually, within the time span of five or six decades, the firm moved from a producing entity to being a learning one alongside production, and learned to face more ephemeral opportunities in fast-growing markets, with large profits awaiting whoever dares make the first move (innovation). Firms tend to congregate in ephemeral groupings and in "galaxies" of cooperation. Competition is accompanied by cooperation, a state of "co-opetition".

The chapter closes with a case study illustrating the evolution of a whole sector under the impact of technological change: the US motion picture industry, with its evolving structure, studios, products, finance, links, value chain and prospects.

Part 4 consists of chapter 7, "Innovation Policy" and chapter 8, "Epilogue". A contemporary definition of Innovation Policy amounts to what government can do to use innovation and progress in scientific and technical knowledge for the purpose of realizing specific national objectives and taking the necessary steps to ensure the meeting of these objectives.

By now, generating growth, raising income and creating employment are the universal objectives but not all countries are able or willing to do what it takes to use Innovation Policy to its fullest extent. Most developing countries do not possess the prerequisites and some of the developed ones, in favor of "small government" or due to ideology, are willing to use Innovation Policy as sparingly as possible and only up to the point of eliminating "market failures", i.e. cases where the effort expended on innovation is less than the situation necessitates.

For those countries that wish to use Innovation Policy beyond the elimination of market failures and wish to further address the realities of our imperfect world, the recipe is straightforward. It consists of possessing (building, ameliorating and maintaining) an NSI consisting of the whole research community, i.e. universities, research institutions and organizations, plus the hopefully enterprising business firms, finance and the state. When accompanied by appropriate policies and regulations this system becomes an ecosystem conducive to the generation, diffusion and commercialization of innovation. Furthermore, it is rendered hospitable when supplemented with IPR and appropriate policies on skilled immigration, steady and low taxation, an efficient civil service, and a low level of corruption, while a balance is sought between variety and selection. The productive sector of the ecosystem is expected to be internationally competitive and able both to withstand external threats such as globalization and to take advantage of the opportunities generated. Diffusion of the benefits from innovation is expected, and steps are taken to guarantee the system's perpetuation.

Each of the four sections of the ecosystem labeled "the building blocks of the ecosystem" are described, followed by a rationale for policy action and policy recommendations. Three topics which stand out as in need of additional explanation are treated individually at greater length: finance, which is part of the NSI; IPR, which helps render the ecosystem more hospitable; and the balance between variety and selection. Other important aspects for the design and implementation of Innovation Policy are also discussed: specificity, flexibility and the nature of the policy, degree of intervention, public funding and governance.

Reference is made to salient characteristics of Innovation Policy in general: England and Germany in the nineteenth century and present-day Switzerland, the EU, Israel, China and the US. The main challenges to Innovation Policy at present are discussed and analyzed: the lukewarm stance towards "real" innovation expressed either as lack of interest or risk aversion or lack of true entrepreneurship. For some among the very large group of countries aspiring to development but clinging to rentier values, the problem of overcoming risk aversion may be met by the encouragement of the creation of venture capital firms, as in Israel. At the world level, we note the lack of interest of policy planners, with the exception of the US, in the regeneration and perpetuation of the system.

Emerging challenges for Innovation Policy are then discussed, such as the waning of the benefits of innovation in terms of employment, income and the rising level of inequality. The chapter ends with action points and brief proposals for future Innovation Policy and some concluding remarks: the spread of the Internet; the need to support some startups but not all – only those that promise to become gazelles; the need to encourage the formation of venture capital firms through lower capital gains taxes and by allowing limited participation of pension funds; the formation of technology rating agencies; the articulation of SBIR-type programs; and turning to the state, the "entrepreneur of last resort".

We also discuss issues we should worry about: the precariousness of the best Innovation Policy – the American one; the need for greater reliance on diffusion; the need to abandon the "Washington Consensus" and alternatives for developing countries; the social impact of innovation; income inequality arising from innovation; the need for intervention; the role of standards; and the need for antitrust laws based on dynamic competition.

The final chapter (chapter 8) serves as an epilogue.

Part I

Innovation and the world economy

1 Innovation and the reshaping of the world economy

This first chapter of the book shows what innovation has been so instrumental in bringing about in the world economy. It has truly reshaped it.

The chapter dwells briefly on some of the most salient consequences due mostly to innovation which have earned it its importance: namely, the use of the rise in productivity as a competitive weapon leading to a new international division of production and employment; the specialization of production; the changes in the mechanism through which competition is carried out; the recognition of the importance of human capital.

It also deals with some of the consequences of innovation, some of the mechanisms it has either set up or supported: globalization, the migration of comparative advantage and the rise of foreign direct investment (FDI); outsourcing, localization, specialization, value chain production. Finally it deals with the "New Economy", the Internet economy, copyright industries – in a nutshell, the world surrounding us.

While the world was changing as a result of innovation, innovation itself was also evolving. The notion of innovation, the process through which it is carried out as well as the environment – including the main actors – in which it is achieved have been evolving. Section 1.3, "Innovation Today" provides a bird's eye view of the present state of innovation. It dwells on the changing characteristics of innovation, only. The remaining dimensions will be treated where appropriate in the book.

The final section of the chapter (1.4) shows the depth and breadth of the subject of innovation by touching lightly upon subjects that will demand our attention in the closing chapters of this work. It raises questions and shows concern as to the sustainability of innovation at the world level in the face of a threatening waning of the benefits of innovation given skepticism regarding both volume and quality of patents in times of jobless growth and rising income inequality.

1.1 Introduction

Many writers on innovation and technology start by reminding us that innovation has become a household word, some sort of present-day mantra. It hasn't always been that way. For thousands of years people, be they serfs, slaves, servants, hired land workers or even rulers and pharaohs, ignored the idea of *progress*. The first four couldn't even dream of change and even if they did dream of it had no time to think about it. Rulers, kings and pharaohs obviously had time on their hands but had no reason whatsoever to change things. So nothing changed until the time period preceding the Renaissance, with the rise of a new social class, the first capitalist class ever, a class bent on profit. The pursuit of profit calls for the use of all available tools of entrepreneurship in the face of the inevitable accompaniment of competition, including innovation.

Later, the accomplishments of New Science were met with enthusiasm, awe, curiosity and admiration. Invention was seen in a romantic light and inventors were the heroes of the day. In the *New Atlantis*, Francis Bacon (1629) described a technological paradise in which inventors were honored by having their busts displayed on pedestals whose size depended on the importance of their invention. Later, Watt was depicted musing over a boiling teapot, a reference to the power of steam. The equivalents of present-day comics were full of drawings of imaginary contraptions.

Western society's conception of technological progress at the time can best be characterized by its main tenets: technological progress contributes directly to the improvement of material, social, cultural and intellectual life, thus accelerating the pace of civilization; it can be measured; its sources, direction and repercussions are under man's absolute control and supervision; it has conquered nature and put it in the service of man; technology and civilization have reached their zenith with the advent of the Western industrialized world.

Such views tend to consider societies that do not exhibit the same characteristics as those developed in the West to be inferior. Speed is a hallmark of our era. But what good would it have done a man in the Middle Ages, a time when people spent all their lives within a fifteen-mile radius of their place of birth, to know that he could travel at a hundred miles an hour?

Gradually the original main tenets of Western society on technological progress have been challenged. Chernobyl, Bhopal, Three Mile Island, Titanic, Challenger, Fukushima, the big oil spills off the coast of Alaska (Exxon Valdez) and in the Gulf of Mexico (BP) and several dozen further catastrophes have demonstrated that technological progress is not always achieved under man's absolute control.

Nowadays it is no longer fashionable to say that man has conquered nature. On the contrary, we wonder how far we are from its annihilation. *Limits to Growth* (Meadows et al. 1972) was one of the earliest, but rather late, warnings of the scientific community. Ever since, we have become more and more aware of the greenhouse effect, pollution and nuclear waste. Nevertheless, beyond any doubt, part of the world population lives longer and enjoys a better life than in the past, and most people associate the evolution of humanity with technological progress. For a good part of the world population the basic needs of sustenance have been met. The history of technological progress is an integral part of the history of humanity.

Some argue that technological progress is associated with progress in defense technologies. There is some merit to this argument though a number of technologies were developed purely for the further improvement of defense-related activities and others produced spillover for civilian use. In the latter category we may cite the laser, GPS, artificial vision, air bags, the modem and the Internet, Tivo's technology in video recorders and many more. Weapons, man's tools for survival or domination, have often expressed his ingenuity. There were many more expressions of this ingenuity that have led to a longer and better life.

Finally, a technology by itself is neither evil nor saintly, as it all depends on the use to which it is put. Computer software generally carries positive connotations, but this is not true when it is used to tinker with other people's bank accounts. Drugs are praised for saving the lives of millions of people; there is less praise for the production of anabolic steroids.

1.2 The importance of innovation: the reshaping of the world economy

Innovation has left its mark on the world for the last several centuries, a fact to which we shall have many opportunities to attest in this book.

At present we are witnessing a paradigm shift that has been taking place ever since the 1980s and whose origins go back to the transistor, something that was recognized only after considerable hesitation. Nobel Laureate Robert Solow's stance reflects the development in the thinking of the economics profession on the matter. In 1987, critical of the scant impact of large investments in information technology, he is reported to have said "you can see the computer age everywhere but in the productivity statistics" (Pohjola 2002: 135). However, 12 years later he stated that "my beliefs are shifting on this subject" and accepted the idea of a paradigm shift by stating "that the story always was that it took a long time for people to use information technology and truly become more efficient" (Lohr 1999: 12). This is essentially David's argument that the period of institutional and organizational adaptation is always long for radical new technologies ushering in a paradigm shift in the way that electricity did (David 1991). As for public opinion, by March 2000, 57 percent of Americans questioned in a poll believed that the US had entered "a new kind of an economy" that is significantly different from the industrial economy (*Business Week* 2000).

Thus, the last three decades have seen the world economy experiencing a major structural change driven by the globalization of business and the revolution in information and communication technology (ICT). This ongoing transformation has been given different names, each emphasizing a different aspect of this structural change: "the New Economy", "the innovation economy", "the knowledge economy", "the network economy", "the digital economy", "the weightless economy", "the e-economy", and in terms of the corresponding change in society "the post-industrial society", "the information society", etc.

For some economists, the New Economy meant primarily the productivity-enhancing effects of ICT. The ensuing acceleration in the rise of productivity would lead to a new kind of macro-economic reality, an economic Valhalla: long-term growth uninterrupted by recession or threatened by inflation. The main actors would be the swarms of firms mostly producing Internet services or other knowledge-intensive products and services.

This version of the New Economy, the "hype" version, did not survive the test of time. What survived is the version of the techno-economic paradigm that stresses the difficulties of change and adaptation of societies, their organizations and institutions. What also survived is the message that "today there is an enormous untapped growth potential that could be mobilized … if our institutional reforms and organizational change promote learning processes" (Lundvall 2004: 1).

The rise of the Internet provided the link between the two broad trends – globalization and ICT – defining the New Economy, thus integrating markets and linking people together across boundaries. Some of the principles that govern it differ from what we have been used to so far, much as its impacts, which may not necessarily be uniform. For instance, there has been accelerated growth in productivity, but this has been less than expected – mainly in the US but also to a lesser extent in Europe and the rest of the world. In addition, people in richer countries with the required infrastructure and skills are in a much better position to benefit from this phase of globalization than people in poorer countries (Pohjola 2002: 138).

In terms of development, it is as if an additional requirement for growth is being added to those that already exist: "the digital divide". As of June 30, 2000, 29.9 percent of the world's Internet users resided in North America, accounting for only 5 percent of the world's population, while residents of Africa, the Middle East and Latin America, all together representing 26.9 percent of the world's population, accounted for only 7.1 percent. Fortunately, thanks to falling prices due to relentless innovation in both software and hardware, the "digital divide" seems to be eroding rapidly over time. By June 30, 2014, North America accounted for 10.2 percent of total users while Africa, the Middle East and Latin America accounted for 24 percent (Internet World Stats 2014).

Another point is that the production of a good is usually credited with an impact on employment, income and exports. Not so in the case of ICT. What is of most importance in terms of benefits from impact is not production but use. This is possibly the first time ever that being able to produce the means of production is not a prerequisite or a main factor in achieving success. The Asian countries, which account for more than 40 percent of world production of electronics, have not benefited as much as countries which have hardly any such production but are heavy users of ICT, such as Australia (OECD 2001: 9).

Some or even all of the reasons above have enabled the US economy to benefit much more than the rest from the diffusion of ICT. Baily and Lawrence (2001) suggest that in the US, ICT innovations have been driven by the demand of globally competitive service industries for improved technologies in order to further increase their competitiveness. The US is the country that has experienced the greatest change in its economic structure. In fact, the change that took place has been so significant that for almost three decades now it has become imperative to create new statistical series in the US for what have come to be known as "copyright industries". This involves, among others, software of all kinds, publications, journals, magazines, movies for cinema and television, video and music. To a considerable extent this industry overlaps with and is in close relationship with the New Economy. According to "The 2013 Report, Copyright Industries in the US" (Siwek 2013) this is the fastest-growing sector in the US, with rates of growth at least double the average for the whole economy, and a pillar of the economy in terms of production, employment and exports. In 2012, total copyright industries accounted for $1.7 trillion of US GDP, or 11.25 percent of the US economy; and employed more than 11.1 million workers or 8.35 percent of all US employment at an average annual compensation of $75,926 exceeding the national average by 18 percent. During 2009–2012, the industry grew at an annual rate of 5 percent, while the entire US economy grew at 2.14 percent. In terms of exports, the industry reached $142.2 billion in 2012, which is more than exports provided by aerospace, agricultural exports, exports of food or pharmaceuticals. Only a handful of countries possess such an index.

A consequence of the change in the structure of production in the US is that the composition of trade has changed dramatically. For instance, a representative basket of US exports worth $1m (at constant prices) in 1970, 30 years later made up only 30 percent of the original basket in terms of weight. Exports of consulting or management services or software have no weight. It is also interesting to note that despite the fact that the US, which hardly produces any of the equipment in the audio-visual sector, enjoys a huge positive balance of payments with the rest of the world in audio-visual services.

Another consequence of the New Economy in the US has been the successful entry of innovative new firms that have grown rapidly and are now among the largest in the world. These dynamic firms had either to grow in totally new fields or at the expense of incumbents: "almost 40 of the top 200 R&D performing firms in 2005/06 were founded after 1980 … while 32% of the top 200 R&D performing firms in 1980 had exited by 2005" (Hall and Mairesse 2009). The New Economy, the e-economy, the digital economy, or whatever we choose to call it, is characterized by increased entries and exits of firms – what economists call "turbulence".

Another aspect of the effort over the last several decades to increase productivity through technological change is that it has been turned into a competitive weapon. While raising the technological intensity of products, it increases the need for R&D. At the country level, studies have confirmed that the higher the level of investment in R&D per person employed, the higher the level of productivity. Over time, productivity increases with the rise of investment in R&D per person employed. The positive impact of R&D has been confirmed even at the level of firms.

Raising productivity through technological change did not simply turn it into a competitive weapon but had two further consequences: the creation of a *hierarchy of products* according to their technological intensity and the change of *the rules of the competitive mechanism* itself. The products with higher technological intensity and faster technological progress no longer compete on the world markets according to price but according to innovation. The winner is whoever first reaches the market with a new product or service. And demand for products of high technological intensity is not very sensitive to price.

Of course, it is not a coincidence that among countries such products are produced by the technological leaders, whether our analysis is at the firm or country level. As a result of these developments an international hierarchy of products was formed, with the technological leaders concentrating their production on high-technology products relying on technological intensity and technological change, and the technology laggards more and more crowded into the production of low-technology products facing relatively limited demand, and competing on the basis of price. Higher value added for finished products produced by the technological leaders means that these countries can afford to pay their workers higher wages than the technological laggards can pay theirs.

The hierarchy of products has a corresponding hierarchy of producers. At the top, there is the group of technological leaders consisting of the US, Japan and the EU. Within this group, also known as the triad, the US leads the way, followed by the other two. A second group of countries composed of technological laggards has been marginalized. It includes Africa, most of Latin America and most of Asia, which seem to belong to the international economic system only nominally.

There is a third group of newly industrialized countries which have managed to break away from the technological laggards and are striving to shorten the distance that separates them from the triad, if not to catch up with them. This group consists of the four "Asian tigers": South Korea, Taiwan, Hong Kong and Singapore. Finally, the last two decades have seen the emergence of a very special fourth group comprising Brazil, Russia, India, China and South Africa (BRICS). This group is led by China and India, whose economic weight has increased and whose roles are bound to be vastly expanded in the not so distant future – China first and India later (Kaplinsky 2006, 2007).

Neither of these countries is still a technological leader but they have managed to attract foreign investment, initially due to cheap labor. China, specializing in manufacturing and already a world power in that domain now accounting for over 12 percent of world exports, at first benefited from foreign investment from Taiwan and Hong Kong, and then beginning in the 1990s from investment from the triad. It has been doing well enough to reach $85 billion in external investment of its own on a yearly average for the period 2008–2011.

India does not specialize in manufacturing (yet) and its exports account for only 1.5 percent of world exports. However, it does specialize in services (informatics-related services, accounting, call centers). Its informatics-related exports accounted for $76 billion in 2011.

A parallel development is the emergence on the world scene of Chinese and Indian multinationals. Huawei, a Chinese firm, has already displaced Ericsson of Sweden as the world's largest producer of telecom equipment, and Lenovo of China is a power to reckon with in PCs. Tata industries, an Indian multinational firm, is already one of the world's largest manufacturers of steel, producing in several localities.

According to the most recent ranking for the 500 largest global companies in 2013, China accounts for 95 and India 8, versus 128 for the US, 57 for Japan, 28 for Germany, 31 for France and 27 for the UK (*Fortune* 2014). In a period of just two years, China gained 22 places on this list, the US lost 4, Japan lost 11, Germany lost 4, France and India were unchanged, and the UK gained 1.

Until the 1970s, internationalization was expressed mainly through international trade. Technological change had been instrumental in raising the level of international trade over the nineteenth century and the beginnings of the twentieth through important innovations in transportation – railways and the iron ship – and freezing, and again in transportation in the 1960s and 1970s due to another innovation – the container. By the mid-1980s, liberalization of capital movements together with technological change mainly through innovation in informatics and telecommunications allowed internationalization to be expressed, this time mainly through FDI, thus contributing to what has been referred to since the mid-1980s as *globalization* and actually enhancing its further rise. The process of globalization has also been expressed through portfolio investments seeking fast high returns, which have proved volatile and destabilizing for some of the recipient countries.

In general, the term "globalization" refers to a dynamic and multidimensional process in which national resources are becoming internationally mobile and national economies are becoming increasingly interdependent. It is also used to describe the increasing internationalization of just about everything, from goods and services to the financial system, competition, technology, firms and industries.

Within globalization, an important development was that MNCs (multinational companies) at first accelerated the setting up of production units in low-wage countries, as they have done traditionally for several decades, as the new technologies multiplied the choices for the location of many economic activities by rendering immaterial the distance from headquarters. Originally, extraction of raw materials and, later, exporting activities requiring low skills were the main beneficiaries. Then MNCs proceeded, albeit still timidly, to more sophisticated ventures intended to exploit the locally available trained human resources, and moved to the production of technologically more sophisticated goods, producing exports to third countries. In some cases, they replaced or complemented domestic production with production in the recipient country – *outsourcing*. Later on, they moved to still embryonic R&D intended to support or adapt these products to the local market. Finally, MNCs realized that there are places with locally developed intellectual capital and dexterities ready to be tapped, with the result that their production there could be combined with genuine innovative research – a process known as *localization*. Both China and India, as well as other countries, strive to upgrade their research potential, and have managed to attract research centers set up by world-famous technology multinationals, such as Cisco Systems, Microsoft, Sun, Oracle and many more.

The reason for the existence of localization is that information and knowledge (the reader will discover the difference between them in chapter 2) have been treated asymmetrically by globalization. Very simply, knowledge does not travel as easily as we once thought. In a world where distance has been obliterated by electronic communication superhighways, proximity and location have become immaterial for the transfer of information but not so for the transfer of knowledge. While the relative cost of obtaining information has been drastically reduced, the cost of obtaining knowledge remains largely unchanged.

> R&D and other sources of knowledge not only generate externalities, but also … such knowledge spillovers tend to be geographically bounded within the region where the new economic knowledge was created. That is, new economic knowledge may spill over, but the geographic extent of such knowledge spillovers is limited.
>
> (Audretsch and Thurik 2001: 6)

Specialization, the long-established currently intensified process through which international trade has always risen, whereby each country uses its comparative advantage and concentrates on the production of a few goods and relies on imports for the rest, turned out actually to be complemented by localization, not substituted by it. Specialization has in fact been pursued beyond the level of products – down to the level of components and parts.

So another development within globalization is that the share of imports in a country's exports rose, or, to put it differently, intermediate inputs rose faster than final goods. This development in globalization is the rise of *"value chain" production* or vertical production, intended to take advantage of comparative advantage within the process of production. In practical terms, this means that a product is produced in steps, each by a different producer, sometimes with each producer in a different country.

Globalization has been on the rise for several decades. Each further step in globalization – starting from trade, then FDI, outsourcing, localization, specialization and, finally, value chain production – has raised the level of interdependence among economies. And this is precisely the reason why the world economy, which has now reached its highest level of interdependence, was shocked as never before by the financial crisis of 2008–2009, which happened in a synchronized way, with the crisis transmitted from country to country and activity to activity almost instantaneously. It started with the housing sector in the US and soon reached exports, which affected imports, especially through value chains, and then reached the international economy. World trade fell and soon everything else did too, including investment, FDI, and even innovation efforts, mergers and acquisitions and patents.

The financial crisis has slowed down the world economy unevenly. The West has been strongly affected, while the Asian countries and some BRICS, though also affected, have escaped relatively unscathed. China, Korea and some of the BRICS, impelled by their own internal drivers seem to have been relatively oblivious to the aftermath of the financial crisis. As a result, China, whose rate of growth of R&D spending has not been affected, is expected to become the largest spender on R&D by 2019 (OECD 2014: 59).

Other developments in the wake of the financial crisis include the slowing down of patenting activities, which are proving slow to recover (OECD 2014: 58). In addition,

Table 1.1 Share of foreign affiliates in value added, employment and R&D

	Share of value added under control of foreign affiliates % 2010	*Share of employment by foreign affiliates % 2010*	*Share of R&D expenditures by foreign affiliates % 2009*
Ireland	54.4	23.5	69.9
Israel		8.9	62.0
Belgium			53.8
Hungary	49.2	24.7	52.6
Estonia	44.6	37.4	
Czech Rep.	42.9	26.9	58.0
Luxembourg	42.2	41.1	
Slovak Rep.	36.5	22.4	37.5
Poland	33.7	24.3	50.5
UK	28.6	18.2	40.7
Sweden	28.4	21.4	29.6
Netherlands	25.7	15.1	30.2
Denmark	25.3	18.9	
Austria	25.2	18.8	52.0
Norway	25.1	20.0	30.5
Germany	20,3	9.9	27.3
Finland	19.7	15.3	14.5
Spain	17.5	10.6	26.6
Portugal	17.4	9.2	23.1
Slovenia	17.1	12.5	33.2
France	16.0	11.2	28.9
Italy	13.9	7.3	24.4
USA	5.8	4.7	14.8
Canada			35.4
Australia	29.3		
Switzerland		10.3	14.4
Japan			6.3

Source: Based on data from OECD (2013).

Notes: Share of value added and employment for UK, Denmark, Israel, 2009. Share of R&D by foreign affiliates for France, Italy, 2010.

the landscape of global R&D is changing. Foreign-funded R&D has been rising in Israel and China, and in Europe in Sweden and Germany, while it has been declining in Canada, the UK, the Netherlands and Russia (OECD 2014: 43) (see Table 1.2). Further, technology is accelerating in certain fields: chemistry, nanotechnology, composite materials, new materials, 3D printing, laser technology.

The US followed by Japan have held on to their positions as technological frontrunners. This is true both in absolute as well as relative terms of the technology balance of payments, which consists of receipts minus payments for the use of technology such as royalties, fees, licenses for the use of patents, or their purchase (see Table 1.3). They are followed by the EU, which is now very slowly gaining some ground in relation to the two frontrunners, thus reversing the tendency exhibited in the 1990s and early 2000s (OECD 2010c: 131).

The US is in a phase of development all by itself. It is the largest producer of technology and the largest supplier of technology, it is the largest investor abroad and constitutes the largest recipient of investments from abroad, but is the least dependent on foreign investment for output and employment. It also produces the most significant technology

Table 1.2 Total stock of foreign direct investment, 2012, US $billions, net outward ranking

	Inward	*Outward*	*Rank*
USA	3922	5191	1st
Hong Kong	1422	1310	
United Kingdom	1321	1808	4th
France	1095	1497	7th
Belgium	1011	1038	
China	833	509	
Germany	716	1547	3rd
Brazil	702	233	
Singapore	682	401	
Switzerland	666	1129	5th
Canada	637	715	
Spain	635	627	
Australia	611	424	
Netherlands	573	976	6th
Russian Federation	509	413	
Sweden	376	407	
British Virgin Is.	363	434	
Italy	357	565	
Ireland	298	358	
Japan	205	1055	2nd

Source: Based on data from OECD (2013).

Table 1.3 Technology balance of payments (current prices), US $millions

	2000			*2012*		
	Receipts	*Payments*	*Balance*	*Receipts*	*Payments*	*Balance*
Australia	1558	2243	–685	4905	8274	–3369
Austria	1687	1235	452	7998	4518	3480
Belgium	4088	3461	627	14507	12586	1921
Canada	3143	1280	1863	2644	885	1759
Czech Rep.	249	435	–186	3085	3232	–147
Denmark	4576	3435	1141	8078	7910	168
Estonia	50	46	4	387	309	78
Finland	1555	1704	–149	10094	8848	1246
France	2742	2644	98	n.a.	n.a.	
Germany	13583	18215	–4632	69082	54727	14355
Greece	139	407	–268	674	814	–140
Hungary	249	396	–147	4396	4058	338
Ireland	15029	19117	–4088	52133	49154	2979
Israel	5584	1011	4573	13237	2451	10786
Italy	2806	3505	–699	13841	12807	1034
Japan	9816	4113	5703	34102	5623	28479
Korea	619	2642	–2023	5310	11052	–5742
Luxembourg	595	423	172	4008	4730	–722
Mexico	43	407	–364	n.a.	1822	n.a.
Netherlands	16367	15888	479	40171	30878	9293
N. Zealand	216	471	–255	1033	1898	–865
Norway	1912	1190	722	6929	3352	3577

Table 1.3 continued

	2000 *Receipts*	*Payments*	*Balance*	*2012* *Receipts*	*Payments*	*Balance*
Poland	188	1097	–909	4473	5933	–1460
Portugal	214	581	–367	1568	1275	293
Slovak Rep.	57	155	–98	948	550	398
Slovenia	12	22	–10	317	660	–343
Spain	2388	3665	–1277	16125	10592	5533
Russian Fed.	204	184	20	689	2053	–1364
Singapore	896	6233	–5337	5924	17075	–11151
Sweden	5281	5635	–354	20464	12437	8027
Switzerland	3407	5336	–1929	28725	27383	1342
UK	19926	9218	10708	36962	12534	24428
US	43233	16468	26765	122586	83957	38629

Source: Based on OECD (2013).

Notes: Starting date: Korea 2001; Luxembourg 2002; Netherlands, Ireland, Estonia 2003; Israel 2004; Denmark 2005.

and specializes in copyright industries, advanced manufacturing, next-generation manufacturing technologies, robotics and advanced materials. By contrast, both Japan and Germany, the largest player in the EU, specialize in medium high-technology manufacturing.

It is also interesting to note what FDI has accomplished. Hungary, the Czech Republic, Slovakia and Poland, that is the largest recent recipients of FDI in the West, have changed their production profile and now owe a large share of their output and employment to FDI. They used to exhibit a revealed comparative advantage in low-technology products but after they became the recipients of FDI, they acquired an upgraded revealed comparative advantage. Ireland, arguably the most successful country in attracting FDI, and very early in the game, has long exhibited a revealed comparative advantage in both high-technology (foreign owned) and low-technology activities (OECD 2005).

The innovative effort by country is very uneven, as witnessed by the gross expenditure on R&D as a percentage of GDP. In general, we would expect that it is easier for a country whose R&D expenditures are mainly incurred by multinationals to achieve high innovative effort. In reality, countries in this category tend to be medium-effort achievers: Ireland, Hungary, Czech Republic, Slovak Republic, Poland. On the contrary, strongest effort achievers seem to fall into two categories. First, the group of small countries whose R&D effort is mostly carried out by indigenous multinationals. This would fit the cases of Finland, Sweden and Switzerland. Second, countries with a strong sense of entrepreneurship, which translates into high shares of funding of R&D by business. This would fit the cases of Japan and Korea. Israel is a special success story. It has managed to achieve the highest rate of innovational effort, which is quite an achievement considering the difficulties for a country at its level of development. Its success most probably depends on acting as a very successful center of excellence. Finally, it is only fair to note China having doubled its effort rate over the last decade, Australia having more than doubled it and Singapore's efforts both in conducting R&D at home and in purchasing technology from abroad. (See Table 1.4.)

Table 1.4 R&D and its funding and conduct, 2012

	Gross exp. R&D as % of GDP	*Funding (%) by*		*Performed (%) by*		
		Govt	*Bus.*	*Bus.*	*Higher Ed.*	*Govt.*
Finland	3.43	26.7	63.1	68.7	21.6	9.0
Sweden	3.28	28.2	61.0	67.8	27.1	4.8
Denmark	3.02	29.1	60.0	66.7	31.6	2.4
Norway	1.62	46.6	44.2	52.3	31.3	16.4
Israel	4.25	12.1	35.6	82.4	14.4	2.1
Korea	4.03	23.9	74.7	78.0	9.2	11.3
Japan	3.35	16.8	75.5	76.6	13.4	8.6
US	2.81	30.8	59.1	69.8	13.8	12.3
EU 15	2.06	32.9	55.6	63.8	23.2	11.9
Germany	2.88	29.2	66.1	68.0	17.7	14.3
France	2.23	35.0	55.4	64.6	20.9	13.1
Belgium	2.24	23.4	60.2	69.0	21.8	8.8
Netherlands	1.97	35.0	48.2	57.8	31.4	10.8
UK	1.62	28.7	45.6	63.3	26.7	8.1
Ireland	1.53	27.3	50.3	72.0	23.1	4.9
Italy	1.26	42.6	44.3	54.2	28.0	14.8
Poland	0.89	51.3	32.3	37.2	34.4	28.0
Canada	1.71	34.3	47.4	51.6	38.7	9.3
Switzerland	2.96	25.4	61.0	69.3	28.2	0.8
Australia	2.13	34.6	61.0	57.9	28.1	11.2
Singapore	2.02	38.5	53.4	60.9	29.1	10.0
Russia	1.12	67.8	27.2	58.3	9.3	32.2

Source: Based on data from Main Science and Technology Indicators, http://stats.oecd.org/Index.aspx?DataSet-Code=MSTI_PUB, in OECD.Stat, http://stats.oecd.org/, accessed on March 23, 2015.

Notes: Funding: Australia 2008, Belgium, Norway 2011, Sweden 2013.

1.3 Innovation today

In terms of competition, innovation has been thought of as a powerful tool for a long time. More recently, an additional reason for striving for innovation and technological change has emerged, as it was realized that innovation determines the rise in productivity and hence income in the long run.

Several studies were carried out in the wake of the "*Abramowitz residual*" – the unexplained part of growth that cannot be attributed either to capital or labor, and is a "measure of our ignorance". These studies confirmed the fact that the rise in productivity is due to the accumulation of physical capital, the accumulation of human capital and technological change, plus a host of other factors (such as infrastructure, economies of scale, market structure, etc.). Tracing the role of these additional factors proved it to be diminutive and intricate (Englander and Gurney 1994). For all intents and purposes, contemporary Theory of Growth rests on the first three factors to explain the rise in productivity, on the understanding that they are interdependent, while technological change is the prime mover in the whole process, creating opportunities for investment in dexterities and equipment.

It is only natural that the rise in productivity has increasingly been seen as depending on technological change. There were points at which the new doctrine was under pressure, as for instance after the first oil crisis in 1973, when growth slowed down in the leading economies but technological change sharply increased. The simultaneous presence of a high rate of technological change with sluggish growth characteristic of that crisis came to be known as "*Solow's paradox*".

Several arguments rushed to the rescue. The most important among them was: erroneous measurements in the measurement of technological change, of the rise in productivity and of qualitative change. Other arguments ranged from the recognition that there is a long time delay for the diffusion of a technology before its repercussions are felt to the argument that product innovation is more and more concerned with products that do not increase productivity, namely products aiming at product differentiation. There is also the argument (Brenner 1998) that employment in manufacturing with high productivity is shrinking, while low-productivity employment in the tertiary sector is rising. All these arguments seem to carry some weight. The moral of the story is that our faith in innovation as the determinant of the rise in productivity and income, in the long run, remains unshaken. For the last five decades the effort to increase the level of innovation in every country has become a top priority.

During the 1990s, the US managed to achieve much larger rates of productivity increases than the EU, which led the Europeans to ascribe this fact to the US's much higher rate of innovation as compared to their own. They decided in Lisbon (2001) and at the Ministerial Committee of March 2002 to increase average spending in the EU for R&D from 1.9 percent of GDP to 3 percent by 2010. A target missed by a mile.

Innovation amounts to producing new knowledge. It results from a number of complementary intangible assets also known as intellectual capital or knowledge-based capital: computerized information (software, databases) and innovative property: scientific and non-scientific R&D, copyrights, designs, trademarks and competences. Some estimates show that intangible investment accounts for up to 1 percent of labor productivity growth in Sweden and just below that level for Denmark, Finland, the UK and the US, which is about a quarter of total labor productivity growth. In recognition of the role of intangibles in further growth, investment in intangibles has risen substantially in most advanced countries, in some cases exceeding investment in tangibles such as machinery and equipment and reaching a stock of up to 12 percent of GDP (OECD 2010b: 11).

While the world was changing as a result of innovation, as described in the previous section, innovation itself was also evolving. Ever since the original enthusiasm for technology five centuries ago, the notion of innovation, the process through which it is carried out, as well as the environment – including the main actors – in which it is achieved have been changing. Here, we shall deal only with the novel characteristics of innovation.

As recently as during the last decade the notion of innovation has broadened beyond R&D to include organizational changes, testing, training, marketing and design. The *Oslo Manual* latest edition (2005) defines innovation as the implementation of a new or significantly improved *product* (product or service), or *process*, a new marketing method, or a new *organizational method* in business practices, workplace organization or external relations. Strictly speaking, for an innovation to be considered as such it has to reach the market. Whether it turns out to be successful or not is a different matter.

According to the *Manual*, *product innovation* is the introduction of a product or service that is either new or significantly improved as to its characteristics or intended uses. Significant improvements in technical specifications are included and so are components, materials, incorporated software. User friendliness and other functional characteristics are included as well.

Process innovation is the implementation of a new or significantly improved production or delivery method including important changes in techniques, equipment or software.

Marketing innovation is the implementation of a new marketing method involving important changes in product design or packaging, product placement, product promotion or pricing.

Organizational innovation, the most intractable and the least studied type, is the implementation of a new organizational method, workplace organization or external relations.

Some writers refer to product and process innovation as *technical innovation* and marketing and organizational innovation as *non-technical innovation*.

Box 1.1 Better Places

Of course, there are combinations of different types of innovation resulting in a whole new package. Better Places is an Israeli company, which serves electric cars. It does not produce or sell the car, nor does it produce or sell the batteries. What it does is to sell kilometers run. The electric car, which has been bought cheaply without the battery, belongs to its owner and the battery belongs to Better Places. The company has set up service stations at which the used battery is exchanged for a fully charged one for a fee and as fast as it would have taken to fill the tank with gasoline. The batteries are recharged at the service stations at night when demand and electricity charges are low. We have a change of business model that has proved so successful that it has been exported to Europe and Japan.

While the term "innovation" refers to all of its types, most writers, including this one, use "*innovation*" and "*technological change*" interchangeably, but strictly speaking the latter comprises only process and product innovation. In this text, as in most others, we shall make this distinction wherever needed. However, as innovation expands as a concept to include several other items besides technological change, the latter represents less and less of it, so the interchangeability of these terms is best avoided.

Novelty, which is a prerequisite, falls into three types: *new to the firm* – innovation generated by some actor may be diffused to a firm irrespectively of whether it has been implemented by other firms; *new to the market* when the firm is first to introduce it in its market; and *new to the world* when the firm is first to introduce it to all markets.

The broadening of innovation means that it is no longer equated with R&D. It is still true that R&D remains the main avenue leading to innovation but it is by now an accepted fact that there are some forms of innovation which do not require R&D, as seen in Table 1.5. Indeed, in this table we observe that services also perform R&D. Note that a substantial percentage of process and product innovative firms in both manufacturing and services achieve their innovations not through R&D but through non-technical means.

Table 1.5 R&D active firms as percentage of product and process innovative firms in each sector (manufacturing and services), 2008–2010

	Manufacturing %	*Services %*
Finland	85.8	77.6
Switzerland	79.5	56.6
Norway	77.8	66.5
Netherlands	70.7	55.8
Sweden	69.5	56.1
Israel	62.7	55.7
Germany	60.8	41.5
UK	60.0	57.6
Japan	59.0	40.3
Italy	54.0	39.3
Spain	42.9	32.6
Turkey	34.8	26.2
France	34.0	30.0
Russia	29.9	9.5

Source: Based on data from OECD (2013).

Innovation is by now thought of as being pervasive throughout the economy, and while we already know that innovation occurs in services too, we use Table 1.6 to better understand what such innovation involves.

Innovation, in its older definition, used to be considered the preserve of manufacturing because this type of innovation used to cover close to the totality of innovative output. This is no longer true as innovation is becoming more and more pervasive throughout the economy. Some decades ago it was said that every time there was an innovation, the innovation clock would also tick in the machinery and equipment sector of manufacturing, which would produce the equipment to manufacture it. Nowadays, an innovation in the service sector or an Internet-based one may probably not need such equipment, and an innovation that may need it might be able to obtain it simply by reprogramming already existing machinery.

Innovation can occur in any sector of the economy including all services, even government services, and is present even in low-technology sectors. Industries with a low R&D intensity usually regarded as less innovative, like printing and paper or textiles and clothing, have turned out in recent studies to have a similar propensity to innovate as those considered to be high-tech or even cutting-edge technologies (OECD 2010a).

Just as innovation can occur practically anywhere, it can also appear in firms of all sizes, a fact known for quite some time. Nevertheless, there are differences between innovation originating in small firms and innovation produced by large firms. Differences range from the fact that large firms tend to produce methods of production while small firms are closer to the market and tend to produce new products. The two types also differ as to their sources. On this score, small firms depend much more on their suppliers and clients than large firms do, and hence depend much less on science.

A small proportion of firms accounts for a large share of both inputs and outputs, thus imparting to the innovation process a pronounced skewness. Such skewness has been confirmed in several studies. In Australia, 10 percent of the firms accounted for 80 percent of the innovation expenditure (Smith and O'Brien 2008).

Table 1.6 Innovation in the service sector, 2008–2010 (as a percentage of all service firms)

	Product or process innov. (only)	*Product, process & organizational innovation*	*Marketing or organizational innovation (only)*
Iceland	17.8	49.2	8.2
Germany	15.5	42.5	18.7
Canada	11.0	39.5	22.0
Sweden	14.5	32.3	12.3
Netherlands	16.6	27.0	11.1
Finland	12.6	28.7	12.2
Italy	8.2	22.6	21.0
France	6.5	22.5	22.5
UK	6.6	21.9	12.6
Japan	9.5	14.6	23.5

Source: Based on data from OECD (2013).

In 2008, the ten top global R&D firms spent about 58 billion euros on R&D, which amounts to about a quarter of the entire R&D performed by the top one hundred global firms. These ten top global R&D spenders spent about 25 percent more than the entire German industrial R&D, even though that country is the third largest R&D spender in the world, following the US and Japan. Skewness at the country level can be illustrated in the case of Canada, where the 25 largest Canadian R&D firms performed 33 percent of all domestic industrial R&D in the country in 2009 (Statistics Canada 2010).

Skewness is an attribute of patenting as well. In 2008, the top ten patenting firms filed about 8 percent of all international patents and the top 20 filed 12 percent of the total. Skewness in patenting, which is a reflection of the skewness of inventive performance, is also present at the national level.

On average, firms in service sectors tend to undertake less innovation than manufacturing firms, although there are exceptions such as the Knowledge Intensive Business Services (KIBS: telecommunications, finance, computer and R&D services) which exhibit similar in-house R&D and innovation outputs to those in high-technology manufacturing (OECD 2010a). But there are wide variations among service industries. On the whole, there are differences between innovation in manufacturing and innovation in services. Manufacturing firms tend to undertake more in-house innovation and are more likely to introduce new-to-market innovations than firms in services. An important point to note is that services, which have become the largest share of total value added in every country, are now the main source of skilled job creation.

A parallel development to the rising pervasiveness of innovation is the rising importance of non-technological innovation. Differences in rates between non-technological and technological innovation seem to depend mostly on the characteristics of specific industries and firms. SMEs (small to medium enterprises) engage much less in non-technological innovation than large firms (OECD 2009c).

"*Lead users*", which are at the leading edge of the market, are an important source of innovation that has not received proper attention until recently (Von Hippel 2005). Innovation by users is usually combined in innovation communities of informal user-to-user co-operation networks in which users help each other to solve problems and innovate. This is one specific type of *open innovation* (Greenhalgh and Rogers 2010: 155). We shall have more to say on lead users in chapter 3 on diffusion.

Constantly rising costs and complexity have forced the process of innovation to become much more open. Firms need to partner in order to share costs and gain access to complementary technologies and knowledge. They become parts of innovative networks, which tend to turn global with an open perspective toward innovation where collaboration and competition coexist (Chesbrough 2006). More recently, however, we have noticed that even firms like IBM, which are large enough and can afford to supply the finance and conduct the research behind the entire innovation project, use open innovation. "The open innovation paradigm is, therefore similar to a flexible and large-scale joint research project" (Greenhalgh and Rogers 2010: 154). Collaboration has become a feature of the contemporary research scene, as shown in Table 1.7.

The degree of openness in innovation differs across firms and industries depending on factors such as the strategy of the firm, the importance that the firm attaches to the technology and the characteristics of the industry. Firms seek to retain their core capabilities but open innovation provides a faster, less risky alternative to in-house development, especially if diversification is the objective (OECD 2008b).

Open innovation offers the benefit of drawing from a much larger base of ideas and technologies, and is a tool to explore new growth opportunities at low cost. It does have some drawbacks too. Managing co-operation with external partners entails costs, there is adverse impact on the flexibility of the company, there is dependence on external parties, and of course there is the danger that one partner may hold the project to ransom.

A relatively recent feature on the innovation scene is that new players have been emerging whose presence is being felt in international patterns of science, R&D and innovation. The presence of BRICS is particularly significant due to the phenomenal rise of China's share of global R&D. The scene with regard to scientific capacity is similar. With the broadening of the concept of innovation and the multiplication of its sources, both technical and non-technical innovations have been on the rise.

In what follows, changes in the process of innovation itself, in the actors involved, the locations in which it is carried out, its means of support, patronage, conduct and collaborations will each be examined in due course.

Table 1.7 National and international collaboration on innovations by firms, 2008–2010 (as a percentage of product and process innovative firms)

	National only	*International*
UK	34.7	31.1
Austria	20.0	31.0
Israel	13.7	27.9
Japan	32.3	9.3
Finland	11.1	28.7
Sweden	14.5	24.0
France	17.9	18.2
Germany	16.2	8.0
Spain	16.3	6.0
Portugal	10.1	9.5
Italy	7.4	4.7
Turkey	3.6	6.9

Source: Based on data from OECD (2013).

1.4 Repercussions of innovation and future prospects of world innovation

One of the results of the intensification of technological change and the subsequent rise in R&D has been a realization of the importance of knowledge and, of course, of the importance of the human factor, the sole carrier and producer of knowledge. The realization of the importance of the incorporation of technology in capital is not recent. What is recent is the recognition of the rising importance of the incorporation of technology in the human factor. There have been various instances in history where technicians were not allowed to leave their place of work lest knowledge should migrate – for instance, craftsmen of crystal and glass on the island of Murano, just off Venice. Human capital and technology constitute two faces of the same coin, two inseparable parts of the accumulation of knowledge.

The specialization of production and the recognition of the human factor as the par excellence carrier and producer of knowledge has further reinforced the separation of the labour force into two parts: specialized labour equipped for a knowledge economy and non-specialized labour non-equipped for such an economy. The first category faces a burgeoning international demand; the second is threatened with a collapse in demand. As a result, the remuneration of the first category, which is the minority in terms of numbers in total employment, has been rising fast over the years while in the other category incomes have practically stabilized at their 1970s level. This explains part of the rise in income inequality between the two groups, which keeps intensifying and which is visible at both the national and international levels. The other part is due to another result of innovation, which has supported globalization and outsourcing with both the losses of jobs to abroad and the loss of labour income in the West. (These facts will be revisited in greater detail in chapter 7, "Innovation Policy".)

At present, the world economy is still slowly recovering from the financial crisis of 2008–2009. Over more than a century, in times of recession or war, that is in times when trade and FDI have been disrupted, globalization has been on the retreat, but then after international links have been repaired it has reached its previous levels again. This time the process of recovery has proved more arduous, as the crisis has turned out to have been deeper and more long-lasting than expected due to the heightened level of interdependency among countries. In comparison to previous crises, drastically higher rates of value chain production in trade imparted the higher interdependency while restrictive macro-economic policies imposed on Europe did not help much either.

With the end of the crisis in view, the prospects for world innovation are rather promising in terms of opportunities. The world is still capable of producing ideas and there are some already in the pipeline. ICT still has a long way to go along the multitude of directions it has moved so far. Biotechnology and nanotechnology are still far from reaching their peak levels in terms of innovation. In addition, there is the prospect of an oncoming industrial revolution ready to unfold: the oncoming third industrial revolution (Gordon 2012) brought about by the digitalization of factory production, new materials and the further surge of roboticization.

However, still being capable of producing ideas and turning them into innovations are two different things.

> The United States has long been, and still is, at the forefront of cutting-edge science, technology and innovation. However, indicators such as business innovation surveys

> and data on growth of multi-factor productivity suggest that the US lead is narrowing in spite of its world-class universities and global technology companies. R&D and patenting by businesses have also grown less rapidly than in the past.
>
> (OECD 2014: 444)

May we add to this picture the declining quality of patents, a fact that may be of some relevance to the growth of the jobless and the waning of the benefits of innovation experienced at present.

China and India, with a few patents among the "highly cited patents", are beginning to make their presence felt in that category, but are far behind, and China, the imminent contender for the title of top research spender, is still very far from reaching US levels of accumulated knowledge and achievement. Above all, the US is still first in all relevant dimensions but, most importantly, still demonstrates the ability and determination to show the way and take the risk in uncharted territories, to open new vistas that may lead to new paths. No one else has tried to do so until now and it seems that it will be rather a long time before anyone else will be in a position to try. If the US, for whatever reason, falters in this most important aspect, then the further pace of development of innovation at world level may be put at risk.

Bibliography

Abramowitz, M. (1956) "Resource and Output Trends in the United States since 1870", *American Economic Review*, 46, 5–23.

Audretsch, D.B. and Thurik, A.R. (2001) "What's New About the New Economy? Sources of Growth in the Managed and Entrepreneurial Economies", *Industrial and Corporate Change*, 10, 267–315.

Baily, M.N. and Lawrence, R.Z. (2001) "Do We Have a New Economy?", *American Economic Review*, Papers and Proceedings, 91, 308–312.

Brenner R. (1998) "The Economics of Global Turbulence: A Special Report on the World Economy", *New Left Review*, 229, 1–265.

Business Week (2000) "Americans see the New Economy All around Them", *Business Week*, May 19, <http://www.businessweek.com/bwdaily/dnflash/may2000/nf00519b.htm>.

Chesbrough, H. (2006) "Open Innovation: A New Paradigm for Understanding Industrial Innovation", in *Open Innovation: Researching a New Paradigm*, ed. H. Chesbrough, W. Vanhaverbeke and J. West, Oxford: Oxford University Press.

David, P. (1975) *Technological Choice, Innovation and Economic Growth*, Cambridge: Cambridge University Press.

Englander, S. and Gurney, A. (1994) "Medium-term Determinants of OECD Productivity", *OECD Economic Studies*, 22, Spring, Paris: OECD.

Fortune (2014) "The 500 Largest Global Companies in 2013", July 21.

Gordon, R.J. (2012) "Is US Economic Growth Over? Faltering Innovation Confronts the Six Headwinds", Working Paper 18315, Cambridge, MA: National Bureau of Economic Research.

Greenhalgh, C. and Rogers, M. (2010) *Innovation, Intellectual Property, and Economic Growth*, Princeton, NJ: Princeton University Press.

Hall, B.H. and Mairesse J. (2009) "Measuring R&D Returns", in Presentation to the Knowledge for Growth Expert Group, Directorate General for Research, European Commission, Brussels, January.

Internet World Stats (2014) <http://www.internetworldstats.com/stats.htm> (accessed 28 July, 2015).

Kaplinsky, R. (2006) "Revisiting the Revisited Terms of Trade: Will China Make a Difference?", *World Development*, 34, 981–995.

Kaplinsky, R. (2007) "The Impact of China and India on the Developing World", CESifo Forum, 1, 15–21.

Lohr, S. (1999) "At Last Economists See a High-tech Payoff", *International Herald Tribune*, 15 April.

Lundvall, B.A. (2004) "Why the New Economy is a Learning Economy", DRUID Working Paper no. 04–01, Aalborg University.

Meadows, D., Meadows, D.L, Randers, J. and Behrens, W.W. (1972) *Limits to Growth*, New York: Universe Books.

OECD (n.d.) Main Science and Technology Indicators, <http://www.oecd.org/sti/msti.htm>.

OECD (2001) *The New Economy: Beyond the Hype*, Final Report on the OECD Growth Project, Paris: OECD.

OECD (2004a) "Multinational Enterprises and Productivity Growth: Insight at the Firm Level", in *OECD Science, Technology and Industry Outlook 2014*, Paris: OECD, <http://dx.doi.org/10.1787/sti_outlook-2014-en>.

OECD (2004b), "Labor Productivity of Foreign Affiliates in OECD Countries: Preliminary Results and Problems of Methodology", DSTI/EAS/IND/SWP, 18, November.

OECD (2005) *OECD Science, Technology and Industry Scoreboard*, Paris: OECD.

OECD (2008b) *Open Innovation in Global Networks*, Paris: OECD.

OECD (2009c) *Innovation in Firms: A Microeconomic Perspective*, Paris: OECD.

OECD (2010a) *Economic Globalisation Indicators*, Paris: OECD.

OECD (2010b) *Synthesis of OECD Country Reviews of Innovation Policy*, Paris: OECD.

OECD (2010c) *Measuring Innovation: A New Perspective*, Paris: OECD.

OECD (2013) *OECD Science, Technology and Industry Scoreboard 2013: Innovation for Growth*, Paris: OECD Publishing, <http://dx.doi.org/10.1787/sti_scoreboard–2013-en > (accessed February, 2015).

OECD (2014) Science, Technology and Industry Outlook, <http://dx.doi.org/10.1787/sti_outlook–2014-en> (accessed 28 July, 2015).

OECD and Eurostat (2005) *Oslo Manual: Guidelines for Collecting and Interpreting Innovation Data*, Paris: OECD.

Pohjola, M. (2002) "The New Economy: Facts, Impacts, and Policies", *Information Economics and Policy*, 14, 133–44.

Siwek, S.E. (2013) *Copyright Industries in the US Economy: The 2013 Report*, prepared for the International Intellectual Property Alliance (IIPA), November 2013, <http://www.iipa.com/pdf/2013_Copyright_Industries_Full_Report.PDF> (accessed July 28, 2015).

Smith, K. and O'Brien, K. (2008) *Innovation in Tasmania*, Australian Innovation Research Centre, University of Tasmania, Australia.

Statistics Canada (2010) *Industrial Research and Development: 2009 Intentions*, Catalogue Number 88–202, Statistics Canada.

Von Hippel, E. (2005) *Democratizing Innovation*, Cambridge, MA: MIT Press.

Part II

Theoretical background of innovation and diffusion

2 Innovation

Relationships, characteristics, origins, drivers

Devoted to innovation, this chapter gathers the essentials needed for its understanding: relevant facts, characteristics, relationships, origins, evolution, drivers, and theory, where appropriate.

As innovation, technology and science are linked with knowledge and learning, it is necessary to discuss these concepts and the relations between them before focusing on innovation itself. Thus, we first discuss the relation between technology and science. Next comes the relation between invention and innovation, and finally a discussion of the relation between knowledge, learning and innovation. This introductory analysis serves as preparation for the understanding of theory and provides some essential definitions. There follow the nature and special characteristics of innovation such as the difficulty of predicting its uses and impact, and the importance of timing, the environment and of interdependencies. Then come the origins and evolution of innovation; variation and selection; combinatorial evolution; innovation by mandate; technology-push and demand-pull; induced innovation; accidental innovation; and reinvention.

Then follows the creation of technological paradigms and trajectories; marginal and radical innovations and technological revolutions. Finally, we address technological opportunities, incentives and differences among sectors. The chapter ends with a discussion on the state of the art on the relationship linking innovation, size of firm and market structure, and new dimensions in research.

2.1 Some basic relationships

2.1.1 The technology–science relation

Technology preceded science by a few thousand years. It is only by the second half of the eighteenth century that science began to have an effect on industry. Since then, people have tended to think that science drives technology and that the latter is nothing more than the application of pre-existing scientific knowledge. In reality, however, science and technology interact in a complex relation of equals. In this relation, mostly in the past but today as well, technology often leads the way and science follows. Taking into account the economic motives driving technological change, it is easy to acknowledge that technological change – based exclusively on technological knowledge – usually precedes scientific understanding. (Rosenberg 1982: 141–159). The market needs a functioning product that meets the requirements of the producer and the user and this objective can be fulfilled without a full scientific understanding.

When scientific knowledge does induce technological innovation, the technology required for the realization of the innovation might well not be the most advanced available. In addition, available science does put physical limitations on an artifact but it does not prescribe its final form, thus leaving room for technological decisions. Maxwell's equations did not determine the design of the electrical circuits of modern radio receivers.

Technology consists of knowing techniques, methods and plans that work in practice. It is not the simple application of knowledge stemming from other fields. Technology works in certain ways and has particular effects, even when we are not fully aware why this is so. We can argue that technology is not a fundamental form of knowledge, but it is a form of knowledge that has led to economic progress for thousands of years. It would be ridiculous to think that invention has humbly to wait for science to open the door. Technology is only about one criterion: does it work? (Cardwell 1994).

Technological knowledge has accumulated over thousands of years without contribution from science. Science would have accelerated technological accumulation, but historically a huge body of technological knowledge has been gained and used, and this trend continues. We use blast furnaces without fully understanding the burning process. We use aircraft that are partly empirically designed, as compressibility and turbulence are not fully theoretically understood.

Technology affects scientific activity in many and varied ways. First, technology functions as a huge reserve of empirical knowledge that is open to examination and evaluation by scientists trying to explain problems. As such, it operates as an agenda for future scientific investigation. Even today engineers in a spectrum of industries are trying to figure out ways to solve problems that have not been scientifically explained. These solutions then induce scientific research that will eventually provide the scientific explanation of the problem (Whyte 1975). Torricelli's scientific proof of the weight of air was a result of his efforts to improve the design of pumps (Coehn 1948: 68–71). Carnot's impressive creation, thermodynamics, was a result of his efforts to comprehend the factors that have bearing on the effectiveness of steam engines, half a century after the steam engine innovation had been introduced by Watt (Cardwell 1971). The development of bacteriology by Pasteur was a result of his efforts to resolve problems of fermentation for the French wine industry.

The science of metallurgy began to develop during the nineteenth century in order to explain some attributes of steel, such as ageing and cracking. Despite the fact that relevant scientific knowledge did not yet exist, new alloys continued to be developed during the twentieth century, through empirical and repeated testing. Duralumin, a new aluminum alloy used since the First World War for the production of airplanes and airships, was used for a long time with no scientific understanding of its attributes. This understanding came about much later, with the use of X-rays and electronic microscopes (Hunsicker and Stumpf 1965: 279).

This sequence from real problems to scientific answers is not characteristic of industries based on scientific research, such as those based on electricity. However, even in such cases practical experience with new technology usually precedes scientific knowledge. A case in point is the study of short-circuits which eventually led to the science of crystals development, something that turned out to be of great importance for the electronics industry.

A second facet of the effect of technology on science that must be emphasized is the fact that technology is not merely a passive pool of knowledge. Instead, technological progress plays a very important role in the shaping of scientific action plans. The natural trajectory of some technological improvements identifies and defines the limitations for future improvements. To a great extent it is on these limitations that following scientific research focuses. For example, in aeronautics, requirements for improvements in performance have always driven technology to limits that could only be pushed back through better understanding of aspects of the physical world. The growth in the use of railways appears to have played a similar role during the nineteenth century in the development of scientific breakthroughs in the fields of metal stress and wear.

One of the basic characteristics of high technology is that it can clearly identify the most promising paths for scientific research to follow. In the case of telecommunications, the endeavor to transmit for ever greater distances and the introduction of new ways to do so also led to much research. In order to improve communications through radiotelephony it was necessary to better understand how electromagnetic radiation interacts with the spectrum of atmospheric conditions (Dinn 1977: 236–242). Research on issues related to the effort for better sound transmission led to a significant improvement in our understanding of the universe. More specifically, research on sources of noise has generated great results. The existence of the "sound of the stars" was realized and this marked the genesis of radioastronomy.

In some industries it might be thought that there is an inner logic guiding research in specific directions. In the development of the transistor the levels of purity that had to be achieved were unprecedented for industrial use. Thus, the development of telecommunications required the achievement of a material purity that could only eventually be attained through intensive scientific research. Telecommunications also required the achievement of extremely high reliability standards and these proved to be of vital importance for the economic performance of the industry (Morton 1971: 25).

Many of the problems related to the use of a new material do not become apparent immediately. It is frequently through fatigue from intensive use, or through the differentiation of the uses of some material, that problems arise. For example, a lot of research went into the use of polyethylene for wiring insulation. Despite thorough research, a whole family of problems arose after polyethylene had been introduced. Research focused on these problems eventually produced a far better understanding of the nature of polyethylene. This "second generation" research led to the understanding that shaping of the material defined many of its mechanical, electrical and chemical attributes, and this understanding led to solutions.

Technology has played one more role in relation to science by affecting, and at times by defining, the direction of scientific research and by determining scientific methodology. For example, in the field of operations research until the end of the 1980s much effort was focused on the development of efficient algorithms for the optimization of large problems. However, with the rapid development of computers the solution of large problems using traditional algorithms (such as Simplex) was feasible. Thus, research was driven towards the development of more complex and realistic problems and emphasis was placed on polycriteria analysis and on the development of interactive methodologies.

The creation of laser had much more noticeable effects, and transformed optics into a vital science with products such as optical fibers. As for solid-state physics, one can argue that its rise to prominence can be attributed to the creation of the transistor, which committed a large body of scientific resources. In this case, the flow of knowledge appears to have been from industry to universities (Braun and MacDonald 1978: 126–127).

To understand the relation between science and technology one has to study the reasons why technological discoveries play such an important role in the direction of scientific research. First, it is obvious that science is spurred in directions that suggest significant economic and/or social benefit. A significant technological discovery signifies the beginning of a series of other important developments, while it offers a totally new framework for technological improvements to come. This framework may direct ensuing research for decades following the initial breakthrough. At present, we are still experiencing the results of research that followed the creation of the transistor, of research on the first nuclear device and of research on heavier-than-air aircraft. Commercial success within this framework requires many complementary discoveries and the development of supporting technologies. These requirements create quite a few new topics for scientific research. A significant portion of research is carried out with the deliberate intention of marginally contributing to knowledge considered crucial for the exploitation of a new technology (Brooks 1968: 399).

There is another explanation for scientific knowledge frequently lagging behind technological improvement. Technological improvement usually changes the economic data relating to specific fields. Expensive raw materials with few applications are not expected to create much interest for further use. When, however, technological improvements reduce the cost of this material, its use may expand. This lower cost may allow the use of the material in other potential applications. For example, important innovations in the production of steel from the nineteenth century onwards created powerful motives for conducting more research on this material. Lowering the price of steel allowed its use in new applications. With the appearance of new applications, there emerged a need for scientific research to identify its attributes, potential and weaknesses under the new conditions imposed by the new uses.

During the last century the relation between technological practice and scientific understanding was formalized through the development of fields of applied science and engineering, that is, fields dedicated to offering understanding of the developing practice of a technology. Nowadays, science has an effect on invention because modern scientific fields are aimed at facilitating technological progress. Examples would include materials science, computer science, electrical engineering and pathology (Klevorick et al. 1995). It is noticeable that all fields experiencing rapid technological progress are closely related to an applied science or to an engineering field (Nelson and Wolff 1997: 205–220). The more extensive the available scientific knowledge on a field, the better guidance it is capable of offering as to the potential outcomes of each possible direction.

In addition, science and engineering offer powerful new ways of experimentation and testing. Thus, extensive scientific knowledge not only allows inventors to identify promising directions sooner, but also to give more accurate and timely forecasts of the probabilities for success. These forecasts are useful as there is no need to reproduce the full future working environment. We thus come to the conclusion that an extensive scientific foundation considerably enhances the effectiveness of efforts for technological progress. It has been observed that the rate of technological progress is related to the power and activity levels of the sciences upon which relevant research and development are founded (Nelson and Wolff 1997: 205–220). "The capability to understand and pursue possible ways for improving technological performance and the interactive obtainment of clear and reliable outcomes contributes to the human capability for developing an applied science capable of effectively supporting technology" (Nelson 2002: 6).

New problems arise following the emergence and use of new technologies, as already noted. The further development of these technologies depends on how effectively science can be brought to bear in order to iron out obstacles to progress. In other words, the successful development of an applied science, or of an engineering field, is often key to the rapid and continuing development of know-how. Electrical and chemical engineering constitute fields of research and education that were created at the same time as the industries using the related technology.

The mobilization of science is a necessary but insufficient condition, however. Most powerful fields of empirical science related to technological fields are essentially of an experimental nature. Thus, empirical sciences used to shed light on specific technological fields are only able to do so when the relevant technologies are open to experimentation.

In short, development of basic sciences such as mathematics, physics and biology that are not directly related to a particular technology or specific problems have an indirect effect on technological progress. Such sciences inform and empower applied sciences and engineering fields that focus on practical empirical problems. To put it more simply, rapidly evolving technological fields are based on powerful applied sciences which are, in turn, based on powerful basic sciences.

In the first part of this section we studied how technology affects science. Later, we saw how science affects technology and how the two co-evolve. We will come to the conclusion that, to this day, probably the best way to describe the technology–science relationship is by arguing that it is a partnership.

This partnership relation raises a question, though. Will this partnership hold true in the future with new and emerging sciences? For instance, quantum mechanics deals with fundamental laws of physics and studies rules applying to the micro world. At this level, the notions being developed are not created through simple eyeball observation. In such cases, technology cannot provide an agenda for science, as the stock of existing technological knowledge is based on a different conception of the physical world, through empirical observation at a different level. For progress to continue, science will need the contribution of a technology that will have to provide new tools requiring very specialized scientific knowledge. It is at this point that the difference between technology and science becomes ever more elusive and a grey zone begins to be formed between them. Within this grey zone, the various forms of engineering created to satisfy specific needs play a very important role and are subject to technological guidance.

2.1.2 Knowledge, learning and innovation

Technique and technology are two distinct concepts that are, however, frequently used interchangeably. Technique actually refers to the way a good or a service is being produced (know-how or hardware), while technology refers to why goods or services are produced in a particular way (know-why or software).

Handbooks of economics usually define "technique" as an abstract combination of capital and labor (as inputs) with the purpose of producing a particular good (outputs). We will define "technology" as the knowledge embodied in various techniques, knowledge of the scientific principles underlying a technique, as well as knowledge required to improve upon and change a technique (Bell and Pavitt 1993: 155–210).

The definition of technique corresponds to information related to a subject, while the definition of technology is related to the knowledge that can be attained on the subject. Information is relatively easy and quick to obtain and is conceptually different from knowledge (Bohn 1994: 61–73). What is arduous, time-consuming and expensive is the acquisition of the knowledge required to analyze and process information. Processing of the new knowledge that may accrue from pre-existing knowledge is also hard. Simply put, information is a raw material that, when properly processed, turns into knowledge.

If technique is about production methods (how) and technology is about knowledge of causality (why), then technological change is about discovering new causal relations that were previously either unknown or unanswered.

Knowledge is the capability or the ability of an organization to comprehend the causal relations related to facts and phenomena of its environment, as well as the ability to assimilate changes to these causal relations. Thus, learning refers to the addition of new causal relations to those already known and understood. Knowledge is then the available pool of causal relations and learning is the rate of understanding of new causal relations. Learning is about expanding and deepening the pool of causal relations.

Generally, innovation aims at solving problems. Solving technological problems implies discovery and creation. In other words, innovation implies the use of pre-existing knowledge, stemming both from experience and from science. That is, innovation requires the use of specific non-codified capabilities on the part of the inventors. In order to better understand concepts such as "codified knowledge" that will be used later on, we will take an explanatory detour.

There are three categories of knowledge: general and specific, explicit and tacit, public and private. All three categories are prerequisites for technological change, while in many cases two categories of knowledge have to be combined.

The fact that technological change is a cumulative process is due to the fact that innovation is based on technological capabilities, that is, accumulated knowledge.

Knowledge constitutes a concept that is broader than information. Information generally only covers parts of knowledge, such as know-what and know-why. These forms of knowledge are closer to the notion of economic goods that can be codified. Other forms of knowledge, such as know-how and know-who, constitute tacit knowledge that cannot be readily codified and that is not easy to measure (Lundvall and Johnson 1994).

These latter forms of knowledge are vague, non-codified, not published, they cannot be accurately described, vary among individuals, but they can be understood by people that share common experiences (Polanyi 1966: 1–18). It is such types of knowledge that require proximity in research, development and may even extend to production.

It is extremely hard to explain precisely to someone how to ride a bicycle as part of the knowledge to ride a bicycle is actually tacit. A cooking recipe consists of two parts: the first refers to a list of the ingredients and the quantities that should be used – this is codified knowledge; the second part consists of an attempt to describe how the ingredients should be used – this involves tacit knowledge and is the reason why cooking is much more than combining ingredients.

A first-year medical student would not know what to do when confronted for the first time with an X-ray. Three or four years later and after having been exposed to several hundred X-rays in the presence of a tutor and accompanying explanatory comments, he is able to diagnose what he sees – or at least he is expected to be able to do so.

"Know-what" is about facts and thus is conceptually close to the concept of information. Many television games use this kind of knowledge and medical doctors and lawyers heavily rely on know-what.

"Know-why" refers to knowledge of scientific principles and of natural laws. Technological development is based on this kind of knowledge that is being created and reproduced within specialized institutions, mainly research laboratories and universities. If a firm requires access to this kind of knowledge, they will probably either have to hire knowledgeable personnel or cooperate with the specialized institutions within which know-why is being produced.

"Know-how" is about a person's or a firm's capability of doing something. The manager of the human resources department of a firm selects and trains staff using know-how. A trainee worker using complex machine tools also uses know-how. Know-how is a kind of knowledge that develops and remains with the firm. One of the most important reasons for the interaction of firms is their need to share and combine parts of their respective know-how.

"Know-who" refers to information as to who knows what, and who knows how something is done. This kind of knowledge is about creating special social interactions that allow access to knowledgeable experts, and about using these interactions effectively. In modern society, where knowledge is dispersed among organizations and experts, this kind of knowledge is extremely important. However, know-who is also the most tacit kind of knowledge of organizations and firms.

The acquisition of the four kinds of knowledge described requires a particular approach for each. Acquiring "know-what" and "know-why" can be done through books, lectures, scientific journals, data banks, etc. On the other hand, "know-how" and "know-who" originate from practical experience. The most typical way of acquiring "know-how" is the relationship between a master and an apprentice. "Knowing-who" is frequently achieved through the technological gatekeeper of a firm or an organization. A technological gatekeeper is an individual with deep understanding of the technological capabilities of his/her own firm, as well as knowledge of the respective weaknesses. This person has an understanding of relevant available technologies and also knows who possesses the technological capabilities needed by his/her own firm, and how to access them.

As we have seen, knowledge is the capability of understanding causal relations and the capability of assimilating changes in these causal relations. Learning is the rate of understanding new causal relations and thus concerns the expansion and deepening of the available pool of knowledge. The way learning is achieved depends on the kind of knowledge to be understood. A firm has to pursue to know as many ways of learning as possible (Bohn 1994).

Learning is cumulative and the rate of learning is greater when what has to be learnt is close to what is already known.

There are several learning mechanisms. The foundation of all mechanisms, however, is learning-to-learn, where learning is defined as the identification and rectification of mistakes.

The importance of acquiring new capabilities and/or of new kinds of knowledge is increasing. The importance of continuous learning of codified knowledge and of capabilities to exploit this knowledge is also increasing. As access to information becomes more straightforward and less costly, so does the capability to choose effectively and eventually successfully use this information to become more important. Tacit knowledge, in the form of the capabilities necessary to process codified knowledge, has become more important than ever before.

Schematically, we can argue that codified knowledge is the raw material requiring transformation and that tacit knowledge (mainly know-how) is the tool used for the transformation. Thus, skills such as choosing relevant (and discarding useless) information, identifying information patterns, understanding and de-codifying information, the ability to learn (and to un-learn), are becoming ever more important.

This learning process is more than conventional training and the role of "learning-by-doing" is significant. One of the main aspects of this kind of learning is the continuous transformation of tacit knowledge into explicit knowledge used in practice, and then back for the creation of new tacit knowledge.

Firms are increasingly becoming aware they have to transform into learning organizations. Firms are also increasingly becoming parts of wide networks of inter-firm as well as user–producer interaction and exchange of information, all of which can produce innovation.

One of the consequences of the existence of inter-firm networks is the rejection of the previous theory of the linear model for innovation. According to this model, innovation is born from new scientific research and gradually continues through development, production and marketing, ending with the successful sale of new products, production processes and services. In contrast to the sequential model, we can now acknowledge that ideas for innovation can come from a variety of sources, including new production capabilities and the identification of market needs (Klein and Rosenberg 1986). The innovation process is not linear and it requires considerable communication between firms, laboratories, universities and buyers, as well as contact between science, engineering, development, production and marketing. This extends to the intense interaction and cooperation between all departments of the firm itself.

2.1.3 From invention to innovation

Nowadays, analyses of invention and innovation are being monopolized by the type of studies recognizing only economic or socio-economic motives for innovation. In reality, things are different, as motives for innovation are quite varied.

It is widely held that the creative process can be attributed to psychological and to socio-economic motives (Basalla 1999). Invention is the first part of the creative process and can refer to conceiving an idea, a plan, a recipe, a chemical formula, etc. An invention might well be a result of an inventor's dream or of playfulness. Many inventors behave as if there is a game that they have to win. Notwithstanding social or economic benefits, finding solutions to difficult problems, facing challenges, demonstrating their intelligence satisfies them.

Box 2.1 Not-for-profit innovation

Claude Shannon is regarded as the father of modern telecommunications. This scientist created the modern concept of information, and defined and quantified information. He proved the unity of information regardless of its source and proved the possibility for transmitting information with total accuracy, as long as the transmission is digital (Shannon 1948). Shannon's principles have been the foundation of a myriad of technological achievements during the last 50 years and they help explain issues as diverse as the sound fidelity of a CD, the capability of transmitting data through modems with noise interference and the way an image can be transmitted from space.

In his private life, Shannon was an idiosyncratic inventor building mechanical toys for self-use, who never thought of promoting them to the market. Among others, he built a mouse robot that could find its way out of a maze, and a doll that looked like W.C. Fields and that could toss balls into a drum. In an interview, he explained that he had always liked building things that moved in a funny way. When asked why he chose to wait for five years from mathematically proving his principles to actually publishing, he responded that his main motive had always been satisfying his curiosity.

There have been signs of this attitude since the Renaissance. A form of illustrated literature flourished between the fourteenth and sixteenth centuries (Keller 1964), describing contemporary technological practice and artifacts and also possible projections. The latter were a product of fiction and were actually proposals for the possible uses of technologies and artifacts with the application of contemporary methods and available technological knowledge (Ramelli 1976).

Patents are another sign of "technological dreaming". Despite the fact that patents undergo rigorous control processes, they are only an indication of technological potential at a given point in time, rather than a technological reality. In addition, there is a myriad of other inventions whose creators never sought a patent.

Box 2.2 Not-for-profit and for-profit innovation

Shannon was not the only one; there were many others. Florence Lawrence was the first star produced by the American movie film industry when movies were still silent (MIT Technology Review 2002). She loved to drive fast cars and invented the brake backlights as well as the little hands designating a right or left turn. She was never interested in patenting her inventions, which were used at once by the automotive industry. In contrast, Lotta, Florence's mother was apparently more enterprising than her daughter. She invented the windshield wiper but was careful in patenting her invention and benefited from it.

Only a small fraction of patents make it to the market and become innovations, while the vast majority are stored in the files of patent offices. In 1869 Samuel Sparks (in charge of the patent office of the USA) argued that only about 10% of patents registered at his office could be commercialized, while a century later Jacob Schmookler (1966) raised this percentage to 50%. Most modern experts tend to concur with Sparks.

The most widely known Renaissance collection of dream machines is the one consisting of the personal notes and drawings of Leonardo da Vinci (1452–1519) that were only revealed to the public at the end of the nineteenth century (Reti 1974). As presented and thought by da Vinci, most of these inventions were not feasible and, most probably, none of these had a noticeable impact on technological evolution. However, one can only admire the genius of a rare technological mind, and appreciate the technological enthusiasm that characterized Western civilization.

Unattainable machines by themselves created a whole class of technological thinking. Among them, perpetual motion machines (Ord-Hume 1977) were the largest sub-category and these troubled inventor-designers for a good 15 centuries. Strenuous efforts and imagination combined to give birth to more than 500 patents during the latter part of the nineteenth century in Britain alone. Since 1911, the US patent office has required that applications for patents for perpetual motion machines be accompanied by a working prototype, and for more than a century, physics has proven perpetual machines unfeasible. Despite all this, efforts towards the creation of such machines are still taking place.

Finally, technological fiction can also be found in arts, such as in science fiction publications and cartoons. Examples of the first case are Jules Verne (submarines), H.G. Wells (time-machine), Carel Capek (robots), Arthur C. Clarke and Isaac Asimov. Examples for the second case are R. Goldberg, Jacques Carelman and more.

Many inventions that came into being from plain curiosity, a game or a dream were then developed into innovation by their original creators or by others. That is, they progressed from ideas, plans, recipes and equations to new products, new production methods or organizational innovations. Of course, not all were successful.

The same differentiation between invention and innovation is true from the economics perspective. Since we are not only interested in invention, but also in innovation and in the successful commercialization of innovation, we focus on goals (i.e. the foreseeable socio-economic benefits). The search for invention may or may not include a socio-economic perspective. However, the search for innovation is based on socio-economic motives.

2.2 The nature of innovation

2.2.1 Innovation: basic facts

According to Dosi (1988a: 221–237) we can distinguish the following characteristics: first, the results of research cannot be known beforehand and this means that technical and economic aspects of research also cannot be known beforehand. Naturally, any innovation activities by private firms are based on the prospect of future economic and technical benefits. However, such expectations are not based on detailed knowledge of all possible facts, situations, input combinations, product specifications, etc.

In other words, innovation is inherently uncertain and this uncertainty is not only due to the lack of relevant information but also due to (1) techno-economic problems, the solutions of which are presently not known and (2) the impossibility of defining the consequences of an action with any degree of precision. These are the main factors constituting the uncertainty embedded in innovation activities.

A second point to be made regarding innovation is that important new technological opportunities are becoming ever more dependent on developments of scientific knowledge.

Third, there is a trend towards ever more research and this is expressed through increasing technological intensity in the search for new production methods, as well as through the efforts for the creation of new products/technologies. This trend is likely to continue in the foreseeable future.

The shrinking of time between scientific discovery and commercial application is the fourth basic fact regarding innovation. It took photography 112 years to appear on the market, radio 35 years, the transistor three years.

A fifth basic fact regarding innovation is that a very large proportion of innovation efforts fail. This phenomenon is widely recognized and has been named "the innovation efforts' graveyard" (Jewkes, Sawers, and Stillerman 1968).

Finally, it appears that technological change does not precisely follow market signals or needs: (1) The direction of technological change depends on the "state of the art" of technologies being used. (2) Very often, the nature of technologies defines the products and production methods that "fit" changing economic conditions. (3) Technological change is cumulative by nature.

2.2.2 Characteristics of innovation

2.2.2.1 The difficulty of predicting the implications of innovation

It is often very hard to predict the technological and economic implications of an innovation. This is true even in cases where an innovation is only being used for the purposes it was designed for, with no inter-sectoral diffusion.

Box 2.3 The Universal Product Code (Bar code)

The Universal Product Code, now known as the bar code, did not make an impressive entrance. The bar code's inventor, Joe Woodland, was from the beginning aiming for an automated cash register (ACR). The ACR was not to materialize for another 25 years, because of the wide consent (between producers, merchants, consumers and the state) such an innovation required. Another hindering factor was that the use of information and laser technology was still prohibitively expensive. Having been used at railway and car industry firms for two decades as an inventory management tool, the bar code was first put to the use with which we are now familiar only in 1974 (Varchaver 2004: 68–76).

The adoption of the bar code system had important implications, much wider than originally expected. The system brought huge improvements in inventory management, led to spectacular productivity increases and imposed radical changes in the field of marketing. It also changed the balance of power between producers and merchants, and it revolutionized the retail industry within the USA and, to a considerable degree, throughout the world.

On aggregate, the adoption of this innovation may have reduced production and distribution cost in the USA alone by some $100 bn per year. A few years ago it was estimated that savings in retail only were at about $30 bn per year. To these savings (that may have by now increased significantly) one should add the significant reduction in manufacturing and assembly errors, the quantum leap in inventory management (by now almost completely automated) and the huge savings in the transport industry. Regarding the use of bar codes in the transport industry, the sight of the bar code printed on railcars, on luggage and on posted items has now become commonplace. The use of bar codes is so widespread that no one takes notice of them any more.

Another implication of the adoption of the bar code system is the transformation of the cash register from a simple cash container to an information system. The recording of each customer's purchases created a vast database. It was then just a matter of time before data mining was used to study consumer preferences, behavior, complementarities between products, customer's loyalty, etc. In other words, the adoption of the bar code system brought along a de facto change in the methodology of marketing that was forced to look into market research through a whole new set of tools. Marketing could then ascertain properties suspected but never proven, and also discover whole new sets of properties.

Those who were to benefit the most were users that adopted this innovation faster, and who managed to better integrate the system into their production processes. Examples are UPS messengers, Procter & Gamble, Cargill in meat processing and Wal-Mart in retail. As a result of the adoption of the bar code system such firms managed to change the balance of power mainly in retail, where Wal-Mart evolved rapidly not only to become number one in the industry, but to become number one in the business world.

There is also a fourth dimension in the seemingly simple case of bar codes. The transformation of the cash register into an information system brought the end of an era. The balance of power between producers and merchants had always been in favor of the producers, who controlled information by controlling their inventory. The adoption of bar codes turned the situation around and it is now merchants who possess more and more valuable data. Bentonville (Arkansas), the home of Wal-Mart, now dictates pricing and specifications even for the largest producers (Brown 1997).

After 30 years in use, the bar code industry is now in its maturity stage and its replacement has begun with the use of RFID (Radio Frequency Identification) systems. These systems can hold and transmit much more information than the present version of the bar code, including the current location of the product, and have the potential to eliminate the checkout counter. Wal-Mart, Target, Albertson's and the US Defense Department were to begin using RFIDs by 2005.

Box 2.4 The Otis elevator brake

Another example of a seemingly trivial innovation that eventually proved extremely important was that introduced by E. Otis who created a safe braking system for elevators back in 1854. This braking system proved to be more important than the elevator itself. Up to 1854 elevators were being used with makeshift props at loading–unloading points. The Otis brake enabled the elevator to be widely used.

The consequences of the elevator brake innovation were huge. Much taller buildings were now usable and skyscrapers became a reality soon thereafter. Building architecture changed because of the Otis brake, as did city planning. Cities could now rise vertically as well, instead of having to expand horizontally. The use and the value of land changed, as did the way of life of modern people.

Box 2.5 Urea

In the early 1960s, urea, a new product, emerged in the oil refining industry in the US. Very quickly it became one of the three basic ingredients in the production of fertilizers and an unexpected gift to the members of the oil industry, who rushed to acquire fertilizer-producing firms. In less than three years, more than 45 fertilizer firms were taken over, including some of the largest. The fertilizer industry has never been the same again.

2.2.2.2 The difficulty of predicting the uses of an innovation

So far, we have discussed the difficulty of predicting the implications of an innovation in cases where the innovation is only being used for the purposes for which it was designed. This is the simplest case. There are, however, cases where an innovation (1) lends itself to other uses within the originating sector/industry, (2) can be used by other sectors/ industries, (3) did not have a specified original use.

To present an example for the first case, the development of the telegraph during the mid nineteenth century could not have given clues to the astonishing eventual development of telecommunications. No one could have predicted in 1847 that telegraphy was to graduate to a cornucopia of services with items like mobile telephony, short messaging system (SMS), Internet accessing, videoconferencing, high-definition television, etc.

The second case, where an innovation is eventually being put to use in sectors or industries other than the originating one, is inherently difficult to forecast. Watt's steam engine was initially used for pumping water out of mines, but was later used to power industrial production and, later still, transportation. The laser was initially developed for military purposes. Later on the laser was applied to a host of uses from medicine to inventory management and commerce (bar codes are read by a laser) to CD players and writers.

The third case, an innovation with no specified use, is a recipe for failure. Failure can only be avoided when an innovation proves to offer something truly different (that is, thought useful by society) that the innovator could not himself identify. This is the case with Edison's phonograph, an innovation with no specific initial targeted use. A year after the creation of the phonograph, Edison himself proposed ten potential uses for the innovation. These suggestions ranged from dictation without a stenographer to the recording of people's final wishes. Music playback was listed only fourth by Edison. This may represent the prioritization given by the innovator, as Edison at first thought music playback was not proper for his innovation. His reaction to the development of the jukebox, which was the most successful application of his invention, was not positive, although he later reconsidered.

2.2.2.3 The importance of timing of innovation

One of the decisive conditions for the success of an innovation is the timing. This is true because both the supply of and the demand for innovation depend on time.

From the supply side, this means that time has allowed the innovation-enabling technologies to develop sufficiently and become accessible. From the demand side, time has allowed social and economic conditions to develop so as to be able to identify and satisfy new needs. Both of these conditions have to be met for an innovation to succeed.

Babbage was a pioneer in developing computer technology. He realized that mathematical operations could be split into smaller steps that could be mechanically accomplished. Despite winning awards that allowed him partially to finance his efforts, he did not succeed in completing his automated computing machine. Babbage was not to blame for this failure, however. To complete his work he would have needed precision engineering. Engineering was at the time not sufficiently developed to provide him with what he would have needed to create a functional machine. If the machine were to be automatic, Babbage would also have to wait for the development of electronics. This is why it is frequently said that Babbage was two generations ahead of time (Clarke 1962: 24).

Box 2.6 The VCR

The development of the video-camera recorder is also indicative. It was originally developed by Ampex in 1957. It was heavy and cumbersome, hence not really practical for consumer use. The price was $50,000. The combination of size, weight and cost prevented this artifact from being used by the public. Only Hollywood studios and television channels operated them in film editing. It was two decades later that the Japanese presented the regenerated product, this time with great international success. Now, size and weight as well as cost were very significantly reduced and the device became widely usable and affordable.

The timing of innovation is becoming more important with the acceleration of the development of innovations. This observed acceleration has led some researchers to argue that we are heading into an era of exponential increase in innovation rates.

The relation between time and innovation is almost always in one direction only. An innovation has a life cycle. The new innovation will have to face the pre-existing and dominating innovation and possibly competing innovations. If an innovation prevails, it will eventually mature and be challenged by a potential replacement.

Two remarks are in order. The first is that, in general, the life cycle of an innovation is not known beforehand. An exception is found in the semiconductors industry, where a life cycle of about three years is being observed. Secondly, when an innovation is being challenged by potential replacements, there is a tendency for improvements in the dominating innovation. Sometimes, these improvements can be quite significant even when the dominating innovation has reached the stage of late maturity.

There are exceptions (rare ones) to this continuous improvement and replacement of older innovations by newer ones. The wheel was developed in Mesopotamia during the fourth millennium BC and was used for over one thousand years before being replaced by the camel. The wheel reemerged some centuries later in North Africa. The main explanation proposed for the replacement of such an important innovation by the means previously used is that climate change helped the desert claim previously fertile areas, making the use of the wheel more difficult.

Firearms were introduced to Japan by the Portuguese in the mid sixteenth century and were widely used for over a century. Eventually they were abandoned because local tradition held it was unmanly to kill an enemy from a distance. Firearms use reemerged when Japan was threatened by the West in the mid nineteenth century.

Typography was a Chinese innovation of the ninth century that begun with the use of wooden ideograms to progress to metal ones two centuries later. During the fourteenth century, however, the Chinese reverted to the use of wooden ideograms. The prevailing explanation is that it was easier to carve a wooden ideogram anew, than to find one among hundreds of available metallic ones.

2.2.2.4 Innovation and the importance of the environment

Protestantism, according to Max Weber, encouraged innovation. On the other hand, religions and societies emphasizing tradition do not seem to encourage change.

Other environmental factors, such as geography, play an equally important role. For more than half a century, the economic development of the American south was based on the textile industry. The industry, in turn, was based on Whitney's machine. This artifact met widespread use in the south, where people used to wear cotton clothing. The machine was also a starting point for a series of more evolved artifacts. If Whitney's innovation had been developed in the north, where cotton clothing was not widely used because of cooler weather, it might have not succeeded, and Whitney might be now known as an eccentric inventor. An invention is only considered important when a civilization chooses to value it as being so (Basalla 1999: 32–34).

Naturally, there are many other exogenous factors contributing to the success or failure of an innovation. The Dutch flute ship was an important development in the design of sailboats. It was much faster than existing types of ship and required half the crew. Such performance warranted huge savings in sea transportation. However, flute ships did not sail in the open seas for a whole century. The reason was that when the flute appeared piracy was a major problem in the Atlantic. In a confrontation with a pirate ship, the deciding factor for survival was the large size of the crew. This factor obviously did not favor the Dutch flute ship (North 1986).

The potential spread of the use of Kamen's ingenious transportation contraption – the segway – depends among other things on whether local authorities will require a license for driving it.

2.2.2.5 Innovation interdependencies

Innovations very seldom function in isolation. Very often, the quest for improvement of the productivity of an innovation creates the need for the development of complementary innovations, as we will see in section 2.3 in the case of the mechanized collector for the tomato as well as for cotton and wheat. The same was true for Babbage's computer and for the video camera, where complementary technologies and additional time were required for success.

Technologies interact in ways that are not always evident, sometimes not even to experts. One of the difficulties in the effort to assess the total implications of an innovation is the fact that these interactions are so numerous, and so diverse. For example, an innovation that led to the reduction of transportation cost: railroads in the USA in the mid nineteenth century (Fishlow 1965). Part of the benefits of railroads was the increase of productivity in agricultural production, which was now able to specialize and enjoy economies of scale with lower production and transportation cost. As a result, it was possible to achieve higher geographic specialization of production, and this eventually led to even greater productivity. This need for complementarities might explain why some important innovations initially only bring marginal productivity increases.

Sometimes, the realization of a particular innovation has to wait for the development of a particular input. In the past, the realization of many innovations had to wait for the development of high-quality and high-performance metal parts. The steam engine was held back by the lack of high-quality steel. The high pressure required to produce energy required steel alloys of high strength and heat resistance properties. Such alloys were difficult to create because of lack of proper metallurgy processes.

Lastly, there is the case where increases in productivity are realized in activities not directly related to the innovation. This kind of innovation is not particularly visible because it mainly concerns intermediate goods (something partially reflected in input-output tables (Carter 1970)).

From this point of view, not all activities are equally important. There are activities where technological progress significantly affects the technological progress of other activities. There are also activities that have absolutely no effect on the development of other activities. An example of the first case is the chemical industry. Technological progress in the chemical industry has affected the technological progress of many other activities, such as the textile industry (synthetic fibers) and metallurgy. The new generation of bar code was actually imposed by the largest user, Wal-Mart (see section 2.5.2.1). New labels, software and hardware were developed to use RFIDs so as to be able to locate a product and hence recognize reference points. The latter task is a prerequisite for the safe use of robots in the home. Eventually, the emergence of the new bar code may turn out to be instrumental for robots to be safely introduced for home use (Mann 2004).

2.3 Origins and evolution of technology

There is still no satisfactory theory on the origins of innovations. As inventiveness is more plentiful than wants and needs, a selection process is required so that some innovations are chosen to be developed and incorporated in a society while others are rejected. Those selected are reproduced and become predecessors to a generation of variants, which in their turn will later be put to the test of the selection process. Those rejected have very little chance of affecting future developments – unless special effort is expended in that direction.

In nature, the natural selection process allows a few of the variants to survive, reproduce and pass on their genetic information. Selection will depend on the prevailing conditions – environmental, biological and social – at the time the variants appear. Those that are to survive carry a survival value that is determined by the conditions they will happen to encounter and not by any absolute criterion of superiority. Therefore, evolution by natural selection has no preordained goal, purpose or direction. This is not true in the case of innovation or artificial selection, where criteria are established to select characteristics considered worthy of preservation in the pursuit of some biological, technological, psychological, social, economic or cultural goal.

Variation and selection, Darwin's evolutionary mechanism for living organisms, could explain quite a lot of what happens in the evolution of technology. It can explain first the multiple versions of a technology (such as the locomotive) at a particular time, as it attempts to address different purposes, operate in different environments and be built by different designers. Then it could explain how some of the better performing versions would be selected for further development, and thus pass on their small differences to future designs. Accumulation through selection of such differences would in the end lead to important modifications of structure. But this is as far as Darwin's theory can take us because it cannot account for the appearance of radically novel technologies, the equivalent of novel species in biology.

> Some technologies – the laser, the jet engine, radar, the Quicksort computer algorithm, the railroad locomotive itself – just appear, or at least they seem to appear, and unlike biological species, they are not versions of earlier objects. The jet engine is not a variation of the internal combustion engine or anything else that preceded it, and it did not come into being by the steady accumulation of small changes in its predecessors.
>
> (Arthur 2009: 17)

Radar "descends" from radio but no matter what we change about radio circuits we will never get radar. It rests on a different principle than radio.

Arthur proposes a mechanism for the evolution of technology that is different from Darwin's. It does not dismiss variation and selection, and in the absence of heredity its crucial element becomes combination.

> If you open up a jet engine ... you find components inside – compressors, turbines, combustion systems. If you open up other technologies that existed before it, you find some of the same components ... Technologies inherit parts from the technologies that preceded them, so putting such parts together – combining them – must have a lot to do with how technologies come into being.
>
> (Arthur 2009: 19)

In combinatorial evolution, where technologies come into being as new combinations of what already exists, the abrupt appearance of novel technologies no longer seems abrupt. Besides the fact that combinations are anything but random, three conditions have to be met: all technologies are combinations; each component of technology is itself in miniature a technology; and all technologies harness and exploit some effect and phenomenon, usually several.

Innovation by mandate encompasses all defense technologies as well as other technologies the existence of which the state sees fit to encourage and which the state helps diffuse. Almost all radical innovations belong to this category, for instance electricity, the telegraph, the iron ship, the Internet.

We have already encountered the case of invention for its own sake and invention which does not have profit as its target. There is a further category encompassing several important innovations, which originated with no specific potential purpose. When Xerox came up with the photocopying machine in 1957, the firm was searching for potential uses. Edison suggested 10 uses for the phonograph, and music reproduction was only fourth in his order of priority – and apparently he was unhappy when his innovation finally became a success through a "trivial" use, the juke box (Basalla 1999: 140).

There is a whole category of inventions that reached the stage of innovation without having, or at least without initially having, the goal to address wishes or needs, that is, without the contribution of demand – electricity, the telegraph, nuclear power, etc. Additionally, the fact that there are needs, which have always existed yet only some of which have induced the materialization of innovations to meet such needs, shows that the relationship between demand and innovation is distant and peculiar. There has always existed a potential market for the heart pace maker, the contraceptive pill, and high-performance contemporary grains, but such needs have only lately been satisfied.

So far, the types of innovations we have referred to are usually described as autonomous or supply innovations – "technology-push". On the other hand, a large category of innovations is on the demand side – "demand-pull". Within this second category there is a large group of innovations that attempt to cover needs or provide solutions to problems. For instance, the need for greater car safety has led first to safety belts, then to air bags. The need to economize on time has led to pre-prepared or partially prepared food and to fast food chains. A similar logic prevails in the case of the provision of solutions to problems. For instance, in a production system the non-correspondence of output speed between parts of the system (such as between a printing machine and a perforator next or further down the line of production) creates a problem to be solved. The attempt to face needs and problems falls within the category of induced innovation.

It used to be thought that a mere change in the relative price of inputs in a production process by itself calls for induced innovation – more precisely an innovation that would economize on the use of the input which has now become relatively expensive (Hicks 1932). More recent thought as well as empirical evidence on this matter have led us to conclude that what is relevant is the economizing on the use of the input with the largest share in production cost. Improvements in parts of a production system result in their increased performance, leading to an imbalance, a non-correspondence, between the improved and the unimproved parts of the production system, thus begging for an induced innovation. An additional eventuality calling for induced innovation arises in the case of shortage of raw materials, a situation most common in wartime. Synthetic alkaline in France, nitrogen in Germany, nylon (to replace silk for parachutes) in the US, are a sample of such induced innovations.

There are instances where the outcome of an innovation (whether induced or not) brings about a chain reaction leading to further innovations. In the early 1970s in the US, the realization that oncoming changes in legislation would limit the flow of seasonal migrants from Mexico employed in the harvesting of tomatoes in California, led to the reappearance of an older innovation: the mechanical harvester developed in the land-grant colleges. When it was realized that natural tomatoes were too soft to be collected by the mechanical harvester the land-grant colleges went back to work this time not to produce some kind of a mechanical improvement but a genetic one. The hard tomato was born. It looked like a tomato but didn't smell or taste like one (Hightower 1972). Such chains of innovation are often met in agriculture – corn, cotton – where the genetic redesign of a plant allows for the easier use of mechanical equipment.

Although cases of accidental innovation – or innovation by chance – are rather few, they are so impressive as to create the feeling that they represent an important share of innovation. In reality accidental innovation accounts for a very small share of innovation; an important point as to its occurrence is that it does not happen at any time or in any place. It usually happens during the research phase and more specifically during the applied research phase, often when aiming at a different target. Sir Alexander Fleming's efforts to explain the sudden death of microorganisms he was studying led to a very important innovation – the first antibiotic, penicillin. Paul Muller, a Swiss chemist, attempting to protect woolen clothes from moths came up instead with DDT which helped eradicate malaria by eliminating the mosquito – although with serious side effects

(Gladwell 2001: 42–51). Warfarin originated from research to combat cows' stomach hemorrhage and ended up as the most effective rat poison. Viagra was originally intended to deal with certain heart conditions but ended up offering its services to better uses.

A case that could more rightly claim the title of accidental innovation is the innovation of the microwave oven. Apparently, the melting of a chocolate bar in a researcher's pocket at Raytheon when he came into the proximity of microwave-emitting equipment led to this innovation (Schwartz 2004: 32–40).

Lead users are another source of innovation (Von Hippel 1988). They develop an innovation up to the level of the prototype and then convince an industrial producer to produce and sell it. In many such instances the ensuing innovation is the result of cooperation between the lead user and the industrial producer. The development of the microprocessor is a case in point. Busicom, a Japanese electronics firm, contacted Intel, a US firm, in the late 1960s so as to be provided with 12 semiconductor chips to be used in its new calculating machine. Dr Hoff, researcher at Intel, had the inspiration to combine the 12 chips into one, which in turn drove him to think of placing the central processing unit for the computer on a single chip. Thus was born the microprocessor, the heart of the personal computer. For the development of the microprocessor Busicom and Intel cooperated over a period of several months. Busicom used it on its calculator while for Intel microprocessors have proved to be the main line of business ever since (Noyce and Hoff 1981).

There is still a similar instance of the lead user where the emphasis is on user rather than on innovativeness, as the lead user uses it extensively and presses for its faster evolution so as to increase even more the advantages this technology is offering vis-à-vis his competitors. Wal-Mart, to which we have already referred, is a case in point. It is the heaviest user of bar code readers (several hundreds of thousands). By demanding that its suppliers use the next generation of bar code using RFID, it has not only encouraged but even hastened the emergence of this innovation (Hen 2004: 90).

An important source of innovation that has not received much attention until recently is reinvention. This term stands for the degree of diversification, change and evolution of an innovation from its initial state during the process of its adoption and application.

As the product of an advanced stage in the process of diffusion it easily passed undetected in studies, which did not differentiate between adoption and intention to adopt, and of course never reached the application stage. So in the past, whenever reinvention was detected, it was ignored or considered peculiar or in any event a rare phenomenon. Its existence and the realization of its importance dates from the time studies started studying the application stage of innovations.

Innovations that fall within the category of reinvention are the effect of a number of causes. Complex or difficult-to-understand innovations offer themselves to reinvention for the sake of simplicity. Reinvention may also happen as a result of limited information, ignorance or insufficient learning. An innovation, which constitutes a general principle or a tool with many potential applications, is a candidate for reinvention. The same holds when an innovation is used as an answer to a wide range of problems. Reinvention is also encountered when an innovation needs to adapt to the structure of the institution that adopts it. Finally, as reinvention arises in the late stages of diffusion, it tends to be more useful to the laggards in terms of adoption than it does to lead users (Hays 1996: 551–566).

Before closing this section, and in fairness to the impatient reader who pines to hear about the buzzword of our times, "disruptive innovation", the truth is that this is not another type of innovation like product or process innovation. It does not characterize an innovation itself but rather its impact on markets in terms of dynamism, structure and market dominance: it either creates new markets or reshapes new ones. We shall visit the concept in more detail later (chapter 6).

The question of supply-push versus demand-pull is a longstanding debate of which here we may retain only a few points. For instance: successful innovations originating from the "market pull" tend to be marginal with smaller long-term value than innovations originating from supply push. Another point is that innovation is a very complex procedure, which cannot be explained by a single factor, as success requires the simultaneous presence of both. Rather than competitive, the two forces are complementary. The majority of commercially successful innovations may originate from "demand-pull" but they still rest on the existing knowledge base.

Finally, the most important result of the debate: the question of demand versus supply does not arise but instead there is a complementarity through time as one precedes the other (Walsh et al. 1979; Walsh 1984). In the first stages of development of an industry "technology-push" tends to be more important while "demand-pull" gains importance as higher levels of maturity are approached.

2.4 Technological paradigms, trajectories, marginal and radical innovations

A "technological paradigm" in the sense of an era, defines the needs that have to be addressed, the scientific principles to be used in finding solutions and the technology that will be used. The present technological paradigm defines today's car or the smartphone, but also pauses questions as to what should be done next – which area of research should be investigated – and sets technological dictates such as "drive for more and more miniaturization", "digitize everything", "re-innovate everything in a mobile rendition". In other words, technological paradigms define technological opportunities and some basic procedures on how to go about doing so. Thus, they channel efforts in specific directions and not towards others. In some instances we have strong paradigms, which generate fast and protracted rates of technological change. Other paradigms are weak, offer few opportunities and hints on how to exploit them.

A technological trajectory describes the activity of technological progress on the economic and technological points of compromise defined by the paradigm. Technological change happens in the form of trajectories defined by specific groups of knowledge. Over time, the origins of new paradigms relate more and more frequently to basic progress in science and related technologies – electricity, informatics.

The "techno-economic paradigm" referred to by Freeman and Perez concerns whole groups of paradigms, in the sense described here. For instance, the techno-economic paradigm (or regime) covers the common characteristics, complementarities and internal links between "small" paradigms: semiconductors, computers, automata.

Marginal innovations (Freeman 1982: 498–499) emerge almost constantly albeit at different rhythms among activities and concern mostly improvements in products or methods of production. The impact of each of them on productivity may be imperceptible but cumulatively their impact may turn out to be significant. It is to this type of innovation that changes over time in coefficients in input-output tables are due.

Radical innovations are much less frequent than marginal ones and they seem to be randomly distributed over time. They are discrete events whose diffusion often presents cyclicality, possibly associated with the large economic cycles. Naylon and polyethylene are good examples of such innovations.

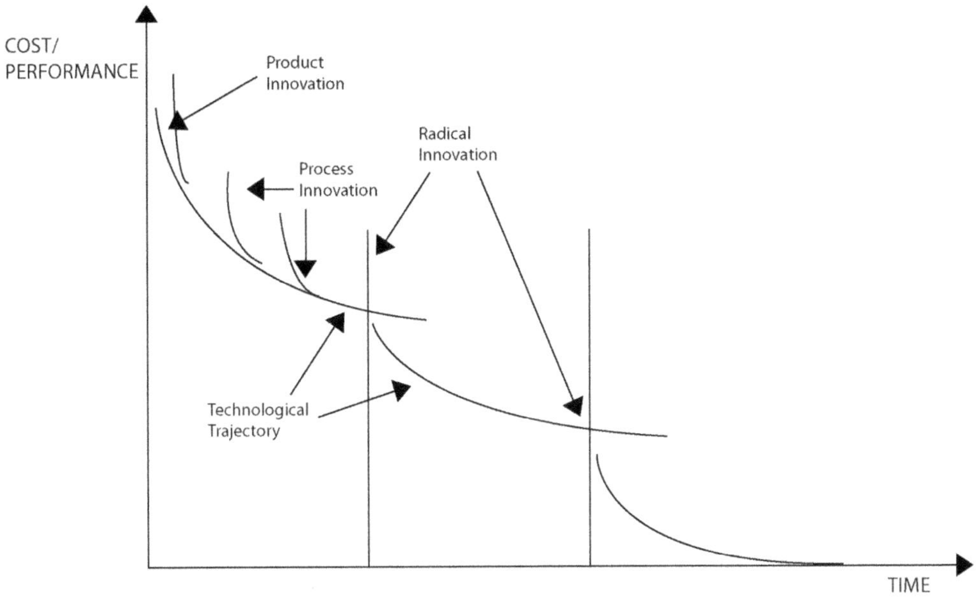

Figure 2.1 The typology of innovation

"Technological revolutions" are what Schumpeter calls the "gales of creative destruction" upon which his theory of large cycles rests. Electricity and the computer are good examples. Such changes entail many groups of radical and marginal innovations, and have diffuse consequences for the entire economic system. They do not lead simply to the emergence of products and services in the activity in which they were first generated but additionally they bring about important changes in other sectors of the economy by changing the structure of their inputs and the conditions of production and distribution.

By definition, a technological revolution should satisfy five basic characteristics: (1) to be able to bring about a substantial cost reduction in many products and services thus allowing for the ability to discern opportunities in new lucrative investments; (2) to be able to bring about substantial improvement in the technical characteristics of many products and services, i.e. reliability, speed, precision, etc.; (3) to be socially and politically acceptable as "a change of paradigm" might entail institutional, educational, legal and regulatory changes as well as changes in management and labor practices; (4) to be environmentally acceptable; (5) to have pervasive consequences on the entire socio-economic system.

A technological paradigm consists of an innovation and its improvement(s), such as the car with an internal combustion engine – and a search procedure – and what is to be done next: that is, in which direction we should investigate, what kind of knowledge we should be looking to acquire. From the moment the technological paradigm has been selected, the innovative effort concerns a specific trajectory. At first, the internal combustion engine was only one of the four technological alternatives for the car, until it established itself. Of course, a change of paradigm signals a change of technological trajectory along with everything that such a change entails in terms of goals and capabilities.

The capabilities and the searching procedure, both of which are involved in the searching process, are to a great extent differentiated among technologies. In other words, for each technological paradigm there is a different "technology of technical change". For instance, in organic chemistry the search is carried out in a totally different logic and rests on different capabilities than in the case, let us say, of mechanical engineering. A crucial consequence of the type of technological knowledge within the paradigm is that innovative processes are very eclectic; they concern specific directions and are cumulative in relation to obtaining the required capabilities.

From the moment we recognize that innovative activity is very eclectic, specific, cumulative, not accessible to all, and that the research activity of the firm (or institution) takes place in specific areas of technology not far from its activities, it becomes clear that the technology the firm will develop over time is not at all random. It depends on and will be realized in technological zones close to its activities.

To recapitulate, the meaning of the term "technology" as we are using it here differs significantly from the one held by orthodox Economic Theory, where the meaning of "technology" is information with a general application, easily produced and reproduced, that is, readily accessible to firms which simply absorb information at no cost from a reservoir of knowledge. In the picture we compose here firms produce products through methods that are technologically differentiated from the methods and products of other firms. Their innovations are based to a great extent on research performed within the firm but also on some contributions from other firms and from public knowledge. The search processes of firm do not (necessarily) cover the totality of technological knowledge and usually do not stray far beyond the firm's technological base. Given the differentiated nature of technology, the firm attempts to improve and broaden its technological base, searching for technological zones that will allow it to use and build upon it. The technological search process is cumulative and what the firm can hope to achieve technologically in the future is narrowly defined by what it has achieved in the past.

2.5 Technological opportunities, incentives and inter-sectoral differences in innovation

For the last two centuries or so, scientific knowledge has been offering a growing reservoir of potential technological paradigms. Each technological paradigm entails technological opportunities, which differ from those encountered in other paradigms in terms of their density and their amplitude. Scientific discoveries play a crucial role especially at the initial stage of development of new technological paradigms. Selection among these paradigms depends on the nature, the interests and the capabilities of the institutions that intervene between research and economic applications. From the moment a technological paradigm has been established, the goals of research become endogenized, that is, they focus on the investigation of the technological trajectory of the paradigm which has been selected (Dosi 1988b: 1120–1171).

The requirements for specificity, cumulativeness and tacit knowledge lead to the fact that opportunities for innovation and the possibilities for their realization are of a "local type", that is, they lend themselves to appropriation in specific innovative firms – they are firm-specific. Another implication is that the characteristics of each technological paradigm and its level of maturity specify opportunities for technological progress. Opportunities are much more numerous at the initial stages of the paradigm than at or when approaching maturity. The result is that at each moment in time, technological opportunities differer depending on the sector and the level of development of the paradigm under which they operate. To a great extent, this is the reason for the differentiation in the level of innovative investments among sectors (Gort and Wall 1986: 741–757)

Changes in relative prices, energy prices, prices of raw materials or changes in demand conditions are important as they have repercussions on the level and the direction of technological change but within the framework defined by the nature of the technological paradigm. However, such changes will also influence the selection and investigation of potential paradigms and consequently their possible potential emergence and establishment. The observed sectoral structure of technological change is the result of a two-way influence of opportunities and appropriability on the one hand, and market incentives on the other. The first of these are decisive in the short run, that is, from the moment the technological paradigm has been selected. The latter are decisive in the long run. They will play an important role in the technological search but they will do so somewhat indirectly through encouragement, support and the focusing of the search for new technological paradigms.

Innovation is not a reaction to relative prices and demand. On the contrary, it is a process to a great extent endogenously defined by competition, where firms constantly attempt to improve their technologies and artifacts independently of whether market signals change or not. Every paradigm and its particularities delineate the opportunities and boundaries within which the incentive mechanism can be exerted by the market, while the conditions for appropriability call to action the possibilities for the use of opportunities.

Finally, to understand the sectoral structure of technological change it is important to realize a number of facts:

- Very often innovation leads to new techniques, which are superior to those previously used irrespective of relative prices. For instance, in the case of electronics, they save on labor, capital, working capital, energy, and improve on quality. So, there is improvement in relation to electromechanical systems irrespective of relative prices – which means that they would have prevailed even at the old prices. This is non-reversibility, an often-seen phenomenon in technological change.
- The incentive mechanisms may involve a large number of factors: (1) technological difficulties (bottlenecks) in related activities; (2) lack or abundance in critical inputs; (3) abundance of specific inputs – energy, raw materials; (4) composition, changes and rates of growth of demand; (5) level and changes in relative prices – especially capital/labor; (6) industrial confrontations – strikes.
- There are some elements in technological change, which remind us of free goods and derive from the stream of free information.
- In addition, there are common characteristics of technology, which relate to non-exchangeable interdependencies at the level of two or more firms. These characteristics may be the result of incentives or owe their presence to chance.
- Over time, the differences in the rates of innovation among sectors and the way innovation is generated, diffused and used are due to inter-sectoral differences. Beside the opportunities for innovation that each paradigm entails (as seen above), the effort for innovation will also depend on the extent to which innovations can reap and hold on to benefits to the firms that attempt to innovate, that is, the capability to appropriate. The latter can be achieved through: (1) patents; (2) secrecy; (3) lead time; (4) time and cost required for perfect imitation; (5) gains reaped from the learning curve; (6) superiority in service and in sales effort.

In a pioneering article (Pavitt et al. 1987: 297–317) the authors have come up with a matrix that relates production characteristics to innovative behavior – a grouping that is akin to the "nature of sectors".

- *Technology-dependent sectors*: Textiles, clothes, leather, printing and publications, wood products. Innovations are mostly new methods of production: opportunities are usually embedded in new machinery and intermediate goods, which originate in firms that do not belong to this sector. What counts here is the efficiency of the firm in the use of the innovation it acquires and on organizational innovations. Firms find it difficult to keep to themselves the innovations they have acquired. Firms in this category tend to be small except in cases involving economies of scale.
- *Sectors with strong economies of scale*: In this group, innovations consist both of methods of production as well as new products. The economies of scale are important and appear at several levels of activity: production, research, design. Product complexity and lead times provide ways for the appropriability of profits within the firm. The firms in this sector tend to be large; they produce a substantial part of the technology in methods of production they use. They spend a large percentage of their resources on R&D and they tend to integrate vertically as to their technological equipment. Transportation goods, consumer electronics, metal production, the food industry, part of the chemical industry, glass, cement all belong to this group.
- *Specialized producers*: Here innovation covers mostly new products. Firms are small and work in close collaboration with users so as to embed specialized knowledge in design and construction of equipment. They comprise mechanization and production of measurement tools. Opportunities are large and derive from changes in design and use of new constituent elements. Gains and their safeguarding depend mainly on accumulated knowledge.
- *Sectors that depend on science*: They consist of electronics and most of the chemical industry. Their innovations are usually directly linked to new economic paradigms, which become accessible through scientific knowledge. Opportunities are very important. Gains are safeguarded through patents in the drug and chemical industry and through lead times and learning curves for the rest. A substantial part of their innovations penetrate several other sectors. Firms tend to be large except for the "Schumpeterian" startups and some very specialized producers. The origin of crucial incentives depends on the nature of the technologies involved and on the economic, institutional and regulatory framework in each country.

Box 2.7 Paradigms, technological opportunities and 3D printing

A short story popular with the media for the last few years may be instructive with respect to terms and principles met so far and the extent of repercussions of an innovation:

The first industrial revolution brought the mechanization of production towards the end of the eighteenth century in Britain and the second industrial revolution brought mass production in assembly. Now after three decades in the making, the stage seems set for the third industrial revolution: the digitization of manufacturing leading to mass customization and increased flexibility in production.

The main actor in this unfolding revolution is the 3D printer (or three-dimensional printer). Referred to as additive manufacturing, 3D printing is a revolutionary manufacturing technique allowing the building of solid objects through the successive application of layers of material. It consists of a contraption, a product innovation almost a decade old. It is a radical innovation. It emerged in small firms in the US, Israel, China and lately in Germany. These firms belong to the category of what we have called "specialized producers" in Pavitt's taxonomy.

The 3D printer can reproduce an item, using a range of sizes, materials and colors, both at close quarters or from a distance. This renders it ideal in producing prototypes for architectural, mechanical, industrial designs, jewelry, sculpture, etc. It is also ideal in producing spare parts and its usefulness is amplified by becoming mobile. It can then provide spare parts anywhere: from remote oil refineries in the desert to ships stranded in the middle of the ocean, but also to the cinema near you, or even to your kitchen sink where your garbage disposal is in need of a spare part. The latter, under normal conditions may be produced in Dayton, Ohio or stocked in the Haidian district of Beijing – or may be so old that it is not produced or stocked anymore. The important thing here is that the distance between the point of availability (usual production or stocking) of an item and the point of its use is rendered immaterial since it can be made available (almost) instantly.

The 3D printer will by itself produce a number of technological opportunities: at first, a number of designs leading to a dominant design for every type of use (see chapter 5). There will also be a number of opportunities regarding inks – and as their price has recently soared, even paper has been tried.

Then of course there is the coming convergence of emerging technologies, which together with the 3D printer will constitute the core of the so-called third industrial revolution. First, a number of investment and innovation opportunities will emerge, for as we have argued earlier in this chapter, a paradigm creates its own technological opportunities. We should expect a large number of opportunities at first (as we have seen that we should expect more opportunities at first and then fewer towards maturity) in the activities of the protagonists of this revolution: specifically, besides 3D printers, new materials, and specialized robots. Then successive waves of (second-order) improvements should be expected, hence more opportunities.

There are still more additional opportunities to be expected through convergence with other technologies and the rise of new activities. For instance, in the examples we used above, once the needed spare part is identified, its design could be found, downloaded and fed in digital form to the mobile 3D printer. However, the possibility of downloading the design of the spare part in digital format was taken for granted. This presupposed the existence of a CAD (computer aided design) file of the item to be produced as well as software to direct the printer on how to lay down the layers of material. Either the producing firm will have to produce such files or firms may emerge that specialize in keeping archives and making designs available through web services, or both.

Technological change does not only create new activities, it destroys some too. The stockpiling of inventories of spare parts will definitely be affected adversely as an activity. At the level of producers, however, reduced levels of needed inventories will reduce costs in general, not just for assemblers.

Additive manufacturing means that several manufacturing activities become redundant when referring to the same level of application, and the real test is mass production, which, given present capabilities, is still at some distance. The available evidence, however, suggests steps have been taken along this line as undeniable advances towards mass production have taken place in some specialized areas of manufacturing. Nevertheless, molding, cutting, boring and stretching are activities that will not necessarily disappear altogether but will be reduced in importance. This means that the relevant equipment and its producers will also lose some of their importance and so will the intellectual capital accumulated in such activities.

Among existing producers those specializing in the manufacturing aspect of production will be the most affected. Adoption of the new methods of production or flexibility in combining them with existing ones will become determining factors of survival.

Depending on how the relevant technologies will evolve will determine the degree of substitutability between the processes of production intended to serve the traditional economies of scale and the new approaches aiming at mass customization. Some of the existing facilities will be rendered obsolete; for others symbiosis will reign. Location of production facilities will have to be reexamined for some US and European firms, which have invested in low wage countries.

Of course, the extent and speed with which all these changes will take place will depend on the degree to which the innovation or rather the bundle of innovations accompanying the original innovation is allowed to diffuse. Incumbents are certain to react vigorously with claims of infringement of copyrights on design – rather than patents which last less long (see chapter 7). Piracy, which up to now has relied on cheap labor costs but the same means of production, will now have added advantages. Using an additive method of production rather than the traditional subtractive methods, it will experience tremendous economies in material costs, plus of course the advantage of lower labor costs.

Another development that seems unavoidable is the reshuffle of comparative advantage in the production of a number of goods, which will now fall within the reach of many more producers – goods, which at present require expensive equipment and accumulated production experience. Whether such changes will prove sufficient to have a significant affect at the country level will certainly become an interesting question.

2.6 Innovation: location (sectoral), size of firms and market structure

Ever since the 1960s when interest in innovation emerged, there has been an impressive amount of research on issues raised by Schumpeter's work. Issues like "where within the economic space does an innovation emerge", or "what is the right size for innovative firms", or "what is the best market structure for innovation to occur" were examined at length. Having read "Innovation today" (section 1.3), the reader must have realized that such questions are now less relevant, as the definition of innovation has broadened to include other forms of research besides R&D, and we now refer to technological and non-technological innovation; innovation is generated practically anywhere, including in small firms; appears in any type of activity including services, and irrespective of the level of technological intensity exhibited by the activity.

This body of studies is not wrong; it has simply lost some of its shine. It provides answers to questions that are no longer in fashion or better, provides answers only to part of the problems posed now. It refers only to a subset of innovation. If our interest centers on issues concerning specifically technological innovation, then we shouldn't look elsewhere. If on the other hand we are looking for answers concerning the entire set of innovation, then we may need to know whether the properties of the subset extend to the whole set. So, it would be unwise to discard the whole lot for they may include some truths which may still hold or from which we may still benefit.

2.6.1 Innovation and sectoral location

One of Schumpeter's well-known arguments was that innovation could be met in only part of the economic space (Schumpeter 1939). Schumpeter's argument was empirically verified in many studies for a variety of countries and economic development stages (DeBresson 1997), such as Great Britain (DeBresson and Townsend 1978: 48–60), Canada (DeBresson and Murray 1984), France (Kaminski et al. 1997), Italy (DeBresson et al. 1994, 1997), Greece (Vernardakis 1993, 1997) and China (Xu et al. 1997; DeBresson et al. 1997). Of course, the verification provided by these studies rested on the definition of innovation as resulting only from R&D. Now we think differently.

Many of these studies at the sector level bring up some new observations on innovativeness. Some focus on the important role that specific sectors play in terms of innovation such as the sector producing capital goods, which affects productivity twice – once at the level of its own sector and a second time through the use of its products in other sectors (Rosenberg 1976). The study on the UK expands on the notion of donor and recipient of technology, stressing the vulnerability of strong innovative sectors that depend on other sectors for the dissemination of their innovations (DeBresson and Townsend 1978: 48–60).

The findings in this study on the UK, plus another one also on the UK (Geroski 1991: 1438–1451), in addition to DeBresson et al. (1997) and the studies mentioned above, lead to a couple of observations that deserve re-examination lest they might still hold under the new definitions.

First, there exists an ordinal ranking on innovativeness among sectors that is practically very similar across countries – meaning that innovativeness derives more from the nature of sectors than from a country's level of development. For instance, software and machinery are found to be at the top of the ranking in every country.

Second, the more backward and forward linkages in an economy, which means the more developed and integrated its production system, the higher is its innovative activity. Linear at first, the relation turns exponential for high link densities.

2.6.2 Innovation, market structure and size of firms

In his early work, Schumpeter spoke of the virtues of competition among small enterprises run by active entrepreneurs (Schumpeter 1911). Three decades later he had gravitated to a position with which he is usually associated and which states that large firms operating in a concentrated market are a necessary support to innovation (Schumpeter 1942). No doubt, this reversal must have been to a large extent due to Schumpeter's witnessing of the drastic change that had occurred in the US with respect to the principal sources of innovation during those three decades.

Schumpeter "wanted to understand the impact of capitalist competition on economic growth and how competition might critically affect growth through its impact on innovation, which he viewed as central to the growth process" (Cohen 2010: 132).

> He was primarily impressed by the qualitative differences between the innovative activities of small, entrepreneurial enterprises and those of large, modern corporations with formal R&D laboratories. Nonetheless, the empirical literature has interpreted Schumpeter's claim for a large firm advantage in innovation as a proposition that innovative activity increases more than proportionately with firm size.
>
> (Cohen 2010: 132)

A number of hypotheses have been advanced to support Schumpeter's positions. (Symeonidis 1996: 3). Innovation increases more than proportionately with firm size because: (1) R&D projects involve large fixed costs which can be covered only if sales are sufficiently large; (2) there are scale and scope economies in the production of innovations; (3) unforeseen innovations can be better exploited by large diversified firms; (4) many projects can be undertaken simultaneously by large firms which spread the risks of R&D; (5) large firms have better access to external finance than smaller firms. In addition, innovation is higher in concentrated industries because: (6) firms with greater market power are better able to finance R&D from own profits; (7) firms with greater market power have better incentives to innovate as they can more easily appropriate the returns from innovation.

Galbraith's view was somewhat similar to Schumpeter's only with a narrower focus. Due to the resources at their command he felt that large firms were uniquely endowed to take advantage of innovative opportunities (Galbraith 1956), and their advantage over smaller firms would rise over time due to the constant rise of R&D costs.

This section examines the relationships between innovation, market structure and firm size; in the end it will be found that all three are determined endogenously.

2.6.2.1 Innovation and size of firms

In search of a "Schumpeterian" advantage to firm size, research has focused on the relationship between R&D and firm size.

In every country, a few large firms carry out a large proportion of industrial research while the majority of medium and small firms perform no research at all. In early empirical studies the overall relationship linking research activity and size of firms was found to be weak, non-existent or even negative depending on whether the large number of firms that perform no research are included or not (Freeman 1982).

All such studies face limitations of some sort. First, they are subject to sample non-randomness as they usually pay attention exclusively to R&D-performing firms and to the top level of firms, i.e. the largest 100 or largest 500. In addition, most studies fail to control for firm characteristics (besides size) despite their likely colinearity and their importance in explaining R&D intensity. Firm characteristics such as cash flow, degree of diversification, complementary capabilities and the ability to spread R&D costs over output are obviously correlated with firm size. Failing to control for their presence leads to biased estimates. Equally difficult to control for are industry effects as large firms are aggregations of business units engaged in different industries.

Variations in innovation intensity are more pronounced among rather than within industries, and thus depend more on the sector than on firm size. Innovation may demand large firm size in some sectors – chemicals, car industry – but small firms may still be innovative in sectors such as tools and machinery and increasingly so in activities related to informatics. There are sectors with high research activity and others with low research activity. There are sectors where even the largest of the firms perform little or no research and sectors where even small firms display considerable research activity.

Cohen et al. (1987) found that it is the size of the business unit rather than the size of the firm as a whole which affects the likelihood of performing R&D. Cohen and Klepper (1996b) found that business unit size alone (or its correlates) explained two thirds of the variance while firm size (or its correlates) explained only 15 percent.

Gradually, and especially since the late 1970s, there has been a rise in the importance conferred on small firms. Since then, the question as to whether small firms differ from large firms in terms of innovation has gained considerable importance. All indications confirm that this differentiation exists and is particularly pronounced. It is expressed through the effort expended towards R&D, its efficiency and the frequency of innovation at the sectoral level and during the technology's life cycle. It is also expressed in terms of advantages and disadvantages vis-à-vis innovation, its sources, its type and importance and in relation to its recipients. In other words, the generation, use, absorption, diffusion and transfer of innovation differ significantly between small firms and the rest.

Until recently, it was widely accepted that large, established firms were the main generators of innovation, economic growth, exports and employment. Studies on the resources spent on R&D confirmed that large firms were the ones to spend more on this aspect. The overall picture began to change in favor of small firms when the input to innovation, that is the effort or expenditure towards R&D, started being challenged as an index of the volume of innovation, and instead output measures such as patent counts were used. The overall picture that emerged during the 1980s was that the role of large firms in the generation of innovation continued to be significant but less so than used to be the case, and correspondingly that the role of small firms was more significant than it had been considered to be until then (Acs and Audretsch 1990). Ever since, the image of the role of small firms in innovation has been significantly enhanced in view of their newly recognized importance as well as their rising numbers. Technological achievements in informatics (especially the Internet) have led to the spawning of myriads of small firms. On the other hand, we have witnessed the tendency of large firms to reduce their size either due to outsourcing, or to diminished scale economies at the level of producing units due to technological change, as in the case of steel and aluminum production.

Yet the odds do not seem to favor small firms in the role of important innovators. After all, they devote less effort to research than larger firms because they face more stringent conditions. The difficulty and cost of accessing capital for research, economies of scale in research, market power, capability in accepting risk, economies of scope in the production and distribution of new products all seem to combine against them (Rothwell and Dodgson 1994).

Would the above justify espousing Schumpeter's and Galbraith's position that large firms are uniquely endowed to take advantage of innovative opportunities?

There are three reasons which need to be attended to in relation to the innovative dynamism of small firms vis-à-vis larger ones. First, the research effort made by small firms is usually underestimated. One reason for that is that research in small firms is dispersed and missed in the census. Another reason is because we usually refer to formal research, that is the effort spent by the research department of the firm, while in most small firms there is no distinct research department, and research efforts taking place informally throughout the firm are usually not registered as such. Efforts to avoid this problem in the Netherlands via a more systematic census revealed that small firms' share of the national total effort in R&D was much larger than suggested by the official census: 43 percent instead of 16 percent (Kleinknecht and Reinen 1992).

The second reason relevant to the dynamism of small firms is that using R&D effort (meaning funds expended), i.e. an input measure, as a substitute for the size of R&D itself amounts to favoring large firms. If instead an output measure were to be used – such as innovation counts, patents, new products and market shares – the outcome would be different. For there is a voluminous body of empirical results supporting the fact that, based on patents, smaller firms present a much more dynamic image than is suggested if only input measures are used.

The use of patents as a measure of innovation carries its own problems. Sectors differ in terms of their propensity to acquire patents – the defense sector has no use for patents while pharmaceuticals have a tendency to patent even very minor results. There is an ongoing debate as to whether large firms differ from small ones in terms of their propensity to acquire patents. In addition, there is the issue that innovations differ vastly between themselves in terms of importance, a trait reflected in their patents.

The third reason to be considered is that in a comparison that tends to favor large over smaller firms the innovative activities of small firms are qualitatively different from those of larger ones. Compared to large firms, smaller firms' innovations concentrate more on new product innovation, and are more focused on the market, the consumer and the user (Auzeby and Francois 1992).

Scholars have also examined the relationship between firm size and the type of innovation pursued. It was found that large, incumbent firms tend to pursue relatively more incremental (Mansfield 1981) and relatively more process innovation than smaller firms (Cohen and Klepper 1996a; Link 1982). More revealing, Cohen and Klepper (1996a, 1996b) confirm the prediction that R&D cost spreading, which is the link between firm size and R&D, will be weakest for those industries in which innovations are most saleable in disembodied form (i.e. pharmaceuticals) or where prospects for rapid growth due to innovation are greatest (i.e. high-tech firms).

Additionally, small firms' sources of innovation tend to be their customers and suppliers and information gathered from exhibitions, in other words sources that have little in common with R&D, and in contrast with the large firms, which are more focused on official research. As for the smallest among firms, they seem to depend mainly on customers and unofficial technical and entrepreneurial knowledge and not on official R&D (Kaminski 1996). Applications of scientific discoveries seem to belong to large firms as a result of their privileged relationship with official research. Nevertheless, small firms often spend relatively more than large firms in innovative investments in production so as to hasten their marginal product innovations to market (Cesarato and Sirilli 1992).

A salient difference between small and large firms has to do with efficiency in innovation. Several studies (Gellman Research Associates 1976, 1982; Pavitt et al. 1987; Acs and Audretsch 1988, 1990, 1991) confirm our reported earlier observation in the literature that smaller firms tend to account for a disproportionately large share of innovative output relative to their size, and so does Lerner (2006), but also add that R&D productivity tends to decline with firm size. A similar result was found with respect to patents as the number of patents per R&D dollar for smaller firms is considerably higher than for larger firms

(Bound et al. 1984), with the pattern varying across industries (Acs and Audretsch 1990, 1991). Geroski (1994) dwells on the clear negative correlation between firm size and R&D productivity. Thus, the predominant pattern is that R&D productivity appears to decline with size.

Efficiency in R&D is still another criterion, only this time the picture changes dramatically in favor of small firms. Wyatt reports that large firms with over 10,000 employees accounted for 43.4 percent of innovations and devoted 80 percent of total funds spent on R&D; that is a relative efficiency of 0.54. In contrast, small firms with 100–499 employees produced 20.6 percent of innovations and used 2 percent of total funds spent on R&D, which equates to a relative efficiency of 10.3 (Wyatt 1984). These findings for the UK have been confirmed by similar results for the US (National Science Foundation 1987).

In all the studies mentioned here it was assumed that firm size is exogenous, but in reality there is a two-way causation: firm size affects innovation but innovation also affects firm growth and hence firm size. To put it differently, innovation affects firm growth and hence firm size, so that firm size in a particular year will be influenced by innovative activity in the previous year.

In short, large firms still account for the majority of innovations but the share of small firms is steadily rising and accounts for a significantly larger share than used to be thought. Additionally, small firms have proved to be much more efficient in their innovative activity than larger ones. Qualitatively, their innovations differ from those arrived at by large firms, they concern marginal innovations, are closer to the market, and have originated from sources which have little to do with formal R&D. Finally, at this point, discussions on the link between firm size and innovation are rather outmoded, partly due to the fact that the boundaries of the firm have become fuzzy in recent decades as the need for collaboration and "strategic alliances – constellations of bilateral agreements among firms – are increasingly necessary to support innovative activities" (Teece 1992: 447).

2.6.2.2 Innovation and market structure

It has long been recognized that the relationship between market structure and innovation is not a one-way causal relationship running from market structure to innovation, since market structure is itself affected by innovative activity. What we have is a two-way causal relationship as both variables are endogenously determined within a complex system of interactions with numerous other variables.

In the early studies Scherer (1965) found that the effect of concentration on patents was not statistically significant. In a subsequent study, using R&D employment as the dependent variable, Scherer (1967) thought his results provided some support for the Schumpeterian hypothesis, although the positive coefficient on concentration was significant in some cases and insignificant in others. In a number of other studies, some found evidence of a positive relation between concentration and research intensity while in others there was no such evidence.

Some studies have explored the possibility that causality may run in the opposite direction and have attempted to investigate how innovation affects market structure. Once more, the results of this type of research were inconclusive. Levin and Reiss (1984, 1988) and Farber (1981) found a positive effect of R&D intensity on concentration, Mukhopadhyay (1985) found a negative effect of R&D intensity on concentration change, and Geroski and Pomroy (1990) found a negative effect of innovation counts on concentration change.

Both Scott (1984) and Levin et al. (1985) found an inverted U relationship between R&D intensity and concentration but when other variables meant to capture inter-industry differences were introduced, either the coefficients became statistically insignificant or the relationship broke down. Aghion et al. (2005) also found a similar relationship but their analysis did not control for the cost-spreading incentive effects on R&D of business unit (as opposed to firm) size. This amounts to a potentially very significant omission (in the sense that it might steal the show) in the light of the very tight relationship noted earlier between business unit size and R&D. Besides, the assumptions imposed (only step-by-step innovations and only one type of rivalry) are so stringent as to render this approach very restrictive in its application. The only study which explicitly recognized that market power cannot be fully captured by one variable, that is concentration, is the study by Geroski (1990), in which he used six different measures of market power. He concluded that the overall effect of market power on innovation was negative.

In fact, the salient characteristic of the empirical literature on the link between innovation and market structure is its inconclusiveness. Nevertheless there are three items worth keeping in mind. First, industry characteristics such as technological opportunity account for much more of the variance in R&D intensity or innovation than market structure. Second, in general there is scant evidence of a positive relationship between R&D intensity and concentration, although there may be circumstances where such a relationship exists. Third, there is hardly any evidence of a positive relationship between innovative output and market structure.

So it is only fair that Sidak and Teece (2010: 588) sum up the situation by saying that "Despite 50 years of research economists do not appear to have found much evidence that market concentration has a statistically significant impact on innovation." Cohen (2010: 194) notes "The relationship between market structure and R&D remains, however, problematic. No clear theoretical rationale for the relationship has emerged. And empirical results remain weak, with measures – albeit poor ones – explaining little of the variance in R&D spending." Equally damaging is the quotation from Scherer and Ross (1990: 642): "through astute choice of assumptions, virtually any market structure can be shown to have superior innovative qualities".

Besides, the aim of the research performed is open to criticism. Sidak and Teece observe that when testing for market power what matters is rivalry. However, market concentration, on which research has focused, does not necessarily determine rivalry. In addition, they note that what should be investigated is dynamic, not static, competition.

In fairness to Schumpeter, we take note of the fact that the research conducted has actually tested a proposition that is different from the Schumpeterian hypothesis, which stipulates that innovation is higher in the presence of market power. Instead, research conducted has investigated whether innovation is higher in concentrated markets (implying that the profit margin is higher in concentrated markets). However, concentration is at best an imperfect proxy for market power in cross-industry studies.

Contemporary wisdom holds that the existence of a variety of mechanisms relating innovative activity and market structure accounts for a good part of the reason why cross-industry econometric studies have been inconclusive. In addition, the results do not easily lend themselves to interpretation, as innovation may impact market structure directly or indirectly and in various ways.

An interesting point to note is that evidence does not support a strong conclusion that competition is uniformly a stimulus to innovation. While there are conditions that warrant a presumption that competition promotes innovation, there are also circumstances for which it is reasonable to assume that competition does not affect, or possibly reduces, innovation incentives (Gilbert 2006: 24). This means that there is no rule of thumb and each case should be investigated on its own merit. A direct implication is that great care should be taken in the application of Innovation Policy as there are clearly limits to its use through the regulation of competition. It goes without saying that equal care should be exerted in both the drawing and implementation of Antitrust Policy.

To make a very long story short, the most striking finding was that the variations in research intensity between industrial sectors could be larger than the variations between firms of the same sector. This casts a shadow of doubt as to the importance of both the size of firms and of market structure as explanatory factors of research intensity.

The variations between industrial sectors, which can be greater than the variations between firms within the same sector, lead to the suggestion that sectors differ among themselves in terms of the level of technological opportunities they possess that determine research intensity and the rate of technological change. In fact, technological opportunities could be thought of as an additional reason, if not the main reason, besides size and structure for the different trends in innovation.

Recent literature has treated market structure and innovation activity as endogenously determined within a complex system in which technological opportunity, strategic interaction, the characteristics of demand and even chance play an important role (Symeonidis 1996). In view of the inconclusiveness so far, contemporary research has moved on beyond market structure and innovation, and attempts to shed light on specific parts of the complex mechanisms piecemeal. One strand of research investigates the determinants of the complex system underlying the relationship and investigates technological opportunity or the characteristics of demand, such as the degree of differentiation. Another category involves the investigation of stochastic patterns of firm innovation with implications for the evolution of market structure over the industry life cycle. Yet another strand of research investigates the sources of self-reinforcing feedback mechanisms such as network externalities, learning by doing, learning by using, R&D cost spreading, etc. and their implications on innovation.

Bibliography

Acs, Z.J. and Audretsch, D.B. (1988) "Innovation in Large and Small Firms", *American Economic Review*, 78, 678–690.

Acs, Z.J. and Audretsch, D.B. (1990) *Innovation and Small Firms*, Cambridge, MA: MIT Press.

Acs, Z.J. and Audretsch, D.B. (1991) "R&D, Firm Size and Innovative Activity", in Z.J. Acs and D.B. Audretsch (eds.), *Innovation and Technological Change: An International Comparison*, New York: Harvester Wheatsheaf.

Aghion, P., Bloom, N., Blundell, R., Griffith, R. and Howitt, P. (2005) "Competition and Innovation : An Inverted-U Relationship", *Quarterly Journal of Economics*, 120, 701–728.

Arthur, W.B. (2009) *The Nature of Technology, What it is and How it Evolves*, New York: Free Press.

Auzeby, F. and Francois, J.P. (1992), "Technological Innovation in the French Industry", *STI Review*, no. 11, Paris: OECD.

Basalla, G. (1999) *Evolution of Technology*, Cambridge: Cambridge University Press.

Bell, M. and Pavitt, K. (1993) "Technological Accumulation and Industrial Growth: Contrasts between Developed and Developing Countries", *Industrial and Corporate Change*, 2(2), 157–210.

Bohn, R.E. (1994) "Measuring and Managing Technological Knowledge", *Sloan Management Review*, Fall, 61–73.

Bound, J., Cummins, C., Griliches, Z., Hall, B.H. and Jaffe, A. (1984) "Who does R&D and who patents?", in Z. Griliches (ed.), *R&D Patents and Productivity*, Chicago: University of Chicago Press for the NBER.

Braun, E. and MacDonald, S. (1978) *Revolution in Miniature*, Cambridge: Cambridge University Press.

Brooks, H. (1968) *Science*, April 26, 399.

Brown, S. (1997) *Revolution at the Checkout Counter: The Explosion of the Bar Code*, Cambridge, MA: Wertheim Publications.

Cardwell, D. (1994) *History of Technology*, London: Harper Collins.

Cardwell, D.S.L. (1971) *From Watt to Clausius*, Ithaca, NY: Cornell University Press.

Carter, A.P. (1970) *Structural Change in the American Economy*, Cambridge: Cambridge University Press.

Cesarato, S. and Sirilli, G., (1992) "Some Results of the Italian Survey on Technological Innovation", *STI Review*, no. 11, OECD, Paris.

Clarke, A.C. (1962) *Profiles of the Future*, London: Pan Books.

Coehn, I.B. (1948) *Science, Servant of Man*, Boston: Little, Brown and Co.

Cohen, W.M. (2010) "Fifty Years of Empirical Studies of Innovative Activity and Performance", in B.H. Hall and N. Rosenberg (eds.), *Handbook of the Economics of Innovation*, Amsterdam: North-Holland.

Cohen, W.M. and Klepper, S. (1996a) "Firm Size and the Nature of Innovation within Industries: The Case of Process and Product R&D", *Review of Economics and Statistics*, 78, 232–243.

Cohen, W.M. and Klepper, S. (1996b) "A Reprise of Size and R&D", *Economic Journal*, 106, 925–951.

Cohen, W.M., Levin, R.C. and Mowery, D.C. (1987) "Firm Size and R&D Intensity: A re-examination", *Journal of Industrial Economics*, 35, 543–563.

DeBresson, C. (1997), "Why Innovative Activities Cluster", in C. DeBresson (ed.), *Economic Interdependence and Economic Activity: An Input-Output Analysis*, Cheltenham: Edward Elgar, 149–164.

DeBresson, C. and Murray, B. (1984) *Innovation in Canada: A Report to the Science Council of Canada*, New Westminster: CRUST.

DeBresson, C. and Townsend, J. (1978) "Note on the Interindustrial Flow of Technology in Postwar Britain", *Research Policy*, 7(1), 48–60.

DeBresson, C., Sirilli, G., Hu, X. and Luk, F.K. (1994) "Structure and Location of Innovative Activity in the Italian Economy, 1981–85", *Economic Systems Research*, 6, 135–158.

DeBresson, C., Sirilli, G., Kaminski, P., Xu, S. and Vernardakis, N. (1997) "Some Surveys of Innovative Activity", in C. DeBresson (ed.), *Economic Interdependence and Economic Activity: An Input-Output Analysis*, Cheltenham: Edward Elgar, 35–46.

Dinn, N.F. (1977) "Preparing for Future Satellite Systems", *Bell Laboratories Record*, October, 236–242.

Dosi, G. (1988a) "The Nature of the Innovative Process", in G. Dosi, C. Freeman, R. Nelson, G. Silverberg and L. Soete (eds.), *Technical Change and Economic Theory*, London: Pinter, 221–238.

Dosi, G. (1988b) "Sources, Procedures and Microeconomic Effects of Innovation", *Journal of Economic Literature*, 26, 1120–1171.

Farber, S. (1981) "Buyer Market Structure and R&D Effort: A Simultaneous Equations Model", *Review of Economics and Statistics*, 62, 336–345.

Fishlow, A. (1965) *American Railroads and the Transformation of the Ante-Bellum Economy*, Cambridge, MA: Harvard University Press.

Freeman, C. (1982) *The Economics of Industrial Innovation*, 2nd edition, Cambridge, MA: MIT Press.

Galbraith, J.K. (1956) *American Capitalism: The Concept of Countervailing Power*, Boston: Houghton Mifflin.

Gellman Research Associates (1976) *Indicators of International Trends in Technological Innovation. Final report to the NSF.* NTIS Document PB–263–738. Jenkintown, PA.

Gellman Research Associates (1982) *The Relationship between Industrial Concentration, Firm Size and Technological Innovation. Final report.* Glastonbury, CT: US Small Business Administration, Futures Group.

Geroski, P.A. (1990) "Innovation, Technological Opportunity, and Market Structure", *Oxford Economic Papers*, 42, 586–602.

Geroski, P.A. (1991) "Innovation and the Sectoral Sources of United Kingdom Productivity Growth", *Economic Journal*, 101, 1438–1451.

Geroski, P.A. (1994) *Market Structure, Corporate Performance and Innovative Activity*, Oxford: Clarendon Press.

Geroski, P.A. and Pomroy, R. (1990) "Innovation and the Evolution of Market Structure", *Journal of Industrial Economics*, 38, 299–314.

Gilbert, R.J. (2007) *Competition and Innovation*, Berkeley: Competition Policy Center, Institute of Business and Economic Research, 24.

Gladwell, M. (2001) "The Mosquito Killer", *New Yorker*, July, 42–51.

Gort, M. and Wall R. (1986) "The Evolution of Technologies and Investment in Innovation", *Economic Journal*, 96(393), 741–757.

Hays, S.P. (1996) "Patterns of Reinvention: The Nature of Evolution During Policy Diffusion", *Policy Studies Journal*, 24(4), 551–566.

Hen, C.Y. (2004) "Wal-Mart Drives a New Tech Boom", *Fortune*, June 28, 90.

Hicks J.R. (1932) *The Theory of Wages*, London: Macmillan.

Hightower, J. (1972) *Hard Tomatoes, Hard Times: The Failure of America's Land Grant Complex*, Cambridge, MA: Schenkman.

Hunsicker, H.Y. and Stumpf, H.C. (1965) "History of Precipitation Hardening", in Smith, C.S. (ed), *The Sorby Centennial Symposium on the History of Metallurgy*, New York: Gordon & Breach Science Publishers, 279.

Jewkes, J., Sawers, D. and Stillerman, R. (1968) *The Sources of Invention*, revised edition, London: Macmillan.

Kaminski, P. (1996) "How Can Very Small Technology Firms Be Helped?", in *Innovation, Patents and Technological Strategies*, Paris: OECD.

Kaminski, P., DeBresson, C. and Hu, X. (1997) "The Clustering of Innovative Activity in the French Economy: An Estimation", in C. DeBresson (ed.), *Economic Interdependence and Innovative Activity, an Input-Output Analysis*, Cheltenham: Edward Elgar, 179–189.

Keller, A.G. (1964) *A Theatre of Machines*, London: Chapman and Hall.

Klein, S.J. and Rosenberg, N. (1986) "An Overview of Innovation", in R. Landau and N. Rosenberg (eds.), *The Positive Sum Strategy: Harnessing Technology for Economic Growth*, Washington, DC: National Academy Press.

Kleinknecht A. and Reinen J.O.N. (1992) "The Experience with the New Innovation Data in the Netherlands", *STI Review*, no. 11, Paris: OECD.

Klevorick, A., Levin, R.C., Nelson, R.R. and Winter, C. (1995) "On the Sources and Significance of Interindustry Differences in Technological Opportunities", *Research Policy*, 24, 185–205.

Lerner, J. (2006) "The New Financial Thing: The Origins of Financial Innovations", *Journal of Financial Economics*, 79, 223–255.

Levin, R.C. and Reiss, P.C. (1988) "Cost-reducing and Demand-creating R&D with Spillovers", *RAND Journal of Economics*, 19, 538–556.

Levin, R.C. and Reiss, P.C. (1984) "Tests of a Schumpeterian Model of R&D and Market Structure", in Z. Griliches (ed.), *R&D, Patents and Productivity*, Chicago: University of Chicago Press for the NBER, 175–208.

Levin, R.C., Cohen, W.M. and Mowery, D.C. (1985) "R&D Appropriability, Opportunity, and Market Structure: New Evidence on some Schumpeterian Hypotheses", *American Economic Review*, 75, 20–25.

Link, A.N. (1982) "A Disaggregated Analysis of Industrial R&D: Product versus Process Innovation", in D. Sahal (ed.), *The Transfer and Utilization of Technical Knowledge*, Lexington, MA: Lexington Books.`

Lundvall, B. and Johnson, B. (1994), "The Learning Economy", *Journal of Industry Studies*, 1, 2.

Mann, C.C. (2004) "Humanoids for the Home", *Technology Review*, May, 25.

Mansfield, E. (1981) "Composition of R&D Expenditures: Relationship to Size, Concentration and Innovation Output", *Review of Economics and Statistics*, 62, 610–614.

MIT Technology Review (2002) "Drive, She Said: A Silent Movie Ingenue – "the first movie star" – Was an Automotive Innovator", July 1.

Morton, J.A. (1971) *Organizing for Innovation*, New York: McGraw-Hill, 25.

Mukhopadhyay, A.K. (1985) "Technological Progress and Change in Market Concentration in the US: 1963–77", *Southern Economic Journal*, 52, 141–149.

National Science Foundation (1987) *Science and Technology Indicators*, Washington, DC: National Science Foundation.

Nelson, R. (2002) "On the Uneven Evolution of Human Know-how", *Research Policy*, 1442, 1–14.

Nelson, R. and Wolff, E. (1997), "Factors Behind Cross Industry Differences in Technical Progress", *Structural Change and Economic Dynamics*, 18, 205–220.

North, D. (1986) "Sources of Productivity Change in Ocean Shipping, 1600–1850", *Journal of Political Economy*, 76, 953.

Noyce, R.N. and Hoff, M.E. (1981) "A History of Microprocessor Development at Intel", *IEEE Micro*, 7, 8–21.

Ord-Hume, A.W.J.G. (1977) *Perpetual Motion: The History of Obsession*, New York: Barnes and Noble.

Pavitt, K., Robson, M. and Townsend, K. (1987) "The Size Distribution of Innovative Firms in the UK: 1945–1983", *Journal of Industrial Economics*, 35(3), 297–317.

Polanyi, M. (1966) "The Logic of Tacit Inference", *Philosophy*, 41, 1–18.

Ramelli, A. (1976) "Le Diverse et Artificiose Machine" (first published in 1588), *The Various and Ingenious Machines of Agostino Ramelli*, ed. E.S. Ferguson, Baltimore: Johns Hopkins University Press.

Reti, L. (1974) *The Unknown Leonardo*, New York: McGraw-Hill.

Rosenberg, N. (1976) *Perspectives on Technology*, Cambridge: Cambridge University Press.

Rosenberg, N. (1982) "How Exogenous is Science?", in *Inside the Black Box: Technology and Economics*, Cambridge: Cambridge University Press, 141–159.

Rothwell, R. and Dodgson, M. (1994) "Innovation and Size of Firm", in M. Dodgson and R. Rothwell (eds.), *The Handbook of Industrial Innovation*, Brookfield: Edward Elgar, 310–324.

Scherer, F.M. (1965) "Size of Firm, Oligopoly and Research: A Comment", *Canadian Journal of Economics and Political Science*, 31, 256–266.

Scherer, F.M. (1967) "Market Structure and the Employment of Scientists and Engineers", *American Economic Review*, 57, 524–531.

Scherer, F.M. and Ross, D. (1990) *Industrial Market Structure and Economic Performance*, Boston, MA: Houghton Mifflin.

Schmookler, J. (1966) *Invention and Economic Growth*, Cambridge, MA: Harvard University Press.

Schumpeter, J.A. (1911) *The Theory of Economic Development: An Inquiry into Profits, Capital, Credit, Interest, and the Business Cycle*. Piscataway, NJ: Transaction.

Schumpeter, J.A. (1939) *Business Cycles: A Theoretical, Historical and Statistical Analysis of the Capitalist Process*, vol. 2, New York: McGraw-Hill.

Schumpeter, J.A. (1942) *Capitalism, Socialism and Democracy*, New York: Harper & Row.

Schwartz, E. (2004) "Sparking the Fire of Invention", *Technology Review*, May, 32–40.

Scott, J.T. (1984) "Firm versus Industry Variability in R&D Intensity", in Z. Griliches (ed.), *R&D, Patents and Productivity*, Chicago: University of Chicago Press for the NBER, 233–248.

Shannon, C. (1948) "A Mathematical Theory of Communication", *Bell System Technical Journal*, 27, 379–423, 623–656.

Sidak, J.G. and Teece, D.J. (2010) "Dynamic Competition in Antitrust Law", *Journal of Competition Law and Economics*, 5, 581–631.

Symeonidis, G. (1996) *Innovation, Firm Size and Market Structure: Schumpeterian Hypothesis and Some New Themes*, Paris: OECD,.

Teece, D.J. (1992) "Competition, Cooperation and Innovation: Organizational Arrangements for Regimes of Rapid Technological Progress", *Journal of Economic Behavior and Organization*, 18, 1–25.

Varchaver, N. (2004) "Scanning the Globe", *Fortune*, May 31, 68–76.

Vernardakis, N. (1993) "Input-output Tables and the Location of Innovation among Small Firms in a Semi-industrialized Economy", *Proceedings of the 10th International Conference on Input-Output Techniques*, Seville, Spain.

Vernardakis, N. (1997) "Locating Innovative Activities in a Semi-industrialized Economy: Greece", in C. DeBresson (ed.), *Economic Interdependence and Economic Activity: An Input-Output Analysis*, Cheltenham: Edward Elgar, 216–224.

Von Hippel, E. (1988) *The Sources of Innovation*, Oxford: Oxford University Press.

Walsh, V. (1984) "Invention and Innovation in the Chemical Industry: Demand-pull or Discovery-push?", *Research Policy*, 13, 211–234.

Walsh, V., Townsend, J., Achiladelis, B. and Freeman, C. (1979) *Trends in Innovation and Innovation in the Chemical Industry*, Brighton: University of Sussex.

Whyte, R.R. (ed.) (1975) *Engineering Progress through Trouble*, London: Institution of Mechanical Engineers.

Wyatt, S. (1984) "The Role of Small Firms in Innovative Activity" (Mimeo), Science Policy Research Unit.

Xu, S., DeBresson, C. and Hu, X. (1997) "The Location of Innovative Activity in China (1992)", in C. DeBresson, *Economic Interdependence and Innovative Activity, an Input-Output Analysis*, Cheltenham: Edward Elgar, 179–189.

3 Diffusion

Diffusion is the process that allows innovation to spread its impact to users and activities throughout the economic system, thus contributing to economic growth. In terms of impact, it is diffusion rather than innovation that determines the rate of economic growth and the rate of the rise in productivity.

Yet innovation and diffusion are equally important, each drawing its importance from the presence of the other. In the absence of diffusion, innovation would have little economic or social impact. On the other hand, without innovation there would be nothing to diffuse. Finally, in comparing diffusion and innovation we may conclude that diffusion is the easiest of the two to study since it is more predictable from observable factors.

We first discuss intra-sectoral (within the same sector) diffusion, starting with the five stages of adoption, a process wrongly assumed to be instantaneous by economists; then we discuss a few stylized facts and the determinants of the rate of adoption. Next we talk about the diffusion curve and give a critique of existing diffusion models – epidemic and probit. Then there is some discussion of direct and indirect network effects, feedback mechanisms, path dependence, "success brings success" and standards.

Subsequently we discuss inter-sectoral (across sectors) diffusion, a subject usually absent from textbooks. DeBresson's theory based on circulation and dominant technological requirements covers some distance but has its flaws.

We then talk about technology transfer, which is a special form of technology diffusion. There follows consideration of conditions for successful transfer and the role of the nature of sectors. Technological espionage, an unwarranted type of transfer of technology, closes the chapter, together with a few words on the modes of technology transfer and diffusion of embodied technology.

3.1 Diffusion: within sectors

3.1.1 Introduction

As latecomers to research on diffusion, economists and business researchers have adopted established terminology, definitions, methodology and an analytical framework.

The most recent research tradition on diffusion is the marketing one, particularly influenced by Bass, whose model was intended to predict the adoption of new products (Bass 1969: 215–227). This tradition emphasizes how perceived characteristics of an innovation, in collaboration with advertising and distribution tools, influence the rate of adoption of a product.

Forty or so years ago, it had already been observed "that diffusion is a slow process" – in some cases very slow – "and that there are wide variations in the rates of acceptance of different innovations" (Rosenberg 1976: 191).

> For many, the question of why things diffuse slowly has become very focused on a single stylized fact about that slowness, namely that the time path of usage usually follows an S-curve: diffusion rates first rise and then fall over time, leading to a period of relatively rapid adoption sandwiched between an early period of slow take-up and a late period of slow approach to satiation.
>
> (Geroski 2000: 604)

Four alternative explanations have been offered for this fact. By far the most commonly encountered model is the *epidemic model*, which is based on information, or the lack of it, and more specifically on the premise that what limits diffusion is the lack of information about the innovation, what it can accomplish in terms of use, and how one can use it. A leading alternative model, the *probit model*, is also based on information. Its main premise is that firms differ as to goals and abilities and need to adopt new technology at different times, and that diffusion occurs as different types of firms adopt it. The third type of models that generate the S-curve, and favored by population ecologists, consists of models of *density dependence*, whereby legitimation and competition assist the establishment of new technologies and ultimately limit their adoption. The fourth type of models is based on phenomena of *information cascades* assisted by network externalities in order to explain the herd-like adoption behavior when a particular variant of a technology is finally selected.

Our plan is to discuss the diffusion process rather than analyze the peculiarities of modeling diffusion, so we shall present at some length the popular epidemic model, then comment briefly on the other models. First, however, we shall examine how each entity or firm decides on whether to adopt an innovation.

3.1.2 The stages of adoption

The adoption process of an innovation "is essentially an information-seeking activity in which an individual is motivated to reduce uncertainty about the advantages and disadvantages of the innovation" (Rogers 2003: 14). This process consists of a series of choices and actions over time during which a person or a decision-making group evaluates a new idea and decides whether to integrate it in its current practices. This means adopting a new idea, an alternative to the one in use or a behavior that is intended to meet the inherent uncertainty of the adoption decision. If the innovation has actually been adopted then the new user is added to the stock of already existing users and the level of diffusion of the innovation is augmented by this last addition.

Adoption is a complex stepwise process consisting of five consecutive stages: (1) information gathering and realization of the existence of the innovation; (2) opinion formation as to the innovation; (3) decision making for or against adoption; (4) implementation of the decision in case it is favorable to the adoption; (5) evaluation of the consequences of adopting the innovation or, to put it differently, evaluation of the wisdom of the decision to adopt.

Not all adoption stages are equally time-consuming and there are overlaps between them. The second stage, where an opinion is formed on the innovation, is the one most liable to overlaps both from the first stage of information gathering and from the third stage of decision taking.

At the start, that is, before the entire diffusion process has started, the person or decision-making group knows of current practices used, of difficulties that are to be met, of market needs in methods of production or new products.

For the passive recipient, the *first stage* of adoption consists of the gathering of information and the realization of the existence of the innovation, the gathering of information about the preconditions for its correct use, and collection of information about how it works. For the active recipient, whom we shall meet later on, there is a further effort-demanding and time-consuming search process requiring knowledge – which may extend to the next two diffusion stages.

This is the stage where change agents exert most of their influence. As for the recipients, the pioneers as to the acquisition of knowledge concerning the existence of the innovation as compared to the later adopters, have a higher level of education, enjoy a higher social level, have higher exposure to information media and interpersonal channels, and are better connected with environments other than their own. Of course, this does not mean that the pioneer in awareness will also turn out to be a pioneer in adoption, for there are two more stages until the decision on whether to adopt. As for the active recipient, what will be at play at this stage is his/her technological knowledge and the realization of his/her capabilities and weaknesses and the "know-who"-type knowledge s/he possesses – in a sense s/he is playing the role of the "technology gatekeeper" (chapter 2).

During the *second stage* of adoption the person or decision-making group forms an opinion on the innovation. Information is sought for the evaluation of the innovation and evidence is sought as to the uncertainty accompanying its adoption. What is at stake is the determination of the relative advantages and disadvantages of the innovation and the main *criterion* is the *economic* one, although cases of status-seeking cannot be written off. *Compatibility, complexity*, and the possibility of *trialability and observation* are also important criteria.

The economic criterion is concerned with the overall cost of adoption, which may turn out to be greater than the cost of acquisition relative to the economic implications, with the rate of depreciation and with the degree to which previous investments to be replaced by the new adoption have already depreciated.

Compatibility is concerned with the degree to which an innovation seems to fit with current values, previous experience and the needs of potential users. Needless to say, an innovation which renders redundant a major part of a production process, stands against current values and routines employed by the firm and does not seem to cover the needs of potential users has practically no chance of being considered compatible.

Complexity refers to how difficult to understand and to use the innovation seems. This criterion may be very important for some innovations as in the case of the first personal computers, where the users, besides deciphering manuals, also had to write their own programs.

Finally, the possibility to try out, which in some cases may allow the partial adoption of an innovation, and the possibility to observe the consequences of an innovation are important criteria as to the formation of a positive opinion with respect to the adoption of an innovation.

The *third stage* of the adoption process concerns taking the decision as to whether to adopt an innovation or reject it. At this point, the person or the decision-making group has at their disposal all the information needed. The decision taken, whether for or against adoption, is not necessarily due to the objective criteria cited already but may be the result of other considerations. These are several. If a person takes the decision alone, personal characteristics may play a role. If a group takes the decision, then most of the factors that make the firm a non-monolithic entity start playing a role. Hierarchy and influence groups within the firm, interest groups in the sense of who stands to benefit even if it is just status, are the usual key drivers. Lack of sufficient communication between the different departments of the firm is another driver.

It is a pity that most accounts of adoption limit themselves to success stories because we would stand to learn a lot from the stories of rejected innovations. Many such cases would fall into the category where the firm has cultivated among its staff a sense of excessive pride, which may or may not be overdue. Many large and very large firms known for their innovative prowess, IBM for instance, are believed to have suffered at least at some time in the past from NIH (not invented here) syndrome – a definite subjective stance against adopting other firms' innovations.

Box 3.1 Kodak: "invented here" but rejected

There is substantial distance between NIH syndrome and rejecting innovations that have been developed in-house. Kodak, by far the largest film producer in the world, collected billions for several decades by producing and selling film and cameras. The emphasis was on the film, the main profit earner, not the camera. But as far back as the 1970s, Kodak had already been a pioneer in experimenting with the digital camera and actually earned seventy-five patents on the digital camera though never used any of them. We all know the end of the story. By the 1990s a swarm of producers around the globe were producing digital cameras but Kodak was still clinging to film. For more than a decade Kodak has been belatedly trying to cover the lost ground and play a role in digital camera innovation and production.

Some management analysts have blamed Kodak's behavior on what is referred to as the "milking cows" strategy, whereby a firm tries to squeeze profits for as long as possible from a very profitable activity, sometimes overdoing it – RCA and GE have also been blamed for similar behavior with solid state technology (Anderson 2004: 56–59). Other management analysts have pointed out that Kodak simply made a wrong decision by expecting that as income rose in the developing world consumers would first purchase traditional cameras and then turn to digital cameras, when in fact they did move directly to digital cameras, bypassing Kodak's film.

Box 3.2 Ampex and the VCR

Rejecting one's own innovations may also be due to the culture of a firm. Ampex Corporation of Redwood City, California, invented the VCR in 1957. It had the size of a commercial refrigerator, weighed accordingly, was very expensive, and so was sold only to television stations and Hollywood studios for film editing. Researchers at Ampex suggested to management producing a miniaturized version of the innovation so as to make it smaller, lighter and cheaper and so more affordable to a larger audience. Management refused on the grounds that they were not in the market for home appliances. Instead, they sold the rights to the technology of the VCR to Sony Corporation, which subsequently commercialized it into a home VCR (Rogers 2003: 151).

Box 3.3 Rejection and rehabilitation: from Alto to the Macintosh

A firm's culture can definitely affect the scope of its expectations and lead to the rejection of very important innovations created by the firm itself. The Xerox Corporation, at the time the leading document copier manufacturer, invested $150 million in creating Xerox PARC (Palo Alto Research Center) in a research park on the Stanford University Campus. Founded in 1970, in the next five years Xerox PARC developed an astonishing array of five important computer-related innovations: namely, the first personal computer named "Alto"; the means for an individual to interact with the computer, known as "the mouse"; icons and "pull-down" menus; local area networks linking computers; and laser printing. None of the personal computing technologies from PARC, except for laser printing, were adopted and commercialized by Xerox. Why did Xerox "Fumble with the Future" (Smith and Alexander 1988)? Laser printing was a very obvious candidate for adoption by the leading document copier manufacturer. It touched directly on the core business of Xerox and offered it high expectations for future growth and profits. To the button-down organizational culture at Xerox headquarters at Stamford, Connecticut, the company was in the copier business only. The remaining four innovations, produced by a freewheeling and casually dressed group of researchers – a different culture from headquarters 4000 miles away – were alien. Besides, they referred to a future product whose success was not obvious to non-specialists. A certain Steve Jobs thought otherwise. He visited PARC in 1979 and attracted many of the researchers to his skunk works. In 1984, five years later, the Macintosh appeared laden with many of Alto's features.

If the decision taken in the third stage of adoption is favorable to the adoption of the innovation, the firm is called to implement this decision at the *fourth stage*. The duration of this stage is difficult to predict as it varies depending on the type and complexity of the innovation. The implementation of the adoption of a home appliance could be particularly short as it may require only its purchase, installation and understanding of the user manual. However, in cases that are much more complex, such as the adoption of a new method of production, implementation may prove to be much more time-consuming, difficult and demanding, and may not necessarily meet with success. There are many intermediate cases such as the replacement of part of the capital equipment in a production unit with newer equipment in activities which, according to Pavitt's taxonomy (chapter 2), depend technologically on other firms in other sectors (for instance, textile producers, plastic objects producers, printers, etc. who all depend for their machinery on equipment and machinery producers). Finally, there are extreme cases where in order for diffusion to take place, the innovator must create the demand for it through awareness or by satisfying the preconditions for diffusion (setting up schools for telegraphers, establishing firms for the distribution of electricity), or by superseding social barriers by gaining the consent of social and economic agents (remember the bar code case).

There is still the case where the adoption of an innovation is encouraged by a third party. This, according to Pavitt's taxonomy, has to do either with "specialized producers" who produce equipment and have a vested interest in its diffusion, or with lead users who are also innovators and belong to "science-based sectors". The innovator creates a prototype, then turns it over to a producer for production and placement on the market. The case of the lead user who is also an innovator usually has to do with a producer who finds a solution to a problem he has run across, foresees that it may be of use to some other producer who may belong to a different sector than his own, and so decides to sell his innovation for production and distribution. As a type pertaining to innovation we have already described such a case in chapter 2, in the case of Busicom and Intel, where the two firms first collaborated but Intel then realized it need not only solve other firms' problems and decided to make microprocessor production the firm's core business. Here, by being far ahead of its competitors, the lead user-innovator is not threatened by imitators. According to some studies, lead user-innovators account for 77 percent of innovations in scientific instruments and for 67 percent of innovations in semiconductors and printed circuits (Von Hippel 1988).

Re-invention as a type of innovation has already been discussed in chapter 2 on "Innovation". One of the reasons we mention it here is because it takes place during the diffusion process and more specifically during the fourth stage, the one we are currently discussing. The would-be adopter has already passed the stage of awareness, has evaluated the innovation relative to his needs, current practices and expectations, and has decided to adopt the innovation. Except that he does not want to adopt it exactly the way it is but rather adopt a better rendition of it, possibly one that fits better his or other would-be users' needs. If he is in a position to provide this better rendition, he is then a very special adopter, an adopter-adaptor, an innovator in his own right. We have already met him when describing the previous stages of diffusion; he is the "active" recipient.

Most innovations at some time during their lifetime are subject to reinvention, and many are reinvented several times. Yet the occurrence of reinvention has only relatively recently been recognized. The reason is that reinvention occurs late in the diffusion process and most studies on diffusion stop at the point of adoption or rejection and actually even before this point as until recently they considered intent to adopt as synonymous with adoption – without ever bothering afterwards to check whether adoption actually took place. Now we know better.

The *fifth stage* of adoption is the last and the least well known. It consists of the evaluation of the consequences of adopting the innovation, that is, evaluation of the wisdom of the decision to adopt. It is usually absent from the goals of research studies on diffusion, as firms are not very willing to divulge how successful they have been with their latest adoptions of technology.

3.1.3 A few stylized facts

In most textbooks on Micro-economic Theory the firm's decision to adopt a new technology is more or less instantaneous, not only in the sense that there are no delays through the stages up to and including the stage of the decision to adopt, but even through the stage of implementation of the decision. In other words, we have a sort of once-and-for-all decision with whatever that implies for the firm. In a split second the firm has been informed about the new technology – how it works, what it does, what its capabilities, advantages and disadvantages are – and has evaluated it. In the same split second, it has taken the decision for or against adoption, as if the decision is being taken by a single, autocratic, decisive and very fast-thinking person. In the same split second the firm also adjusts to the decision to adopt. That is, the innovation is acquired, installed, adapted to the existing surrounding equipment and technologies, new staff has been hired and operates in harmony with existing staff, which has been trained to new practices and routines, and, of course, the firm has been reorganized to better respond to all the above. In a split second, the firm has not only adopted an innovation, it has also been transformed into a totally new firm.

This says a lot about orthodox Economic Theory in which time plays no role; it also tells us a great deal about the monolith that the firm constitutes in terms of that theory.

The description of the stages of adoption should be sufficient to convince us that we are dealing with a process that is time-consuming, a process that cannot possibly be completed instantly. True, an event broadcast on the six o'clock news, such as a report on the Boston Marathon bombings, or on a tsunami in Indonesia, will reach most people in the next few hours and practically all potential audiences in the next two days or so (Rogers and Seidel 2002: 209–219). But this is not what we mean by adoption here. This is mere transmission of raw information and refers only to part of the first stage of adoption, with four more stages waiting to be completed. All it does is create awareness.

Another interesting fact is that innovations that are worth adopting are not adopted simultaneously by all potential adopters. There is always a time lag, sometimes very substantial, between the emergence of an innovation and its adoption by a group of users. There follows another relatively long period during which the innovation is gradually adopted by all the relevant users.

We also know that diffusion can be represented by an S-shaped curve; that the rate of diffusion differs substantially among innovations, sectors, countries and regions; and that the expected profitability of adoption of an innovation and the average size of firms of the sector within which the innovation is diffused constitute the main explanatory factors as to the rate of its diffusion.

Further, we know that there is hardly an innovation that travels unadulterated during its life cycle as adopters come up with more promising variations of the original innovation. We have referred to this phenomenon as reinvention. Yet, with very few exceptions, research studies on diffusion behave as if the original innovation diffuses unchanged and alone, that is with no interference from its variants.

The diffusion of innovations does not simply spread their ensuing rising productivity in a country's economy. It also transmits information concerning their performance both to firms that are considering adoption as well as back to the innovators themselves.

3.1.4 The rate of diffusion of an innovation and its determinants

"Diffusion is the process in which an innovation is communicated through certain channels over time among the members of a social system" (Rogers 2003: 5).

The basic elements in the diffusion process are the attributes of the innovation/technology which is being diffused; the population of the potential entities (firms) of adopters and the way these entities decide on whether to adopt; the information channels concerning the innovation between the innovation producers and the potential adopters; the nature of the social system in which the innovation is diffused; and the extent of change agents' promotion efforts to affect the rate of diffusion of the innovation.

Among these factors, the attributes of the innovation are the most important ones in explaining the rate of diffusion – that is, the attributes as perceived by potential adopters. These attributes consist of: relative advantage, compatibility, complexity, trialability, observability. These attributes were discussed earlier in the section on the stages of adoption, though some additional comments are warranted in the case of relative advantage.

Relative advantage is the degree to which an innovation is perceived to be superior to the one it is meant to supersede. It is the sum of all expected benefits as compared to the cost of adoption. Relative advantage is the most important explanatory factor of the rate of adoption of an innovation, which is natural given the fact that the most crucial point during the process of the decision making as to adoption is the search for information intended to reduce uncertainty surrounding the relative advantage of the innovation. Potential adopters wish to know to what extent the innovation excels in comparison with current practice. As a result, relative advantage very often constitutes an important part of the message content on the innovation.

Relative advantage includes economic profitability, diminished initial cost, and reduced discomfort, enhanced social prestige, saving of time and effort, and immediacy of reward. Innovations that exhibit rapid rates of adoption are those that are perceived to promise the greatest economic rewards, the smallest risk and the least uncertainty. Adoption will be slower the more the reward will not be immediate or clearly visible. Preventive medical treatment belongs to this category. Preventive hygienic practices to contain the diffusion of AIDS are a case in point. The event may happen potentially. In addition, ignoring the preventive hygienic practices may not necessarily imply the contraction of the disease.

The second category of explanatory factors with respect to the rate of diffusion of an innovation has to do with the way in which the decision to adopt is taken. The decision may be considered optional, it may concern individuals or be collective, it may concern a large group of individuals, or it may involve a small group wielding power. It has been observed that the smaller the decision-making group the faster is the decision taken – whether in favor or against adoption.

Older studies considered the decision to adopt to be an instantaneous event and analyzed diffusion at the level of individuals categorized as to their receptiveness since institutions and firms were considered to behave like individuals. More recently, the fact that diffusion within the firm is a complex process has been accepted (Van De Ven et al. 1989, 1999). Its rate depends on the receptivity of the firm to change, which in its turn depends on general characteristics of the firm (Mahler and Rogers 1999), its structural characteristics (Meyer and Goes 1998) and links to its environment.

The third category of explanatory factors of the rate of diffusion has to do with the channels through which diffusion takes place. That is, whether it is accomplished through communication mass media or through interpersonal channels. Mass media have proved to be relatively more important in the very early stages of adoption when the innovation's appearance is first announced, while interpersonal channels have proved to be relatively more important later on in the process of adoption, i.e. that of the persuasion stage. Mass media are relatively more important for the early adopters while interpersonal channels are more important for late adopters.

The last two categories of explanatory factors for the rate of diffusion are the nature of the environment in which the innovation diffuses – its social system – and the extent of change agents' promotion efforts in diffusing the innovation.

3.1.5 The diffusion curve

Traditionally, the determination of the diffusion curve is carried out without reference to the explanatory factors of the rate of diffusion – although their absence does not imply the lack of their existence or their importance – but instead with reference to the oldest diffusion model, the epidemic model.

In an epidemic the disease is transmitted to healthy individuals through contact with patients. As the disease diffuses, the number of carriers increases and so does the rate of diffusion until the number of individuals who remain healthy diminishes to the point that the rate of diffusion is forced to diminish. It keeps decreasing until it stops altogether once it has reached all those individuals that could get contaminated.

In the case of the diffusion of an innovation what is diffused through "word of mouth" is information about the innovation itself. Firms have little information about an innovation, which has still not been adopted by many, and as a result they associate it with a high level of risk. As the level of adoption rises, so does the amount of information to the firms, which potentially could adopt and hence the risk diminishes. The rate of diffusion rises but not indefinitely. As the number of firms that have adopted the innovation out of the total number of potential adopters rises, the number of potential adopters who have still not adopted diminishes, the rate of diffusion keeps decelerating, and in the end diffusion stops.

Unlike the event broadcast, where we expect a curve rising very fast and almost reaching saturation fast, now we get an S-shaped curve representing diffusion over time (see Figure 3.1). All along the curve, at each point, the slope of the curve is the rate of diffusion. In the first part of the curve the slope rises, then it stops rising at a point we refer to as an inflection point, (A), then the slope becomes less steep and finally the curve

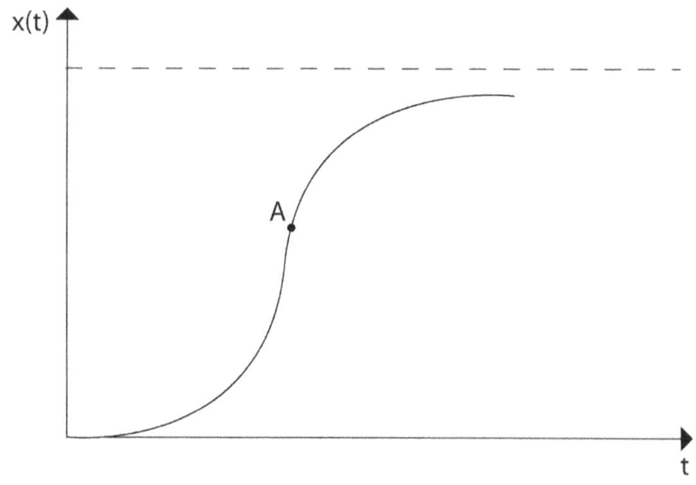

Figure 3.1 The diffusion curve

approaches asymptotically the dotted line, its maximum level or level of saturation, and diffusion stops altogether.

Just like the epidemic where it is difficult to explain how the first contaminations started, the "word of mouth" model cannot explain the diffusion of an innovation from the time it is invented, but only from the time some number of early users have begun using it. Early adopters have evidently chosen to adopt the innovation despite not having benefited from the experience of a previous user – they are somehow different from subsequent users. We have already described this category of early adopters and how they differ from the rest as well as the fact that they are more susceptible to common source information.

Diffusion will be faster for simpler technologies which can be easily learned and transmitted, for dense populations, where early users are active in spreading information, in cases where switching costs are not substantial and in situations where the new technology is clearly superior to the old one. Figure 3.2 portrays two diffusion curves. Curve A describes a diffusion process that has started earlier and rises faster than the diffusion process described by curve B.

One of the earliest and possibly the most frequently portrayed results of studies on diffusion is the diffusion of hybrid corn planted in the US. The figure is very similar to Figure 3.2 with the two curves, except that now we have more curves, one for each state. Each state differs from the rest in terms of initial use of hybrid corn, rate of diffusion and final level of saturation (Grilliches 1957). Similar curves are obtained if we attempt to portray diffusion curves for a piece of machinery across states or countries. We may also portray the diffusion of some characteristic products. For instance, we may portray the diffusion of radio, television, the refrigerator, etc.

For each of these items, the rate of diffusion will depend among other things on whether close substitutes exist. The radio had no close substitute and diffused faster than the automatic clothes washer for which there was a substitute. Tellis et al. (2002) find that the single most important factor in the explanation of the speed of diffusion of

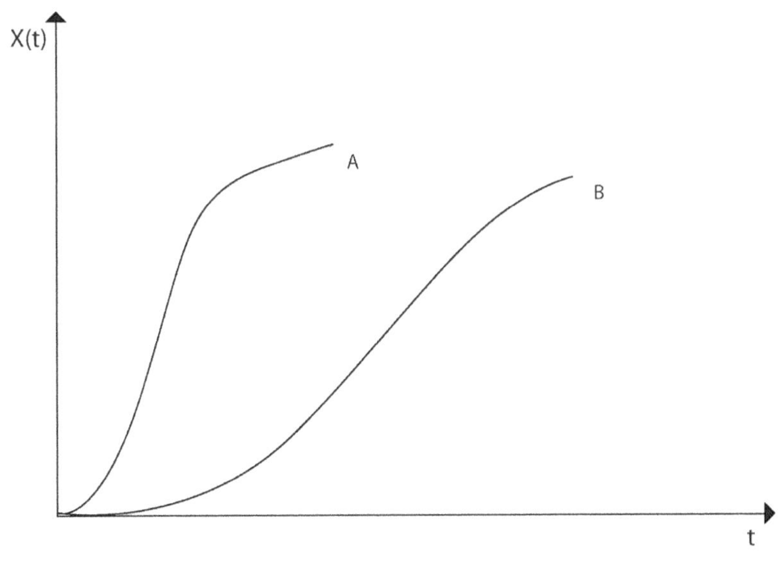

Figure 3.2 Two different diffusion curves

goods across European countries during the latter half of the twentieth century is how status-enhancing the good diffused was considered to be. "Brown" goods, consisting of information and entertainment consumer durables are more status-enhancing and diffuse faster than "white" household appliances, as they are more readily observable to non-members of the household.

Diffusion curves are used extensively in forecasting both for marketing and production purposes. The rate of diffusion, the time interval elapsed from innovation to first adoption and the saturation level can all be measured, and several studies have linked them to the characteristics of the innovation, of the receptor and of the environment. Methods have been developed whereby knowing the level of diffusion at some early stage of the process allows us to forecast the remainder of the shape of the curve – that is, predict in time how the rate of diffusion will vary and what will eventually be the saturation level. Alternatively, based on the characteristics of a product, a forecast as to its diffusion curve can be obtained based on the diffusion curves of products with similar characteristics.

So far, we have portrayed the diffusion curve as the well known S-shaped curve also referred to as the cumulative diffusion curve as it represents at each point in time the level of total adoption achieved. Depending on the use we want to make of the process of diffusion, we may wish for a different representation than the cumulative S-shaped curve, for instance period-by-period adoption. In the diffusion of mobile telephones, a service provider and a handset supplier have different needs in terms of forecasts. The service provider is concerned about the demand on the infrastructure, that is, the level of adoption at a particular time and the eventual number of adopters. It is concerned with cumulative adoption. On the other hand, the handset supplier would like to know the level of demand during a time period, i.e. the rate of adoption at a given time and the magnitude of peak demand that it will have to face. It is interested in period-by-period adoption. The two representations of the diffusion curve are presented in Figure 3.3.

Diffusion also depends on the characteristics of the technology to be diffused. Two technologies, which differ in terms of complexity, will exhibit different diffusion curves.

3.1.6 Critique of the epidemic model and the alternative probit model

In the epidemic model firms are treated as similar to each other with no difference in goals, capabilities or actions. It is also implicitly assumed that neither the new innovation nor the technology it replaces change during the diffusion process and that the new technology is better than the old one. The only difference allowed is that of progressiveness among firms in terms of adoption: "early adopters", "later adopters" and "laggards". If, however, we start from the observation that the decision to adopt is taken by a particular firm, and then if we recognize that there are differences among firms, these differences may have an important role to play in explaining patterns of diffusion. A natural way to deal with this situation is to use a *probit model* to analyze individual decisions to adopt, and to suppose that firms differ in terms of a particular characteristic, which affects the profitability of adoption.

Firm size turns out to be a very commonly explored characteristic in the empirical literature on diffusion, partly due to the relative ease of observation and partly because it is very often used as a proxy for a number of other features. Most of the evidence supports the hypothesis that, for one reason or another, large firms are quicker imitators than small firms.

Suppliers who are responsible for providing the new equipment constitute another important set of agents which through their pricing and servicing policies influence the

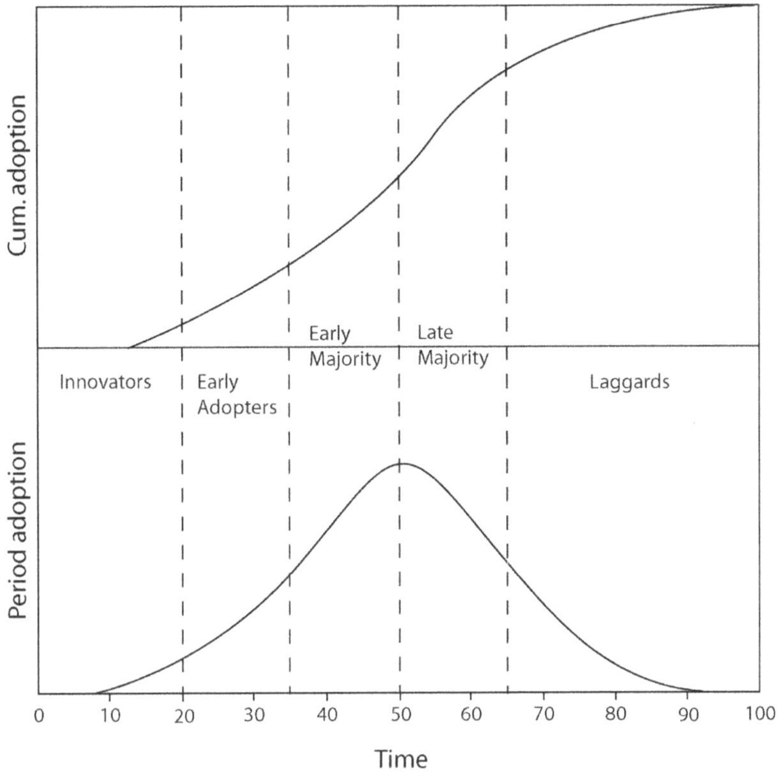

Figure 3.3 Two representations of the diffusion curve: cumulative adoption (up); period by period adoption (down)

cost-benefit calculations on which the decision to adopt rests. Conditions of competition in the supplying industries will affect diffusion in the using industries. Suppliers are interesting for another reason. They are often suppliers for both the new technology and the old technology that is being displaced. One way or the other, the old technology almost always reacts through some incremental innovation, which will slow down the diffusion of the new one. Diffusion of the new technology is also slowed down whenever it stimulates innovation in the older technology (Mokyr 1990: 90, 129). In contrast, the introduction of a complementary good or an application enhances the speed of diffusion. The emergence of a new application enhances the diffusion of an operating system; the emergence of CD software affected positively the diffusion of CD players (Gandal et al. 1999); the impressive speed of diffusion of the mobile phone in recent years in some parts of Africa has been mainly attributed to the emergence in these countries of a money transfer service provided through the phone.

When the new technology arrives, chances are it will not appear in its final form. The decision to adopt now or to wait for the next generation of the innovation is a real consideration which raises the opportunity cost of adopting now rather than later, and hence moving in the direction of slowing down diffusion.

Costs are another type of driver of diffusion. Both learning costs and search costs involved in gaining information on the new technology tend to slow down diffusion especially in the first stages but they tend to become less of an impediment as diffusion proceeds with more and more adoptions, thus increasing the amount and quality of information divulged. Switching costs may sometimes turn out to be a major retarding factor in diffusion. A favored example in textbooks was to mention that purchasing a CD player is no longer a major expenditure but transforming one's vinyl record collection to one on CDs is. Now of course this example is irrelevant as there are better and cheaper ways to reproduce music.

In the case of a process technology, things are more involved. Besides purchasing the new technology, switching means that the firm has to change just about everything from design and production to marketing and distribution. In addition, the firm has to learn to use the new technology, a process which depends on its own "absorptive capacity" – an expensive item to upgrade. The more drastic the breaks with previous activities, the greater are switching costs. Established, incumbent firms, more adept at spotting costs than new sources of revenue, firms "too close to their customers", are unlikely to adopt new "disruptive" technologies (Christensen 1997).

Finally, opportunity costs are important and can be created by previous investments in equipment, which have not fully depreciated. Firms with new vintages of capital stock are less likely to switch to a new technology than firms with older equipment.

The diffusion of an innovation rarely takes place in a stable environment. Very often an innovation is hindered in its diffusion process by one or more of its own variants, the results of reinvention. A superior variant raises the level of saturation and usually affects the rate of diffusion as well. The number of cameras produced worldwide had stagnated before reflex caused the saturation level to rise and so did later the introduction of the digital camera.

An interesting case is when different technologies compete to achieve the same goal (Norton and Bass 1987), or of competing different generations of the same device (Kim et al 2000).

Box 3.4 Competing technologies and market power

Competition was very intense among the six technologies that aimed at recognition of individuals through biometric technologies for digital or Internet transaction use. The problem is more complex than it might at first seem because at the start, the level of development is different among the competing technologies and each has a different development potential. At one point, voice recognition, compared with iris recognition or digital fingerprint recognition, was considered to be the least developed yet the technology with the greatest expected potential. To complicate things even more, this is not a straight comparison among unsponsored technologies. The outcome of which technology will prevail and diffuse will very much depend on the sponsoring firm. In the particular situation at hand, the largest firm among the competitors in the six technologies was taken over by Microsoft and the outcome was the combination of some of the technologies.

Usually unnoticed in diffusion is the fact that the diffusion of an innovation may sometimes accompany the diffusion of some other innovation. The diffusion of the heart pacemaker also refers to the diffusion of a system of innovations including the transistor and the miniature battery.

A positive aspect of the use of the probit model is that it lends itself to policy making. The fact that it identifies observable factors that may trigger an adoption decision makes it possible to identify levers that may be used in policy making in order to speed up or slow down the diffusion of a particular technology. On the other hand, use of the probit model, with its emphasis on the decision-making process by each individual firm (based on weighing benefits versus costs), ignores the social feedback effects, known as externalities, that might result from one individual adopting an innovation and thereby encouraging another user to do the same. This forces us to consider network effects and the related issue of standards.

3.1.7 Network effects and standards

The importance of *network effects* has increased in recent decades with the rise of ICT as *network products* have proliferated. For network products demand, as it does for all products, depends on price, quality, substitutes, etc. but in addition also depends on how many others have already adopted it or on how many are expected to adopt it overall. The reason is that there are benefits to be gained by consumers or firms from the fact that other consumers or firms have chosen the same technology that they have. These benefits fall into two categories: direct and indirect.

Direct network benefits arise because they allow the adopter to communicate with other adopters that have chosen the same technology. The most obvious examples that come to mind are the telephone or the fax. If there were only a sole telephone on earth it would have no value as a communication system as there would be nobody else to communicate with. But as the number of telephones rises so does the value of the service the system offers. In fact there are at least three different theories for the evaluation of a network. They all allot a rising value to the network depending on size, and differ only as to the weight each attaches to size.

Indirect network benefits arise from the fact that adoption of a product that uses a particular technological standard by a larger number of people increases the probability that the standard will survive and that goods compatible with the standard will continue to be produced. The most cited example of indirect network benefits is the VHS/Beta competition which resulted in a single standard for video recorders/players, mostly ascribed to the initial advantage held by VHS in the length of program that could be recorded and to consumer desire for a large volume of pre-recorded tapes.

Consumer demand for network products stratifies consumers in terms of whether they value the network character of the product more highly than the rest of its qualities, such as its technological prowess. If consumer demand by those that value the network character of the product overshadows demand by those that value the rest of its qualities, then an inferior technology may be preferred to a superior one. This also helps explain why a superior technology may not be able to displace established inferior technologies.

Network externalities have generated a *positive self-reinforcing feedback mechanism* thus inducing "*path dependence*" in the diffusion process. In other words, the (equilibrium) point at which the process converges depends on its previous historical path. Contrary to the dictates of orthodox Economic Theory, *history (or initial conditions) count(s)*. Effectively, early accidents in the choice of technologies can lead to the adoption of an inferior standard because of the existence of an installed base which makes that technology more attractive to each successive wave of new adopters.

With demand depending on size or anticipated size of the network, it goes without saying that suppliers of network products will rush from the start to build up a network of users – usually known as a *standards race* to dominate the market – making that network more attractive to potential new users. As the number of users increases, the network becomes more and more attractive to other potential users. We have what is called *positive feedback* where *success breeds success.* (We shall meet these concepts again in chapter 6 in connection with increasing returns.) The outcome in such cases tends to be that one firm captures the market or most of it – *winner takes all or winner takes most.* From the moment a particular technology or variant has gained the upper hand, the market tilts in its direction as its installed base gains new adopters and demonstrates that it is no longer the risky proposition it was assumed to be earlier. Besides, if A has proved to be working, why invest in B, which is still uncertain? We have what some writers have called an *information cascade* where "it is optimal for an individual who has observed the actions of those ahead of him, to follow the behavior of the preceding individual without regard to his own information". This is used to explain *herd behavior,* a *bandwagon* or even *penguin effect* to describe the very rapid diffusion in these cases. (Afraid of sharks and orcas, penguins wait on the ice edge till one of them jumps in the water and survives, and then they all jump in to catch fish.)

Winner takes all, or most, leads to the emergence of *de facto standards* or *dominant designs* as the rest of the industry adopts the winner's standard. These informal or market standards, which have resulted from market success, are not open and are proprietary designs. They may have resulted from a costly standards race – costly to winners, especially to losers, but also to consumers. As a proprietary standard, it may lead to an element of monopoly and bears the risk of leading to a lock-in to an inferior good or technology.

In contrast, formal standards known as *de jure standards* or *institutional standards* are open and public. They consist of documents that are the result of deliberations within standard bodies (CEN, CENELEC, DIN, BSI, ISO) and consortia relevant to specific technologies and are the result of consensus or agreement. On several occasions this has proved to be time-consuming but the resulting open standard may be a significant enhancement to innovation. In many instances the future agenda of the committees on standards may be a precursor of where a particular technology is heading next.

There is a close connection between standards and network externalities. This is due to the fact that standards generate a number of effects that tend to maximize the probability that the particular good in question will exhibit network externalities. First, the compatibility or interconnection standard raises the probability that two products, such as telephones, facsimiles, instant messaging services, CDs and CD players, will be able to communicate with each other. Second, a successful standard expands the size of the potential market for a particular good, leading to economies of scale in its production, hence lower prices and greater availability of complementary goods. Standards to limit variety such as clothing size and standards for measurement and description also work in the same direction. Third, a standard helps create communities of adopters with learning effects among them, such as the users of an operating system like Windows. Fourth, there is the safety standard or minimum quality standard, which is intended to protect the consumer by providing assurance that the product is safe to use and that it meets a certain minimum level of quality. In so doing it helps raise the size of the potential market.

3.2 Diffusion: across sectors

In contrast with diffusion within a sector, diffusion across sectors is a topic the study of which has not generated much interest. Yet several innovations are known to have migrated from the sector in which they first appeared to other sectors and have been applied to new uses. The laser originated in the defense industry, found its way to the surgeon's operating table and then to a host of other uses including the counter at the supermarket. The bar code was first put to use by railway companies for logistics purposes, and was then used by all sorts of manufacturing assemblers, dispatchers, supermarkets, warehouses, retail stores and e-stores. In earlier times the steam engine was put to use for pumping water in mines, in cotton, wool and grain mills before being used by locomotives, threshing machines and steam plows. Semi-conductors followed a similar path. In the form of (germanium) transistors they were first used by the military. Silicon transistors and then integrated circuits were used in aviation, medical devices, calculators and computers. More recently, since the advent of the microprocessor, semi-conductors have been used extensively in an endless array of applications such as cars, computers, watches, appliances and many more.

In diffusion within a sector the existence of the S-curve has been proven in hundreds of studies and the rationale for its path explained extensively. In contrast, in the case of diffusion across sectors, very few explanations have been offered so far to account for its hardly explored path. There is an exception: the work of Chris DeBresson (1991, 1995), which nevertheless does not cover this vast uncharted territory because it rests on very few case studies. Despite this writer's additional reservations on DeBresson's hypothesis – to be spelled out in due course – rather than ignoring diffusion across sectors altogether, as most textbooks do, it was decided to include it because it is an important topic, albeit not sufficiently researched, and because DeBresson's attempts at least set a framework.

Ideally, we would have liked to have an explanation of how technologies migrate from the sector they have originated in to other sectors, how they diffuse to the rest of the economic system. What are the basic rules of how technologies migrate and cross borders across sectors, which sectors they can cross and what are the main determinants and conditions governing their migration? As a tool that would help explain the evolution of technology and, ideally, visualize what the "next game" will be all about, it is valuable to the global long-term player entrepreneur, to the innovation strategist, to the technology analyst and to the historian.

It has been observed that innovations with pervasive repercussions in activities beyond the ones in which they have been generated, innovations that diffuse through the economic system, have followed a migration path that is anything but random.

Some activities are always first in feeling the influence of an innovation ahead of all the rest. Usually, new technologies make their entry in industry when they are still in a rough form. Such activities may entail high costs, which is why they appear in activities supported by public funds such as public laboratories, in research related to defense activities and spending, and in medical research activities. At first, all they have to do is prove their technical feasibility. In these activities low cost is not a priority but achieving a specific task – and some promise for improvement – is. A missile may not necessarily always reach its target on the first trials, and its cost may leave a less lasting impression on a general than a manager. This is why such activities, which are comparatively undemanding, are considered privileged entry points into the economic system and called "demonstration sectors".

"After this entry into the economic system, all technologies seem to follow a predictable path: from instruments to machine tools to machines, and then to transformation industries, utilities, and finally consumer goods" (DeBresson 1991: 244). As these new technologies move along their diffusion path from the demonstration sectors to consumer goods industries, the economic criterion, which was at first insignificant, keeps gaining in importance. Machine producers already compare long-term return to cost. For the materials transformation industry the decision to adopt is often based on costs and benefits.

By the time the new technologies have reached the utilities, transportation and communication sectors, these technologies have already been subjected to some standardization. This allows for the meeting of a basic condition of the existence of such complex systems, namely compatibility among their parts. Standardization will allow for economies of scale to materialize and cost per unit to drop. The drop in cost is a condition for the new technologies to be adopted by the consumer-durable industries whose products are characterized by high elasticity of demand as to price. This means that producers in this sector, in contrast with producers of tools or machines, have every reason to lower production costs and prices, as sales should rise with the drop in prices. Finally, the economic criterion reaches its utmost importance in the subsistence consumer goods industries as the decision to adopt is taken exclusively on cost considerations, even on short-term marginal cost considerations.

Now why should new technologies repeatedly move along a privileged path – a path that is predictable in the sense of "most probable"? According to DeBresson, the path depends on two sets of conditions: conditions on circulation and conditions on technological performance thresholds.

The condition on circulation states that the sector in which the innovation was generated and the sector towards which it is headed have to be linked by a supplier–user relationship or the reverse. If there are no links between the two sectors migration from one sector to the other will not be able to occur along the links, which are missing except in cases of a technology-push innovation.

The technological performance conditions state that the adoption criteria of an innovation differ among sectors. Even if the links exist and do permit the passage from one sector to another, migration may not necessarily take place, as minimum thresholds of technological performance have to be met. These thresholds may not have been achieved despite the fact that diffusion may have reached saturation levels in the sector of origin of the innovation.

The only requirement for the demonstration sector is technical feasibility, which means that the innovation works. For tools, however, there is the additional requirement of reliability – additional to the requirement of technical feasibility, and each threshold requirement presupposes the existence of the previous less demanding ones. Machine tools are still more demanding: they require additionally technical efficiency. The list of requirements gets longer and longer as we descend to the level of consumer goods. Here, the apparent requirements no longer seem technical, as these are taken for granted and increasingly cede their importance to cost and other considerations. New utility is expected from a car but that does not mean that reliability, a quality seen further up the list, is not expected, because every time we turn the key we expect it to start. User friendliness gains importance for luxury goods and mass luxury goods. Personal computers invaded homes only after Apple made them friendlier through the mouse and icons than they were through IBM's voluminous user manuals. Consumer goods rest on low prices hence there are economies of scale in production and a level of standardization to allow

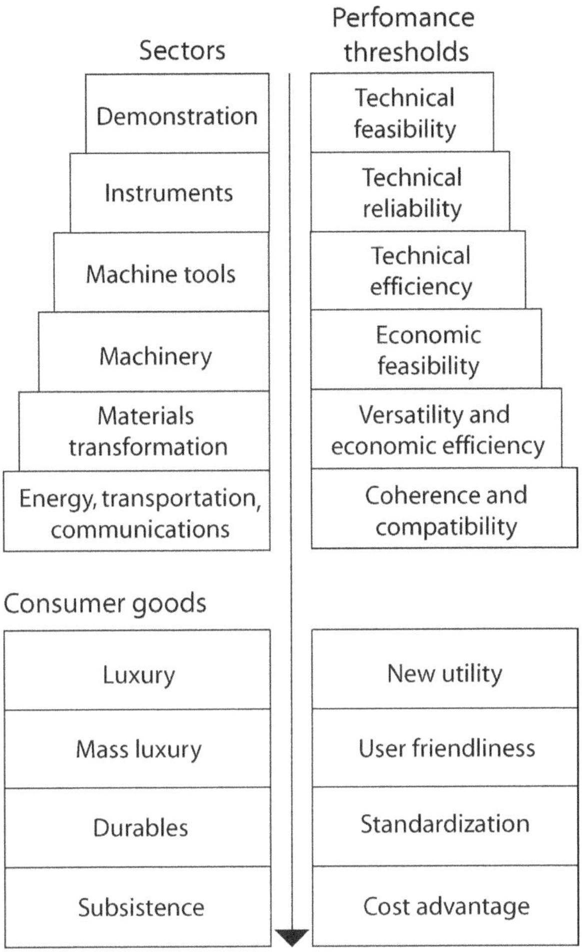

Figure 3.4 Dominant performance thresholds required for acceptance in an industry

for easy replacement of parts and provision of service. In basic consumer goods new technologies become acceptable when, beyond proving their superiority, they also exhibit lower fixed and variable cost.

Beyond the fact that DeBresson's hypothesis rests on very few observations (three to five), it also rests on assumptions that may no longer be valid. For we know by now that diffusion does not occur exclusively through a user–supplier (or the reverse) relationship. This jeopardizes the circulation hypothesis. In addition, we also know that the pressure to reach the market first, even beyond the logic of patent races, is so strong that the innovator rushes to be first to reach the market in a move to preempt competition, despite the fact that he often knows that his product is not as good as it should be and will soon have to be improved – an example is Apple's iPhone 5 experience with its maps. This jeopardizes the hypothesis on technological performance thresholds.

DeBresson has shed some light but we still remain in need of a theory of diffusion across sectors.

3.3 Technology transfer

3.3.1 Technology transfer and the new meaning of technology

Diffusion, as we have seen it till now, consists of the unhindered – by technology producers – flow of technology with no a priori intent on the part of users to adopt it. Technology transfer refers to the situation when there already exists intent to adopt; this is an additional type of diffusion.

Until the early 1980s inherited wisdom was that the opportunities offered by new technologies would hasten the process of economic development in the developing world. It was thought at the time that once an innovation was generated, all that was left to be done was for that innovation to be acquired through technology transfer, a process considered to be simple and straightforward.

These beliefs turned out to be simplistic and above all wrong. The massive technology transfer that accompanied US investment in Europe in the 1950s and 1960s proved to be a special case rather than a lesson to be applied to the rest of the world. Now we know that technology transfer is a much more complex process than we used to think. It turns out to be not only very expensive but also very demanding and complex. In addition its success depends on many more factors than originally thought, among which are the level and direction of local technological efforts and other characteristics of the host country.

For mainstream Economic Theory, technology consists of the required information for the design and production of a product through alternative methods of production. This information is taken as a public good; it is available, accessible, costless and can also be reproduced at no cost. It can be codified in designs and working manuals.

But now we know that technology is accumulated knowledge and hence does not consist of information alone but also of tacit knowledge, which is usually its most important element (see chapter 2). This knowledge is not public but private. It is the result of accumulated experience on the part of the firm in design, production and investment activities, and it is mostly in tacit form. It is acquired through the solution of problems encountered by the firm in the process of the performance of its routines and remains within the firm's confines mostly in codified form. In addition, each firm acquires its capabilities through a time-consuming and strongly cumulative, expensive learning process. As a result, the spectrum of technologies that a firm can muster at any given time is much smaller than the sum total of existing technologies. In addition, each firm is relatively unique as to its capabilities since its experiences, capabilities and learning methods all differ from those of other firms.

3.3.2 Technology transfer: conditions for success

The fact that most of the technology transfer that takes place internationally is through MNCs has never been challenged. Besides, it takes place between the mother company and its affiliates with hardly any spillovers to the host country. Of course, we refer here to official technology transfer in contradistinction to unofficial technology transfer.

Transfer of technology between the mother company and its affiliates refers to the closest possible relationship between donor and receiver, which means that the donor has no reservations and is determined to pass on technology to the receiver. It is in its interest to achieve that goal as best it can and as completely and efficiently as possible. However, there is a part of technology transfer which does not reflect the mother–affiliate relationship, but instead a weaker relation. It is only natural that the closeness of the relationship between donor and recipient should prove crucial for the donor's determination and interest as to the success of the transfer since, the donor's interest varies depending on how close its relationship with the recipient is.

A number of factors have proved important as to assuring commitment and determination on the part of the donor. First, there is the level of competition among donors. If competition among donors is weak, in the sense that there are few prospective donors each with its own technology, technology transfer will usually take place through affiliates, which are fully controlled. If competition among donors is strong and prospective donors all have the same technology, then transfer usually takes place through cooperative business ventures and technology contract agreements.

Second, if the technology to be transferred is new, the transfer is usually towards an affiliate, and as technology to be transferred matures, the number of transfers through cooperative business ventures and technology agreements rises.

Third, high technologies have the tendency to be transferred between firms linked by a strong relationship.

Fourth, the closer a technology is to the core business of the firm, the more its transfer will tend to be between the mother company and its affiliate.

Now, what are the important factors and conditions for the transfer of technology to be successful? First, previous experience of both donor and recipient in technology transfer activities has a tendency to decrease the cost of transfer.

Second, the donor should have the proper organizational structure and capabilities that would allow it to transfer technology.

Additionally, the absorptive capacity of the recipient, its technological level, the characteristics of the environment in which the transfer is attempted, and the duration of the transfer have all been recognized as playing a significant role in the transfer's success. Above all, however, the most crucial factor as to the success of technology transfer is the recipient's amount of learning expected to lead him to higher levels of technological capabilities.

As technology transfer is not a simple transfer of information but concerns learning, it has repercussions for the relationship between the donor and the recipient, which may be at a distance from each other. The importance of distance depends on the type of technology. When the technology to be transferred is standardized, relatively stable and codifiable, the information to be transferred can be performed through standard codes at a distance and at low cost. When the technology is complex, changes rapidly and is tacit, then short distance becomes an advantage. In many cases where the recipient's needs are complex or change rapidly, proximity is compulsory.

In addition to geographical distance there is the distance between donor and recipient in terms of level of technological capability. In earlier times, when what was at stake in a transfer of technology was mostly cost and payment, this issue mattered less. But now we define transfer of technology not only as the acquisition of the technology but also as the ability to use it in production, to assimilate it, to develop it and to be in a position to improve upon it some time in the future. According to this line of thought, if the distance between donor and recipient in terms of technological capabilities is very large, then there could be acquisition but no transfer. If donor and recipient were on the same level in terms of technological capabilities, there would be no need for a transfer. The most promising situation for a transfer is when the recipient is at a lower level of technological capabilities than the donor but still not far away, and thus in a position to understand so as to be able to learn.

We noted above that the donor firm should acquire the necessary organizational structure and the capabilities to transfer technology. By analogy, the recipient should get ready to receive technology, which means developing the necessary technological capabilities through the learning process that the transfer implies but in addition acquiring organizational capabilities and if need be acquiring an organizational structure that would allow the firm to make the most from the transfer. After all, the gain of the recipient firm will depend on how well its choice will materialize, in other words, from the learning process within the firm: organizational learning.

3.3.3 Technology transfer and the nature of sectors

In practice, technology transfer is influenced by the nature of sectors in which firms operate, and so they differ according to the sources of technology, type of use and means at the disposal of the firm in creating technological advantage (Pavitt 1995).

Using Pavitt's taxonomy (chapter 2), for the category of firms that are technologically dependent, the transferred technology is embodied in the equipment provided by suppliers. The firms are small and their technological capabilities limited; the only advantage that can be gained vis-à-vis the competitors who also purchase the same machinery is through better management and use of existing stock. Technology is appropriated through advertisement, marketing, trademarks and esthetic design. Practically all technological change takes place at the level of suppliers in other sectors. Here, transfer and diffusion are basically synonymous, as transfer is unhindered, even encouraged, by the producers of machinery for whom it is a product. For the recipients it is a new method of production.

In the category in which firms are characterized by economies of scale, the main preoccupation of those firms is to take advantage of those economies in order to satisfy price-sensitive users. Here, the producing systems are large, complex and interdependent, and their main goal is reliability. Part of the machinery and equipment is produced within the firm and part is produced in collaboration with small specialized machinery producers. Capabilities and know-how in design, construction and use of large-scale production systems are of great importance and are reflected in payments for know-how in licensing agreements.

Product design and reliability are the success factors of specialized producers but cannot be protected through patents or secrecy. Reverse engineering and links with the mechanics of production departments play a significant role in the transfer of technology.

For the category of firms that rely on science, the technological growth paths are dictated by strong techniques that emanate from science. Here, more than in any other instance, technology depends on industrial research and development, and on public scientific and technical knowledge achieved in universities and other educational and research institutions. Again, more than in any other instance, successful technology transfer, leading to taking advantage of the abundant technological opportunities that emerge, depends on the ability of users to understand and assimilate the underlying scientific and technological principles involved. It is precisely for this reason that industrial R&D is so important in the capacity to assimilate someone else's technology.

3.3.4 Technological espionage

A substantial part of technology transfer occurs without the donor's knowledge or approval. This is referred to as industrial or technological espionage and it has been known to exist for centuries, witness official steps taken to discourage it. A British law of 1781 prohibits the export of (among other things) machinery, machines, tools, models, designs and parts used in the (textile) industry. Despite the prohibition, technology managed to diffuse both to Europe and to the US, either in the form of prohibited exported machinery, or through specialized craftsmen who migrated, or through both simultaneously. Of course, at the time, innovations were simple, hence easily imitated. In addition, people were the only vehicle for knowledge, that is, tacit knowledge, since there was then no codified knowledge, i.e. no scientific journals, books, etc. This is the reason why in several instances craftsmen were not allowed under heavy penalty to leave their place of work. The most cited example is the case of the crystal craftsmen of Murano, the island off Venice. When under Louis XIV the Huguenots were thrown out of the country en masse, France lost its expertise in clockwork technology. At the time, the knowledge of the specialized craftsmen consisted of the empirical knowledge that they had acquired through their experience of work close to an older specialized craftsman plus their own experience in the use of some machinery.

The (even unlawful) possession of machinery without the contribution of the appropriate human input has been proven inadequate for centuries. In 1783, machines were disassembled in London and unlawfully shipped by Americans to the US. After four years of futile attempts to find specialized persons to reassemble them, the parts were shipped back to England.

The fact that technology transfer is a difficult process, whose success is not assured even when the presence of both machinery and codified knowledge have been secured, has also been known for quite some time. Sometimes, resort to technological espionage has attempted to cover the void left by tacit knowledge.

In the late seventeenth and early eighteenth centuries:

> if the infant English silk industry was to flourish, it had to gain access to the water-powered throwing machines used by the Italians to produce superior silk thread cheaply. The Italians regarded the workings of these machines a state secret, and according to the laws of the Kingdom of Sardinia the disclosure of the operation of silk mills was punishable by death.
>
> (Bassala 1988: 84)

In fact, unexpectedly, codified knowledge existed in the form of not one but three publications of detailed engravings of one of the machines between 1607 and 1656. However, interpretation of a depicted machine without the tacit practical knowledge of its construction and operation is close to impossible.

> After an abortive attempt by Englishman Thomas Crochett to mechanize silk throwing in 1702, a London textile merchant decided that information about the machine must be stolen from the Italians. John Lombe, who came from a family of English textile weavers and merchants and who had a good head for mechanics, was sent to Italy in 1715 to do the job.
>
> (Bassala 1988: 84)

During a two-year stay, Lombe "found means to see this engine so often that he made himself master of the whole invention and of all the different parts and motions" (Cipolla 1972: 47). In this case, the transfer of technology could not have taken place without industrial espionage.

Nowadays, industrial or technological espionage or piracy has increased substantially due to a number of factors. Globalization increased the possibilities for industrial espionage, and so did the fact that more people have access to trade secrets than in the past. In addition, nowadays, in the face of decreasing company loyalty, employees have greater opportunities to gain from the knowledge of trade secrets, either by changing jobs or becoming self-employed. Intangible property is easier to steal than tangible property. Advances in communications and the Internet in particular have made the collection of trade secrets easier. Finally, in many cases the new technologies end up as products and services (like for instance Napster and its descendents Kazaa, Morpheus, Grokster) which bypass "lawful ownership" of a service or product and render obsolete existing means of intellectual property protection (Parloff 2003).

Espionage may take the form of either economic or industrial espionage. The key difference between them is that only the former involves a government's efforts to collect information. Cases like British Airways versus Virgin Atlantic ending with an apology on the part of British Airways, the case of Unilever falling victim to "dumpster diving" settled out of court with Procter & Gamble, or Canal Plus' indirect settlement with News Corporation are examples of industrial espionage (Crane 2005: 233–240).

On the other hand, in a case where the government of country A eavesdrops on, let us say, IBM's communications, or supplies the equipment to a company to do so, for the benefit of a company residing in country A, or owned by the government of country A, we have economic espionage. The latter acts as a form of strategic trade policy, shifting profits from foreign firms to domestic firms, and potentially improves national welfare (Whitney and Gaisford 1996). It also has the long-term effect of reducing the incentive for innovative behavior on both sides.

Nowadays, economic espionage is very substantial. Estimates for the US alone, which is the primary target as it has the most sought-after technology and is the largest producer of intellectual property, were estimated at up to $260 billion per year in 1996 – up to $400 billion if overseas operations were included. Jobs lost to economic espionage ranged from one to six million (Tucker 1998: 1121). An FBI study concluded that in the course of a decade, 57 countries had engaged in covert operations against US corporations (Tucker 1998: 1115). "Since the end of the Cold War, foreign intelligence services have increasingly devoted their resources to stealing US technology … Perhaps most surprising about this disturbing trend is that the perpetrators are often long-time United States allies" (Tucker 1998: 1110).

During the Cold War, "The US intelligence community kept economic espionage by our friends secret to ensure that allied intelligence services continued to spy on the Soviet Union. Victimized US companies rarely revealed the theft of their confidential information." "An admission can embarrass the company, lower stock prices, scare away investors and customers, and reduce market share" (Tucker 1998: 1120). Besides, it might destroy the company's chances of dealing with a fast-rising prospective business partner.

At the end of the Cold War there began a transformation with respect to US intelligence concerns, that is, an impetus more towards economic than military objectives. But it is only in 1996 that the Economic Espionage Act (EEA) was signed into law. In contrast with the National Stolen Property Act, which was until then the main enforcement tool, the EEA gave intangible property the same degree of protection as tangible property. Yet, some would state that even such a measure is likely to prove ineffective in the light of the persistent attempts by foreign governments to make cyber intrusions (Lewis 2008). In particular, China stands to gain a lot (Hjortdal 2011) in addition to the great deal that it has already gained.

Some countries exert a sustained effort over decades through education and research complemented by very substantial sums spent on acquiring foreign technology through licenses and other agreements – this is the case for South Korea, among others (Nelson and Pack 1997). But China's rather short record on education, research and purchase of foreign technology would prove insufficient in explaining the country's achievements. In the short period of two decades, China has moved from the level of a producer of cheap products of low technological intensity to the production of technologically sophisticated products. It is hard to believe that China now has an aerospace industry and the seats of its astronauts are molded specifically to the shape and size of their bodies by indigenously developed and produced 3D printers (*Economist* 2013: 14).

Recent studies (Lemoine and Unal-Kesenci 2007; Altenburg et al. 2008) confirm that China, and India to a lesser extent, have already caught up with the OECD countries in terms of accumulation of production capabilities that use and adapt existing knowledge. High-tech goods accounted for 30 percent of China's manufactured exports in 2004 (only 5 percent for India), up from less than 10 percent in the early 1990s. However, China and India's transition from production capability to innovation capability, which requires new knowledge and putting it to use, has progressed rapidly but has still some way to go. China's performance in exports is based on the international segmentation of production processes, as the overwhelming part of high-tech exports comes from assembly of imported high-tech parts – hence the official dissatisfaction with the low value added characteristic of its production. Foreign affiliates play a dominant part in its high-tech export trade, as 53 percent of it was due to foreign fully owned firms and another 27 percent was due to joint ventures in 2003. The same happens in imports of high-tech products where foreign fully owned firms accounted for 50 percent and joint ventures for another 21 percent. This also means that the East Asian production network that results in Chinese exports is to a large extent in the hands of foreign affiliates: they import components and parts from other countries in East Asia, some from facilities they own, have them assembled in their own facilities in China, but also by indigenous firms, and then have them exported abroad.

To put it simply, the road ahead is long, and therefore the incentives for economic espionage on the part of China, so often recently accused in the press, still remain very potent.

3.3.5 Modes of technology transfer

There are many different classifications of technology transfer, each placing emphasis on a different aspect of the transfer process, such as foreign direct investment (FDI), licensing, joint ventures, franchising, marketing contracts, technical services contracts, turnkey contracts and international subcontracting, plus non-conventional channels like reverse engineering and reverse brain-drain. Most attention is devoted to the first three.

While transnational companies (TNCs) were previously identified solely with FDI, the rise of minority-owned investments and new forms of investments led to rather complex patterns of technology transfer. TNCs today engage in diversified types of relationships of which FDIs are only a part: joint ventures, sub-contracting, franchising, marketing and manufacturing.

Fees and royalties for technology reflect flows of disembodied technology taking place within TNCs as intra-firm transfers between the parent and the affiliate. It was estimated that in 1995, "some four fifths of payments of fees and royalties for technology of US and German TNCs took place between parent firms and their affiliates … Over 60% of payments to Japan originated from their own foreign affiliates" (Radosevic 1999: 23). Global payments for disembodied technology apparently quadrupled between 1983 and 1995 and reached $48 billion that year. More recent figures (see Table 1.3 in chapter 1) show that the US alone sold $122 billion of disembodied technology in 2012, Germany $69 billion and Japan $34 billion. These estimates are exaggerations as measures of payment for disembodied technology, as it is known that they are linked to repatriation of profits and transfer pricing – practices which TNCs have long been accused of using to avoid paying local taxes. For some countries such estimates may indeed be close to representing hard-earned income from disembodied technology, for example the $16 billion earned by Israel in 2012. For others, like Ireland, which is far from being a technological hotbed but rather a very successful attractor of foreign investment, the $52 billion earned in disembodied technology sales is, at the least, misleading.

3.4 Diffusion of embodied technology

Transfer of embodied technology means the technology that is embodied in products that are internationally exchanged. This includes all sorts of goods: capital goods, intermediate goods and raw materials. Embodied technology constitutes a large part of total acquired technology worldwide that rises over time. It is estimated that for the period 1970–1995, the percentage of embodied technology to overall acquired technology for the 10 technologically most advanced countries rose substantially (with the exception of Japan for which it has remained constant).

There is a reverse relationship between the amount of embodied technology acquired by a country through imports and the size of that country. It has been observed that the larger the size of a country, the more it exhibits characteristics that lead to increased local production of technology: it tends to be autarkic, to be better developed and to have strong links among sectors and firms. In contrast, small countries depend heavily on the acquisition of technology through technology embodied in imports, which often accounts for more than half of the total. For the US, the greatest producer and exporter of technology, imported embodied technology stood at 12 percent of the total in 1993, 7 percent for Japan, 25 percent for Germany, 35 percent for France, 52 percent for the UK and 71 percent for the Netherlands (OECD 1996: 47). It goes without saying that embodied technology plays a very important role. In particular, a vast literature dwells on the strategic role played by capital goods in economic development, as they are regarded as the goods with the highest technological content. This assessment seems to persist despite the introduction of electronics.

Bibliography

Anderson, H. (2004) "Why Big Companies Can't Invent", *Technology Review*, 197, 56–59.

Bass, F. (1969) "A New Product Growth Model for Consumer Durables", *Management Science*, vol.13, no.5.

Altenburg, T., Schmitz, H. and Stamm, A. (2008) "Breakthrough? China's and India's Transition from Production to Innovation", *World Development*, 36(2), 325–344.

Bassala, G. (1988) *The Evolution of Technology*, Cambridge: Cambridge University Press.

Cipolla, C.M. (1972) "The Diffusion of Innovations in Early Modern Europe", *Comparative Studies in Society and History*, 14, 46–52.

Christensen, C. (1997) *The Innovator's Dilemma*, Boston: Harvard Business School Press.

Crane, A. (2005) "In the Company of Spies: When Competitive Intelligence Gathering becomes Industrial Espionage", *Business Horizons*, 48, 233–240.

DeBresson, C. (1991) "Technological Innovation and Long Wave Theory: Two Pieces of the Puzzle", *Journal of Evolutionary Economics*, 1, 241–272.

DeBresson, C. (1995) "Predicting the Most Likely Diffusion-sequence of a New Technology through the Economy: The Case of Superconductivity", *Research Policy*, 24, 685–705.

Economist (2013) "From Dental Braces to Astronauts' Seats", September 27.

Gandal, N., Kende, M. and Rob, R. (1999) *The Dynamics of Technological Adoption in Hardware/Software Systems: The Case of Compact Disk Players*, CEPR Working Paper, London: CEPR.

Geroski, P.A. (2000) "Models of Technology Diffusion", *Research Policy*, 29, 603–625.

Grilliches, Z. (1957) "Hybrid Corn: An Exploration in the Economics of Technical Change", *Econometrica*, 501–522.

Hall, B.H. (2006) "Innovation and Diffusion", in J. Fagerberg, D.C. Mowery and R.R. Nelson (eds.), *The Oxford Handbook of Innovation*, Oxford: Oxford University Press, 459–484.

Hjortdal, M. (2011) "China's Use of Cyber Warfare: Espionage Meets Strategic Deterrence", *Journal of Strategic Security*, 4, 2, 1–24.

Kim, N., Chang, D.R. and Shocker, A.D. (2000) "Modelling Intercategory and Generational Dynamics for a Growing Information Technology Industry", *Management Science*, 46, 496–512.

Lemoine, F. and Unal-Kesenci, D. (2007) "China and India in International Trade: From Laggards to Leaders?", Centre d' Etudes Prospectives et d' Informations Internationales, 19.

Lewis, J.E. (2008) "The Economic Espionage Act and the Threat of Chinese Espionage in the United States", *Journal of Intellectual Property*, 8, 189–235.

Mahler, A. and Rogers, E.M. (1999) "The Diffusion of Interactive Communication Innovations and the Critical Mass: The Adoption of Telecommunications Services by German Banks", *Telecommunications Policy*, 23, 719–740.

Meade, N. and Islam, T. (2006) "Modelling and Forecasting the Diffusion of Innovation: A 25-year Review", *International Journal of Forecasting*, 22, 519–545.

Meyer, A.D. and Goes, J.B. (1998) "Organizational Assimilation of Innovations: A Multilevel Contextual Analysis", *Academy of Management Journal*, 31(4), 897–923.

Mokyr, J. (1990), *The Lever of Riches*, Oxford: Oxford University Press.

Nelson, R.R. and Pack, H. (1997) *The Asian Miracle and Modern Growth Theory*, Washington, DC: World Bank.

Norton, J.A. and Bass, F.M. (1987) "A Diffusion Theory Model of Adoption and Substitution for Successive Generations of High Technology Products", *Management Science*, 33, 1069–1086.

OECD (1996) *Technology and Industrial Performance, Technology Diffusion, Productivity, Employment and Skills, International Competitiveness*, Paris: OECD.

Parloff, R. (2003) "The Real War over Piracy: From Betamax to Kazaa", *Fortune*, October 27, 54–59.

Pavitt, K. (1995) "Technology Transfer among the Industrially Advanced Countries: An Overview", in N. Rosenberg and G. Frisch (eds.), *International Technology Transfer: Concepts, Measures and Comparisons*, New York: Praeger.

Radosevic, S. (1999) "Technology and Modes of Technology Transfer", in *International Technology Transfer and Catch up in Economic Development*, Cheltenham: Edward Elgar.

Rogers, E.M. (2003) *Diffusion of Innovations*, 5th edition, New York: Free Press.

Rogers, E.M. and Seidel, N. (2002) "Diffusion of News of the Terrorist Attacks of September 11, 2001", *Prometheus*, 20(3), 209–219.

Rosenberg, N. (1976) "Factors Affecting the Diffusion of Technology", *Explorations in Economic History*, 10(1), 3–33, reprinted in N. Rosenberg, *Perspectives on Technology*, Cambridge: Cambridge University Press, 189–212.

Schweizer, P. (1993) *Friendly Spies*, New York: Atlantic Monthly.

Smith, D.K. and Alexander, R.C. (1988) *Fumbling the Future: How Xerox Invented then Ignored the First Personal Computer*, New York: Morrow.

Tellis, G.J., Stremersch, S. and Yin, E. (2002) "The International Takeoff of New Products: The Role of Economics, Culture and Country Innovativeness", *Marketing Science*, 22(2), 188–208.

Tucker, D.S. (1998) "The Federal Government's War on Economic Espionage", *University of Pennsylvania Journal of International Economic Law*, 18, 3.

Van de Ven, A.H., Angle, L.H. and Poole, M.S. (eds.) (1989) *Research on the Management of Innovations: The Minnesota Studies*, New York: Ballinger/Harper and Row.

Van de Ven, A.H., Polley, D.E., Garud, R. and Venkataramam, S. (1999) *The Innovation Journey*, New York: Oxford University Press.

Von Hippel, E. (1988) *The Sources of Innovation*, New York: Oxford University Press.

Whitney, M.E. and Gaisford, J.D. (1996) "Economic Espionage as Strategic Trade Policy", *Canadian Journal of Economics*, S627–632.

Part III

The impact of innovation on Economic Theory, the economy, the firm

4 Innovation and Economic Theory

Innovation, while under constant change itself, has brought about important transformations in both Economic Theory and in the real economy. Economists first understood it as process innovation, then as technological change, then as innovation, and now as innovation in a very broad sense (1.3). In this chapter we argue that Economic Theory with a mission to explain (ex post) how the real economy works has been forced to change as well (to some extent but not enough) so as to keep abreast with the transformations.

At first, Economic Theory hardly took notice of innovation and by the time it finally did decide to deal with it, it did so less than whole-heartedly; by then hypotheses had been adopted that were both rigid and hard to replace without destroying a major part of the entire logical edifice.

The Classics were interested in technological change but in a different, more spherical way than we are at present, with the emphasis on the impact rather than the generation of the process. The Neoclassics had no real interest and Schumpeter's contribution, ahead of his time, is the next notable point in this topic. Several of his assumptions did not withstand the test of time but he left a legacy on the topic of innovation that can hardly be matched by the rest of economists. He is responsible for placing innovation not only within the economic system but at its heart; for the "Schumpetarian entrepreneur"; for the "gales of creative destruction"; and for dynamic competition. He wanted to "answer the question how the economic system generates the force which incessantly transforms it" (Hall and Rosenberg 2010: 4).

The belated advent of the economics of technology covered some but still not all of the ground. Yet it already produces a much better understanding in the less than perfect world we live in than Mainstream Economic Theory with its rigid and simplistic hypotheses. Developments in convergence do not support theory and the same is true for comparative advantage, growth and trade. The only type of innovation that Microeconomic Theory can address is methods of production. Growth cannot be accounted for and neither can trade be explained. The way the firm is portrayed ignores the concept of the firm as the main generator of innovation. Time is ignored while information and knowledge are one and the same.

We should realize that it is getting more and more difficult to live with the concept of equilibrium in a world heading for increasing returns and "winner takes all", or most.

Parts of Economic Theory will have to be rewritten.

4.1 Technological change in the history of economic thought

4.1.1 The Classics

The Classics were definitely interested in economic development and its repercussions but showed much less interest in its explanatory factors. With respect to the most important of these factors, that is, technological change, they recognized it but did not attempt to explain it because they considered it exogenously determined. They saw technology in a light different from ours, which is the reason they chose to fathom some of its aspects in a different way than our contemporary queries. With the exception of Adam Smith and Marx, their interests were in the direction of the socio-economic repercussions of the introduction of (new) technology but not the processes of its generation, use and diffusion. They believed that the economic development taking place all around them was the result of mechanization. The machines, simple artifacts of their times, were the symbols of economic development and needed no further explanation.

Adam Smith emphasized the division and specialization of labor and the concomitant rise in productivity. In his famous example on pin production (Smith [1776] 1937: 4), whereby the entire process, rather than being performed by a single worker moving along successive tasks, is broken down into a series of tasks each performed by a worker, results in higher efficiency. Such productivity gains are explained through specialization of labor and are due to three reasons: the improvement in dexterity; the saving of time; and the use of machinery. Additionally, it is specialization deriving from the division of labor that brings about the benefits of international trade in production and economic development.

For the Classics, innovations are the new methods of production. For Smith, the new technologies derive from the use of production processes, which allow the rise of productivity. As for innovation, that is, the invention of machinery, it is due to the division of labor, and derives from three sources which nowadays we call "learning". More specifically, innovation is the result of: learning by using, on the part of workers who have been entrusted with specific machinery; learning by doing, on the part of machinery producers through production and specialization of machinery; and official scientific learning (Smith [1776] 1937: 10).

For the Classics, as we shall discuss later in this section, stability is of special importance. Some maintained that stability is disrupted by innovation (which in this instance meant mechanization), but that there is a return to stability. *James Stewart* and *David Ricardo* were among those to differentiate the nature of change into sudden and gradual. But while Stewart reassures us that whatever negative consequences innovation has on employment will be temporary (Stewart [1767] 1966: chapter 19), Ricardo is optimistic at first with regard to employment, but later adopts a pessimistic position (Ricardo 1821: chapter "On Machinery"). Another interesting dimension in Ricardo's thinking on technological change is his position that it could be influenced, directed and used as a tool of economic policy. Along this line, he proposed that the state should not attempt to restrain mechanization, lest capital should emigrate to wherever it considers beneficial. This would be much worse for the working class since it is better to lose some jobs than lose all jobs.

For *Malthus*, the extent to which unemployment caused by technological change can be absorbed by the economic system depends on demand factors and more specifically on the extent to which further technological change blends with the preferences of (social) groups and the structure of society (Malthus [1820] 1966: 413).

Finally, *John Stuart Mill* did recognize the importance of innovation but noticed that it does not necessarily benefit all or even most people (Mill [1848] 1923: 751).

4.1.2 *The Neoclassics*

From the last quarter of the nineteenth century when the Classics ceded their position to the Neoclassics, interest in economic development faded and so did interest in technological change. Of course, there were some exceptions such as *Böhm-Bawerk*, *Marshall* and *Rae*, who dealt with technological change but only indirectly. The former referred to the impact of labor cost on the type of innovation. Marshall touched upon the impact of technological change on society and more specifically on employment, referred to the increased opportunities offered by innovations, and the better use of scientific and technical knowledge. In contrast with Smith, for Rae, it is innovation that precedes the division of labor (Brewer 1998).

With the Neoclassics mathematics were introduced in economic analysis and so were new standards of analytical rigor. This approach led to the principle of maximization subject to conditions. Thus, consumers maximize their utility subject to income limitations, while producers maximize their profits subject to cost limitations. Economic Theory chose simplicity over complexity, ignoring the multidimensional character of the real world – a high price to pay for gaining analytical rigor. Schmookler captured the situation perfectly: "The older economists studied the world and knew it was spherical. Only their maps were flat. Many of the economists that came later, studied the maps and mistook them for the real world" (Schmookler 1972: 62). Basic principles such as labor specialization were forgotten. Important subjects, like technological change, were thought of as part of the limiting conditions, and hence not in need of investigation.

Theorizing became especially artificial both in its hypotheses and in its conclusions. It also lost the time dimension as part of the process of economic development. Hence, analysis became static or comparatively static, whereby a parametric change is introduced and we watch the results.

The most important contribution of the Neoclassics to Economic Theory is the production function which links the level of output to the level of inputs of the factors of production, given the level of technological knowledge. For Neoclassical economics there is no theory of technology. Technology is nothing more than productivity.

4.1.3 *Schumpeter's contribution*

Both *Schumpeter* and *Marx* before him shared the Classics' interest in economic development, but they differed on the role of technological change, which they considered to be an endogenous variable, a product of the socio-economic system, and the core of development in the theory that each of them developed.

Schumpeter's main goal, as we have said, was "to answer the question how the economic system generates the force which incessantly transforms it" (Hall and Rosenberg 2010: 4). This goal he shared with Marx as

> this idea and this aim are exactly the same as the idea and the aim which underlie the economic teaching of Karl Marx. In fact, what distinguishes him from the economists

> of his own time and those who preceded him, was precisely a vision of economic evolution as a distinct process generated by the economic system itself.
>
> (Schumpeter [1937] 1989: 166)

Schumpeter also adopted Marx's idea that capitalist evolution is fueled by technological competition between firms. This became the central idea of his proposal and, for him, this was the true nature of capitalist competition and much more important than price competition (Schumpeter 1943: 84).

Starting with Schumpeter's own contributions, and they were many, probably one of the most crucial was to extend the concept of innovation beyond mechanization (process innovation) to which Marx had limited it, by including new products, new inputs, new combinations (of existing resources, equipment, and so on), new markets, new ways of exploiting markets and new organizational solutions.

Further contributions by Schumpeter refer to his definition of entrepreneurship and the distinction between innovation and invention. The "new combinations" constitute the "entrepreneurial function" which consists of "doing of new things or doing of things that are already being done in a new way (innovation)." This activity is distinct from invention (discovery): "as long as they are not carried out into practice, inventions are economically irrelevant" (Schumpeter 1934: 89).

Entrepreneurs may happen also to be inventors (or financiers or something else) but they don't have to be. They could be putting to work someone else's invention that is already in existence but that has not been applied to commercial reality yet. The entrepreneur becomes the champion of the invention, and struggles against all odds, accepts the risks and has to put up with inertia at all levels of the undertaking. For his efforts, he is rewarded with the profits, which are entirely due to his entrepreneurial tasks, contrary to Marx who considered them the result of exploitation based on the social and political power of the capitalistic class and on innovations. This did not restrain him from expressing his admiration for the capitalistic system, which he considered to have the advantage of being able to offer the required incentives for the generation of innovations and hence enjoy huge rises in productivity (Rosenberg 1990: chapter 8).

Profits are expected to be of a transitory nature. After an innovation reaches the market, imitators start appearing, and gradually they start coming in "swarms" and bandwagon behavior sets in. As the number of competitors rises, profits evaporate. The "swarming" of imitators following an innovation provides growth for the activity in which the innovation materializes. This will usually bring about some repercussions in related fields, as some innovations tend to facilitate, induce or complement other innovations. These interdependencies concentrate in particular sectors or groups of sectors, which will exhibit fast growth for a while. Eventually, they will slow down, thus generating a cyclical pattern which may contribute to business cycles of varying lengths, even to "long waves" known as Kondratiev cycles (see chapter 5).

On the question of the origins of innovation, Schumpeter, who considered technological change to be the engine of development, is clearly on the side of supply, with the emphasis on the importance of new products. He also raises both the issue of the ideal market type and the optimal firm size that would encourage innovation. For him, the large firm is the one that has the greatest possibilities for R&D, and therefore is the better choice. If size is such as to lead to oligopolistic or monopolistic situations, this by itself does not necessarily imply negative consequences, provided the large firm does not rest on its laurels but continues to innovate. This was not Schumpeter's early position but rather his later one. Many economists distinguish between the early Schumpeter (Mark I) dating before the First World War and the later Schumpeter (Mark II).

Schumpeter Mark I was criticized for glorifying the entrepreneur, where he is the main agent and champion of innovation as well the actor to bring it to fruition in a fight against social inertia in a "competitive capitalistic" system. The by now mature Schumpeter Mark II, 30 years later, realizes that most of the innovation is taken care of by "giant firms" in a system best described as "trustified capitalism".

> He did not think of large firms as a threat against technological competition ("perfect competition" he had always regarded as pure fiction) ... in Schumpeter's view, technological competition between firms should be expected to continue to drive capitalist evolution "even in the world of giant firms".
>
> (Fagerberg 2003:134)

Schumpeter does not accept the traditional theory of individualistic and static competition, and disagrees with the idea that the aim is to maximize short-run profits with prices as parameters and the level of technology as a given. On the contrary, he asserts that in the real world, competition takes place among a few large firms, where each of them attempts to gain profits but not in a static way, that is, not by choosing the quantity that is to be produced with available technology at a given moment, but in a dynamic way, through the adoption of a long-term innovative strategy. Firms compete among themselves, not by accepting technological limitations but by abolishing them. The competitive process is a process of "creative destruction", which fuels economic development by destroying the old and creating the new.

To a large extent Schumpeter met his main goal by developing an understanding of how innovation shapes economic evolution. His analysis is based on a perspective based in economics and the social sciences and not borrowed from biology – a popular practice at the time of his early writings. Of course, a system that is constantly evolving, with bursts of innovation, shocks to the system, is at odds with the notion of "equilibrium".

> Schumpeter is suggesting that the essence of capitalism lies not in equilibrium forces ... Equilibrium analysis fails to capture the essence of capitalist reality ... Whereas a stationary feudal economy would still be a feudal economy, and a stationary socialist economy would still be a socialist economy, stationary capitalism is a contradiction in terms.
>
> (Hall and Rosenberg 2010: 5)

4.2 The rise of the economics of technology and its main tenets

4.2.1 The emergence of the economics of technology

Several centuries intervened between the deliberate use of innovation in the pursuit of profit and the emergence in the 1970s of a field consisting of the systematic study of innovation and other related subjects, a field later known as the *economics of technology*, or *evolutionary economics*, or, as some prefer, *heterodox economics*. It would be unfair to say that these centuries of seeming inertia are due to the lack of intellectual curiosity in the face of business acumen, for the later centuries gave birth to the "science" of economics – a huge step in the right direction but not a step in the direction of covering the gap concerning innovation.

Since the Industrial Revolution the intensity and the role of technological change has evolved qualitatively. At the time, its presence had been noticed but not in the way we do now. By observing the new machines, economists in the nineteenth century could observe the consequences of technological change, presuming that technological change – or part of it – was the reason for the speeding up of economic growth at the time. Paradoxically, they did not attempt to explain how these machines contributed to economic growth. As a result, the machine became the symbol of economic growth rather than an explained part of the growth of the production process.

Adam Smith and Karl Marx were exceptions among the economists at the time, for their interests centered on the basic characteristics of the mechanisms of the capitalistic system combined with an analysis for the creation of technological change. Besides these exceptions, the general attitude recognized the importance of new machines as an explanatory variable of economic growth but a variable that could be taken for granted with no need for further explanation – in essence exogenous to the economic system. In contrast, economists at the time were ready to recognize and were very interested in studying the repercussions of this exogenous factor on the economic system.

There are several reasons for the emergence of the economics of technology, foremost among which is the rising inability of inherited wisdom provided by Orthodox Economic Theory, that is, Neoclassical Theory, to describe our economic environment. This is particularly true when attempting to explain growth. Starting with the seminal article by Abramowitz (1956) and the studies conducted in its wake, there remained a substantial unexplained part of growth which could be attributed neither to capital nor to labor– between 40% and 60% – and which has since been referred to as "the Abramowitz residual" and "a measure of our ignorance". Still within the realm of growth one cannot but realize that while inherited wisdom expects a convergence among economies things are drastically different in real life (OECD 1996).

Problems with Economic Theory are not confined to Growth Theory, as we shall see later in this chapter. At this point we need only name a few. For instance, microeconomics refers to innovation only as far as process innovation is concerned, as it has still not developed the tools to deal with either product innovation or organizational innovation. In international economics inherited wisdom is of little help as present-day reality is much different from its descriptions and its dictates. In a world where trade is dominated by transactions by MNCs, between the mother corporation and its affiliates, there is less and less room for theoretical explanations based on factor endowments. Besides, the principle of comparative advantage as we know it is becoming less relevant – on top of the fact that it is not only migrating but also mutating. Migrating to wherever labor is cheaper and mutating from a state of being given due to resources to a state of being construed.

Another source of unease with existing theory arises from the realization that technology is almost literally knocking on our door and its presence is increasingly felt by the day, a development for which Economic Theory has not adequately prepared us. It is more and more used as a weapon in the attempt to raise productivity in an environment that is becoming more and more competitive, leading to a changing international division of production and labor. Even the mechanism through which competition is performed has been affected, with the result that only part of the market behaves according to the dictates of Microeconomic Theory, that is, according to price, the result of supply and demand. For there is now a substantial part of the market composed of products of higher technology intensity which no longer compete on price but on innovation – first to reach the market. A new PC, tablet, smartphone, when first on the market and with superior characteristics than existing models, exhibits reduced price sensitivity.

Finally, the treatment by the existing theory of innovation as exogenous to the socio-economic system, which implies that innovation is somehow produced in some other system and is autonomous, is untenable. By now, we know that R&D, most often present in the process of the generation of innovation, do depend on the state, firms and individuals, i.e. the same socio-economic system – and as Schumpeter has argued, innovation is at the heart of it.

The economics of technology can be seen as a reaction to orthodox economic wisdom, a protest against the fact that orthodox economic knowledge refers to technological change as something apart, something that does not affect economic analysis except as an exogenous factor. It is high time for economics to recognize that innovation is an integral part of the socio-economic system and that there is a two-way causation between technology and the rest of the system – not as it is seen today in terms of a one-way causation running from technology to the rest of system.

The economics of technology is a horizontal branch of Economic Theory because it cuts across vertical branches. It thus affects practically every branch of economics, from microeconomics to international economics to Growth Theory.

The economics of technology, or evolutionary economics, is also referred to as the "economics of an imperfect world" and its main tenets are: the differentiation between knowledge and information, emphasis on learning, introduction of the notion of time in theory, the acceptance of the existence and important role of increasing returns, the role of prior experience, the fact that perfect competition is a utopia, the fact that firms do not behave as monoliths, and several more which we shall explain as we meet them in the text.

Some would say that there is a further reason for the emergence of the economics of technology, which arises from the need to reconsider the basic sources of inspiration on which the edifice of Economic Theory was built. As most sciences have done, economic "science", in its effort to gain status and acceptance, initially relied on the "hard" sciences, and more specifically on Newton's physics – as in the case of the notion of equilibrium. A reconsideration of the sources of inspiration for Economic Theory could lead in the direction of the Theory of Evolution (Veblen 1898), several of whose processes exhibit characteristics similar to the corresponding processes of technological progress (DeBresson 1987).

4.2.2 The economics of technology and the advent of Science Policy

As we have already mentioned, for a very long time there was no interest in providing a unifying theory of innovation, not even to explain the reason for its existence or the workings of its processes. Nevertheless, by the end of the Second World War, when interest in growth started materializing, there was Schumpeter's theory, which set innovation at the heart of the economic system. Economists at the time found it interesting but irrelevant to the explanation of business cycles, which was the problem that was preoccupying them, because, ironically, they saw no connection between innovation and business cycles.

In addition there were a number of individual and seemingly unrelated issues akin to innovation that had already been raised, preparing the way for the emergence of the economics of technology. One issue concerned the question of whether technological change constituted a *new method of production* or *a new product*. Most economists perceived technological change simply as a new method of production that led to diminished production costs, as if the understanding of the importance of technological progress could be exhausted by the relation of inputs to output. This oversimplification ignored the birth of new products as well as qualitative improvements, overlooking technological change's most striking contribution to human well-being. As Rosenberg put it, it is "like playing Hamlet without the prince" (Rosenberg 1982: 4).

Kuznets remarked that whether innovation could be considered a new method of production – process innovation – or a new product depends on whether we consider it from the producer's or the user's point of view (Kuznets 1929: 534–560). To the producer of textile machinery a new machine is a new product; to the user who purchases it to produce textiles it is a new method of production.

Kuznets, like Schumpeter, is credited for realizing the importance of new products in long-term growth (Kuznets 1930). An economy relying solely on process innovation would bring down the cost of production but in the long run would be able to do so less and less, thus retarding growth. Additionally, low income and price elasticities for existing products would minimize the impact of further technological change on demand. Without new products the economy would tend to stagnate.

The question of whether innovation is a *continuous* or a *discrete* process is another issue that has attracted some interest. Schumpeter emphasized the discrete nature of the innovation process and the role of the individual. A number of reasons helped maintain this position: one is "conveniently forgetting" the existence of a previous innovation, as in the case of Whitney (who was accused of forgetting an ancient ancestor of his machine). Another is national pride and the allocation of an innovation to a specific person from one's own nation. The inventor of the electric bulb is considered in England to be Joseph W. Swan, in the US, Thomas Edison, and in Russia, A.N. Lodygin. The inventor of telegraphy for Russia is A. Popov and for the West it is Guglielmo Marconi. An additional reason for considering innovation as discrete is the patent system, which recognizes that every invention is specific, and so discrete. Besides, it recognizes (legal recognition) the inventor as an individual irrespective of whether the individual has relegated his/her rights to the firm for which s/he works from the first day of being hired.

The view of innovation as continuous was held by Marx, who minimized the role of the individual and amplified the role of society, and remarked that innovation is the result of a series of contributions and improvements upon an initial invention.

An issue that attracted considerable interest is how the *rate of technological change*, i.e. the pace of innovation, varies over time and geographical region. Some emphasized the role of religion in affecting human behavior, and Max Weber suggested that Protestantism brought about capitalism ([1905] 1930). Others gave weight to the economic system in use. According to Marx, technological dynamism is directly linked to the historical emergence of capitalistic organizations. He believed that capitalism leads to great rises in productivity because the system creates especially strong incentives and institutions that accelerate technological change and the accumulation of capital (Marx and Engels 1951).

What turns a *technological champion* into a *laggard* and vice versa? Francis Bacon remarked that all the great innovations up to his day, such as gunpowder, the compass, and printing, were Chinese and there are many indications that until the fifteenth century China was more advanced technologically than the West. Some writers ascribed the decline to the value system of the mandarin class which was inimical to profit and enrichment (Needham

1969). Others attempted to explain the success of the West through its ability to borrow selectively from other civilizations (Hall 1957) and to its ability to produce and live by a culture based on objectivity rather than magic and prejudice, a culture based on the scientific revolution (Landes 1969).

An issue that created considerable interest was the *direction of technological change* and more specifically its bias or the lack of it vis-à-vis the use of the factors of production. Hicks observed that innovation is directed "naturally" towards the decrease of use of the factor of production that tends to become relatively more expensive (Hicks 1932). This view was supported by economic historians as it advocates for the way in which both the US and Europe developed – bias against the use of labor and in favor of capital which was used more abundantly (Habakkuk 1962).

The *way technological change evolves* and whether it follows a predetermined path was also of some interest. The relative prices among factors of production affect the decisions as to which techniques will be used and which in their turn will affect the further development of technological change. Choices among decisions are thus much more fateful than usually thought as they mobilize a long-term evolutionary process which links the relative prices of the factors of production to the choice of techniques and to the direction of technological change. Put simply, understanding technological change requires knowledge of its history (David 1975: 4).

The issues mentioned above more or less constituted the body of knowledge on innovation at the time when interest in growth was revived after the Second World War and considerable effort was expended in understanding its causes. The studies in the wake of the "Abramowitz residual" contributed to the realization that technological change is an important source of economic growth but also pointed towards the limitations of the ongoing approach in the analysis of such change.

Understanding of these limitations went through a number of phases. The period right after the war was a time of total trust in technology and admiration for its accomplishments during the war. This period was characterized by an unprecedented rise in R&D expenditure. However, little by little it became clear that there is no direct relationship between the level of R&D expenditure and the rate of economic growth. This implied that the links between R&D and economic growth are much more complex than they were thought to be at the time. It was only natural that the mood in the developed world changed and that skepticism towards R&D efforts (expenditure) replaced the original enthusiasm.

As a result science policies were developed during the 1960s and 1970s as an attempt on the part of the state to manage the channeling of resources to R&D for specific economic and social goals. Since R&D was found not to lead automatically to economic growth it was thought that scientific and technological inputs would have to be analyzed and studied further. The same would be true with regard to the way they are used by the economic system. This could not be done under the theoretical conditions as they existed then – in other words, looking at the problem in the tradition of neoclassical economics. So several series of studies on innovation developed that did not belong to any specific framework. These studies, which paid added attention to the intrinsic characteristics of the process of innovation, created a necessary condition for the construction of a different theoretical framework. They were not limited to the consequences of innovation but also dealt with its origins and the internal structure of its processes. They sprang from a number of theoretical fields: from the theory of production to industrial sociology, and from economic history to research on the nature of management. They attest to the existence of a few common characteristics, thus underlining the unique entity of technological change. For instance, it is common nowadays, irrespective of a specific context, to make the distinction between product and process innovation or between radical and marginal innovation or to refer to a strong versus a weak technological paradigm.

The most important point to be made is that innovation and technological change have emerged as processes with a series of specific characteristics, which are up to a point independent of the socio-economic environment in which they have been created. Consequently, they have earned the right to be studied as separate autonomous fields. However, this does not mean that innovation or technological change is totally independent from the socio-economic conditions. On the contrary, the common characteristics of these processes are exhibited in different ways under different environments exactly because innovation and technological change are endogenous to the socio-economic system. To put it simply, we can differentiate analytically between innovation and technological change and socio-economic processes and thus we can study them separately.

This new body of knowledge which arose from the desire and need better to understand innovation and technological change led to hypotheses that explain how the socio-economic system works better than the inherited wisdom of neoclassical economics.

4.3 Inherited wisdom and the economics of an imperfect world

Although Schumpeter was the protagonist of the evolutionary approach to long-run economic development for half a century, a revival of his positions began only in the 1970s, two decades after his death, and well after the resurgence of interest in growth after the Second World War. By that time, a number of individual and seemingly unrelated issues akin to innovation, as described in the previous section, had already been raised. Thus, the way was cleared for the emergence of what came to be known as the economics of technology, and gradually there appeared a number of additional studies in the "Schumpeterian tradition". Depending on the particular emphasis of certain groups of studies, they earned more descriptive titles such as "evolutionary economics" (Nelson and Winter 1982) and "economics of an imperfect world", and are more generally referred to as "heterodox economics".

Gradually there developed a body of knowledge with quite distinct positions from mainstream Economic Theory. For instance, while in mainstream Economic Theory innovation is exogenous, i.e. generated outside the socio-economic system, but impacts upon it, in evolutionary economics, *innovation is at the heart of the socio-economic system* from which it emerges, and the impact goes in both directions. In other words, the socio-economic environment creates, encourages, accelerates, directs, focuses and characterizes technological change, which in turn influences several other factors such as productivity, income, employment and methods of production, as well as the level and quality of living, within this same socio-economic environment.

Possibly the most important difference between the two approaches is their treatment of the concepts of *information* and *knowledge*. In mainstream Economic Theory, knowledge as a concept is the same as information, and therefore is subject to the same rules governing (the rest of the production inputs, that is) capital and labor. But we explained in 2.1.1 that knowledge – which also means technology – is vastly different from information. Knowledge is much more than information, which, if it is thought of as raw material, when properly processed turns into knowledge. It differs from capital and labor as it is not subject to *decreasing returns* in the way that they are, since it is not used up or diminished through use by others (even when used simultaneously) and it is cumulative. In fact it is subject to *increasing returns* as additions to knowledge become cheaper and more efficient. Such a definition of knowledge is tantamount to setting a bomb in the foundations of Economic Theory – which is built on the assumption of decreasing returns – with immediate but not exclusive victims being the theories of growth and trade, as several of the theorems on which they are based are rendered invalid.

Usually ignored or treated as an oddity in textbooks, increasing returns are explained in Box 4.1 along with the more familiar concept of decreasing returns.

Box 4.1 Increasing and decreasing returns

Our aim here is to explain what increasing returns mean by showing how they relate to total product, marginal and average product, and how they differ from decreasing returns. We do this through two sets of curves (X), (W), which are interrelated (Figure 4.1). We start with the production of saffron, the very delicate and most expensive spice, which is grown in very few micro-climates in the world. Production consists of picking the crocus flowers, which are then separated from the stigmas. The latter are dried and stored in sealed containers to avoid bleaching. The flowers can be picked only by hand and it takes about 150,000 stigmas to produce one kilogram (dry) of saffron. Our producing unit has a fixed amount of land, so the amount of labor used is the only variable input.

Levels of production are depicted by TP or Total Product curve, which relates levels of output with changing amounts of input, see (X), in Figure 4.1.

Both curves depicted in (W) derive from the TP curve in (X). The Average Product (AP) curve shows the average output produced by a unit of labor as the amount of labor changes. It is obtained by dividing output at each point of production by the corresponding number of units of labor used. It reaches its maximum level at point (B) where the steepest slope line from the origin becomes tangential to the TP curve in (X), and corresponds to point (E) on the AP curve in (W).

In similar fashion, the Marginal Product (MP) curve depicts at every level of labor use the additional production brought about by the last additional unit of labor in (W). Starting from the origin as units of labor are added, the most recent addition to labor increases TP by more than each previous addition. The TP curve in (X) rises steeply at first till it reaches point (A) in (X), which is a point of inflection, that is it signals the change in the slope of TP from increasing to decreasing. In addition, it corresponds to point (D) in (W). From the origin up until this point, the MP of each additional unit of labor is rising, and we call this space the region of Increasing Returns, also known as stage I.

From point (A) on TP to point (C) also on TP, there is the region of Decreasing Returns or stage II, as each additional unit of labor adds less and less to TP. Point (C) is the maximum level of output achieved by the TP curve, and corresponds to point (F) on (W), where MP is reduced to zero. Adding additional units of labor beyond this point leads to negative MP in stage III as additions to labor decrease TP. While stage III may not be observed in a number of cases in real life, in the case of the picking of such a delicate flower as saffron we may expect that overcrowding could provoke it.

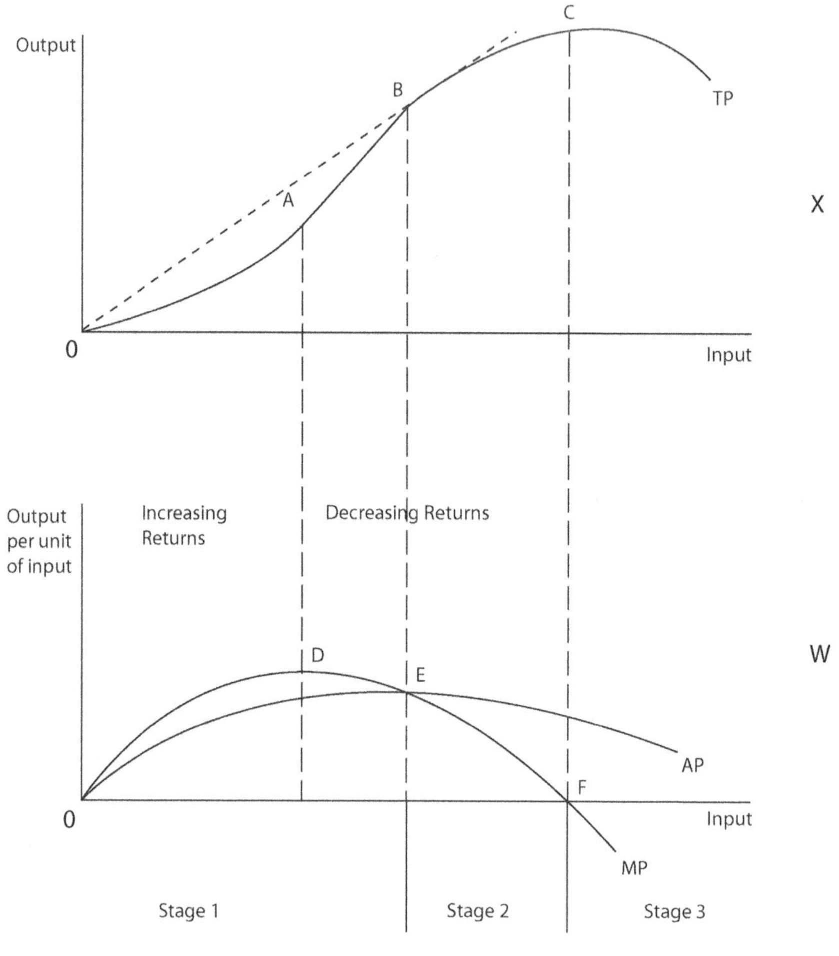

Figure 4.1 Increasing and decreasing returns

Increasing returns play the role of the phantom in Economic Theory as they appear and disappear, or more precisely are evicted from it. The existence of increasing returns or positive feedbacks has long been recognized and has always been part of the literature in economics but was never sufficiently elaborated upon. Adam Smith (1776) emphasized increasing returns to explain both specialization and economic growth. But then they disappear as David Ricardo and John Stuart Mill both conveniently forgot that "division of labor is limited by the size of the market". They made a comeback with Edward Chamberlain and Joan Robinson who analyzed the implications of increasing returns for imperfect competition. Allyn Young and Nicholas Kaldor analyzed the role of increasing returns in economic growth.

More recently, the work of Joseph Stiglitz (Rothschild and Stiglitz 1976) showed that the domain of diminishing returns is much more limited than had previously been realized (Arnott and Stiglitz 1988).

But the most striking appearance–disappearance act by increasing returns in terms of suddenness and swiftness occurs in the work of Alfred Marshall. In the first edition of *Principles of Economics*, his most famous work (Marshall 1890: 452), he not only accepted their existence but also expounded on the implications of increasing returns with respect to economic growth. By the time of its second edition, only a year after the first, any mention of increasing returns had disappeared. They were quietly and suddenly evicted from Economic Theory, and beside the few exceptions mentioned above, they were never to return.

Why such abhorrence at increasing returns? The answer is that they spoil the game because their implications inflict irreparable damage on mainstream Economic Theory. The first implication of increasing returns is their *incompatibility with the concept of equilibrium* (Viner 1937: 475–482), a concept borrowed from Newtonian physics intended to contribute to the foundations of the logical construct of economics and to provide the then new "science" with credibility and prestige.

Conventional Economic Theory is built on the assumption that economic actions generate a negative feedback that leads to a predictable equilibrium for prices and market shares. Such feedback – known as diminishing returns – tends to stabilize the economy because any major changes will be offset by the same reactions they generate. The equilibrium achieved is considered to be the "best" possible outcome given existing circumstances, and represents the most efficient use and allocation of resources.

Incompatibility of increasing returns with the concept of equilibrium means that they allow for the possibility of existence of *multiple equilibria and ex ante non-predictability* as to outcome. At the start there is no way to predict who among the competitors will win or what their market shares will be.

One of the consequences of increasing returns as to product, firm or technology is the tendency of the *frontrunner to increase its lead* from the rest and for the laggards to fall back even further. As soon as the frontrunner moves ahead of the rest of the pack, the *market tilts in its favor*. The reason for getting ahead of the pack may range from appropriate strategy to pure chance. We have seen it happen over and over again in real life with operating systems, first with PCs then smartphones, where the game is played out in terms of who will manage to attract the developers.

After winning the first round, a player's *chances of winning the next one are reinforced*, and more so every time. We have a "*first mover's advantage*" which may take the form of "winner takes all" or "winner takes most", depending on the situation.

Dominating the whole market or most of it implies that *an inferior innovation or technology could very possibly be "adopted"* in a market to the detriment of a better one or cheaper one that might be slower to reach the market. Hence, we may say that increasing returns also allow for the *possibility of inefficiency*, in the sense that selection may turn out not to be the best choice. While under decreasing returns, the market is supposed to be able to choose the best technology, under increasing returns, the market can no longer perform that task and an inferior technology may win – an eventuality that often happens in the real world.

A further dimension to the selection of an inferior technology that lends it even more importance is the eventuality of a *lock-in*, that is, once an innovation or technology has been selected and applied, it is difficult to exit from it. The most prominent example usually cited is QWERTY, the keyboard "somehow handed down to us from the Age of the Typewriters" which "nobody has been persuaded by exhortations to discard" (David 1985: 332).

Finally, a last feature that derives from increasing returns is *path dependence*: that is, history, including small events and chance circumstances, could determine which solution prevails.

More recently, the work of Joseph Stiglitz (Rothschild and Stiglitz 1976) has shed more light on competition and has shown the incompatibility of perfect competition with the existence of asymmetric information, in essence showing that the domain of diminishing returns is much more limited than had previously been realized.

Asymmetric information is not the only source of increasing returns. In fact, they derive from *four generic sources*: namely, *large fixed costs* – which give the advantage of falling unit costs as output increases; *learning effects* – which act to improve products or lower their cost as their prevalence increases; *coordination effects such as network effects* – which confer advantage to "going along" with other economic agents taking similar action; and *self-reinforcing expectations* – where increased prevalence in the market enhances belief of further prevalence.

The reader may recall that knowledge in the case of heterodox economics is subject to increasing returns and is cumulative. So far, we have investigated the implications of increasing returns. As for *cumulativeness*, it does not simply imply a dimension of gradual change as to whatever relates to knowledge but also introduces the concept of time, which is missing from conventional Economic Theory. For instance, growth can only exist as a gradual process, but history also counts in the sense that what I know today depends on what I knew yesterday. The introduction of time is in disagreement with conventional Economic Theory as the latter considers some processes to be instantaneous, such as the taking of a decision by a firm.

Time, here in the sense of history, adopts different interpretations under different regimes in returns.

> Under constant and diminishing returns .... ultimate market shares are built in a priori to the endowments, preferences, and transformation possibilities that describe the economy and small events cannot sway the outcome. Here the dynamics of the market reveal the superior choice. But while this is comforting, it reduces history to the status of mere carrier – the deliverer of the inevitable. Under increasing returns, by contrast, the process becomes path-dependent ... many outcomes are possible, and heterogeneities, small indivisibilities, or chance meetings become magnified by positive feedbacks to "tip" the system into the actual outcome "selected". History becomes all important.
>
> (Arthur 1989: 127)

Another dimension that sets apart conventional Economic Theory and heterodox economics derives from the latter's tendency to realism – the economics of an imperfect world. In this regard, it sets among others a number of serious caveats as to assumptions governing practically the totality of the firm's activities. By banishing increasing returns from Economic Theory, Marshall allowed it to be inhabited by clones of the "representative firm" (Reinert 1994), which behave as if they maximize, minimize or optimize. The firm is conceived as an agent that applies optimal technological practices received from the exogenous system of science and technology in which they have been developed. This conception ignores completely the notion of the firm as the main generator of technological change. "In textbooks firms have one goal, profit maximization, one strategy, price competition, and a single organizational means, cost minimization" (Best 1990: 139).

The world is too complex, there is a massive amount of information and humans are unable to behave in the unrealistic way neoclassical Economic Theory assumes they do when taking a decision. What they do instead is not "maximizing" but "satisficing" (Simon 1959, 1965; Cyert and March 1963), that is, they stick to a behavioral rule as long as it leads to a satisfactory outcome. Only when this stops being the case does the actor attempt to move to another alternative. Satisficing behavior belongs to "bounded" or "procedural" rationality, and has been applied to the firm's behavior in evolutionary economics through the evolutionary model of Nelson and Winter (Nelson and Winter 1982).

4.4 Evolutionary economics and Economic Theory

The criticism of mainstream Economic Theory brought about by evolutionary or heterodox economics has not affected all strands of Economic Theory evenly. Most affected are: Micro-economic Theory, International Trade Theory and Growth Theory.

4.4.1 Micro-economic Theory

Micro-economic Theory of Production is the main product of Neoclassical Theory. It analyses the relationships between quantities of inputs and outputs at the level of the producing unit, and in doing so it adopts important simplifying hypotheses on the nature of firms and markets. In other words, it combines a theory of production with an implied theory of the firm and a price theory or market mechanism.

The multiplicity of goals of the Micro-economic Theory of Production in conjunction with the way the firm is portrayed, i.e. in a way that is too general and imprecise, have earned it the description of unrealistic. In fact, we have a theory of markets. The theoretical framework is usually covered by the following hypotheses: the firm produces a homogeneous product; it has perfect information concerning inputs and outputs; it is a price-taker, that is, it accepts the ongoing price and its actions do not have an impact on either inputs or outputs; demand and supply are in equilibrium in the markets; the firm is managed by its owner; the firm maximizes profit.

As a result of its assumptions Neoclassical Theory of Production is unrealistic. In real life, the firm cannot choose among an infinite number of techniques as assumed but has a limited choice among the few techniques available (see Box 4.2). In addition, the theory allows only for limited capital–labor substitution, which means that beyond a certain point, changes in the economic environment do not bring about technological change. But we do know that existing methods can be changed and new ones can be invented, i.e. methods that use smaller quantities of expensive input. In other words, a change in the economic environment may bring about a change in the selection of the chosen technique among available and accessible choices or it may induce an innovation that could produce a series of more suitable techniques (Box 4.2).

Box 4.2 Microeconomic Theory and technological change

Production within the firm is considered to be a combination of inputs, capital (K) on the vertical axis, and labor (L) on the horizontal axis, so as to produce output, (Q): Figure 4.2. Every point on Q like C or D, represents a specific combination of (K) and (L), that is a specific method of production to produce the fixed amount of output Q. Theoretically, the firm can choose to produce at a point among an infinite number of points all along Q, not just C or D. In real life, however, the firm will be lucky to find more than a few alternatives. This constitutes the first weak point in this analysis.

Among all the infinite supposed alternatives along Q, the firm will choose to produce at the point representing the method of production that minimizes the cost of production, given the relative price of inputs. AB stands for the relative price of capital for labor as every point on it combines relatively OA of capital with OB of labor. Applying this criterion, point C, which is the point of tangency of Q and AB, and belongs to both Q and AB, is the only point that satisfies both conditions: it is one of the (supposedly) infinite alternative production points on Q and the guarantor that this is the cheapest method of production, given the existing relative prices of (K) and (L). Point D for instance, as any other point on Q, does not fulfill the second condition of also belonging to AB.

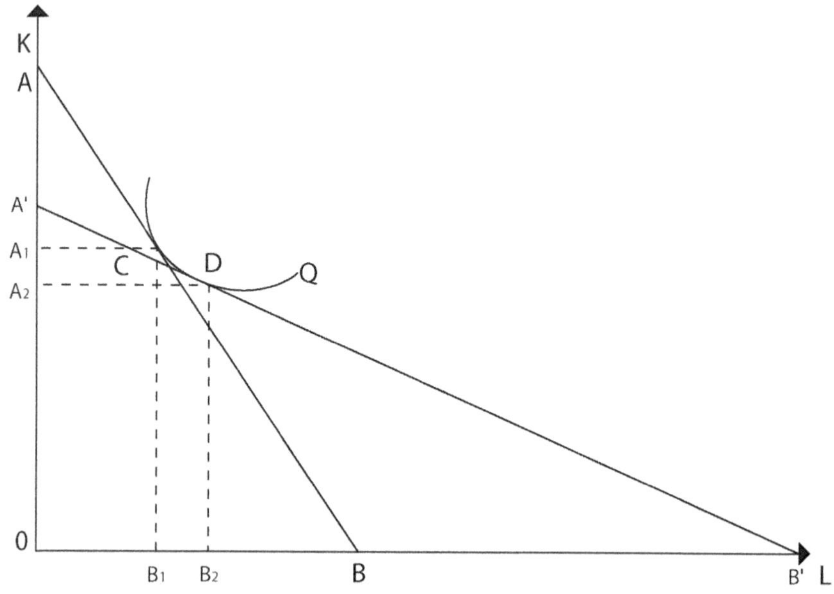

Figure 4.2 Most efficient production method and change in relative price of inputs

Now suppose that relative prices of inputs change in favor of capital (K), that is capital becomes more expensive with respect to labor (L) than it used to be when represented by AB. The new relative price of inputs line A'B' is now tangential to Q at point D, which represents the new cheapest method of production using OA2 of capital and OB2 of labor – instead of OA1 of capital and OB1 of labor for production method C.

But D cannot move much more to the right as there are theoretical limitations, due to the convex shape of Q. This amounts to saying that substitution of one input for the other is not infinite, and becomes increasingly more difficult until we reach the limits of capital–labor substitution and that beyond this point, changes in the economic environment cannot bring about technological change. This constitutes the second weakness of this analysis.

We know from historical evidence that existing methods could be changed, or new methods could be found using relatively lesser quantities of expensive input. To put it differently, a change in the economic environment could either cause a change in the choice of the selected method among existing and accessible methods, or induce the emergence of an innovation, which could generate a set of more suitable methods of production.

Yet the same tools as the ones used above may not necessarily lead to wrong conclusions, as for instance in portraying technological change, rather explicitly, in Figure 4.3. Indeed, we need no explanation in understanding that the graph depicts how technological change takes place through the movement from Q to either Q1, Q2 or Q3, which all produce the same amount of output as Q. The only difference is that Q1, Q2 and Q3 are all more advanced sets of technologies than Q as they use fewer inputs. The technological change from Q to Q2 is known as neutral technical change since the combination of inputs is the same in both cases; from Q to Q1 it is biased as it is labor saving; from Q to Q3 it is biased and capital saving.

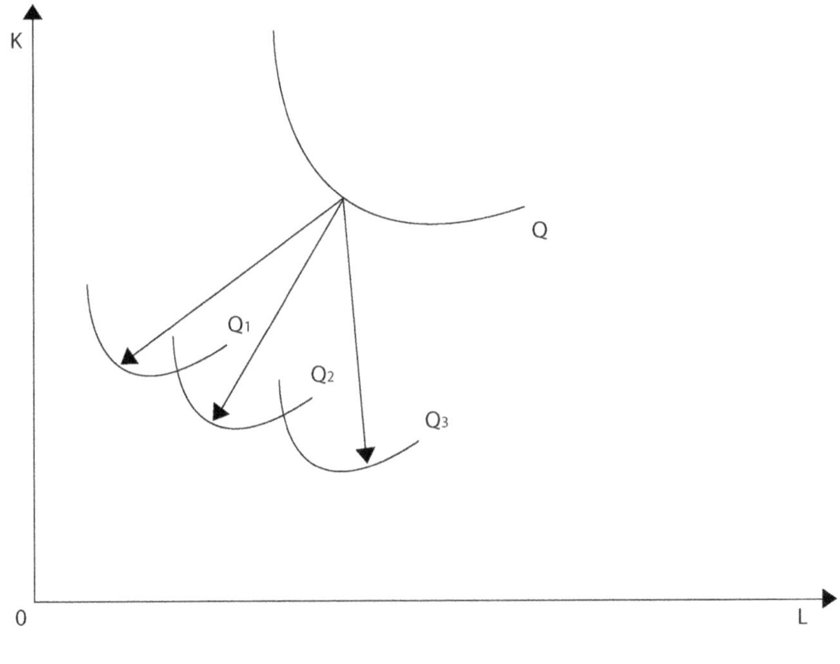

Figure 4.3 Technological change: labor saving, neutral and capital saving

However, what is most damaging to Microeconomic Theory as a framework for the analysis of technological change is the fact that it can address only one type of technological change, namely new production methods. It cannot deal with new products or services, or for that matter any other type of innovation, for it has not yet developed the means to do so, and hence constitutes only a tool of limited use for the explanation of innovation.

4.4.2 International Trade Theory

The idea that differences in technology or that differences in productivity explain the differences in specialization among countries is not at all recent. In fact, it dates back to David Ricardo (1821), the most illustrious economist on Trade Theory of all times. Yet for more than a century, the Neoclassical position on this matter has won the day. According to the Heckscher–Ohlin theorem, comparative advantage is not determined by the technological gaps between countries, but by the unequal distribution between countries of the relative factors of production (K/L in country 1 compared to K/L in country 2). Factors of production K (capital) and L (labor,) are considered immobile at the international level.

The dominant position of the Heckscher–Ohlin theorem certainly cannot be attributed to its meager prowess in the explanation of international trade transactions. On the contrary, it is understood that it has served "political correctness" well as it leads to results proving the existence of (total overall) gains from trade, hence advocating for free trade. In addition, it also served as the foundation on which an impressive series of further theorems were erected, as for instance the theorem on relative specialization based on the relative abundance of the factors of production and the theorem on price equalization of the factors of production. In return, Neoclassical theory had to accept a tight straightjacket of assumptions: same technology and same preferences among countries, perfect competition, profit maximization and the "clearing" of markets.

The first time that the Neoclassical position on trade was questioned was through the empirical work of Leontief (Leontief 1953), who found that the US, a country with abundant capital (also in relation to labor), is biased towards labor-intensive exports and capital-intensive imports – in other words, exactly the opposite of what conventional theory would have anticipated, hence referred to ever since as "Leontief's paradox". Despite the irritation caused by the paradox, it was ignored for some time; then, as usually happens, further theoretical support came to the rescue of embattled Neoclassical Theory through the relaxation of some assumptions in order to render the theory more realistic.

One such strand of renewal involved the relaxation of the unrealistic assumption that all countries have the same production function. Another way of relaxing assumptions on the production function was through the introduction of economies of scale and imperfect competition. There were many such propositions which became known as "new" theories of trade (Dixit and Norman 1980; Krugman 1979; Krugman 1980, 1986; Markussen and Melvin 1988), whose models are susceptible to multiple solutions many of which are in open disagreement with the Heckscher–Ohlin theorem (Markusen and Venables 1988; Grossman and Helpman 1990).

Another attempt at supporting beleaguered conventional theory within the framework of "new" trade theories came from a synthesis of theories of monopolistic competition, inter-sectoral and intra-sectoral trade. Intra-sectoral trade – that is trade between two countries, such as Germany and France, whereby both produce and trade the same commodity (cars) – is explained through monopolistic competition and economies of scale. Inter-sectoral trade, let us say involving trade of agricultural goods for industrial goods between two countries, is explained through traditional Neoclassical Theory (Krugman and Obstfeld 1988). Intra-sectoral trade takes place between countries with similar structures, which is the reason we meet it between developed countries. Inter-sectoral trade takes place between countries with dissimilar structure and is met in exchanges involving developed and less developed trade partners.

Contemporary theories of the technological gap between trading partners and theories of the product cycle question mainstream theory and attempt to redress comparative advantage on the grounds of the technological gap rather than on the factors of production. In most of its versions, contemporary theory of international trade based on the technological gap assumes that technology is not directly available and accessible, that it is not accessible to all, that it has a cost and that there are important advantages to being first in having access to it. Technological change generated in a country induces the generation of trade between this country and the rest of the world, and this trade will last for as long as it takes for imitators abroad to satisfy foreign needs.

The frontrunner innovator develops an important advantage, which he defends by using different means such as secrecy or patents (Freeman 1963: 22). Once imitation has taken place, trade flows depend once more on the traditional factors of adaptation and specialization. Such a position explains short-term trade flows but they may last longer due to static and dynamic economies of scale. In addition, when joined by new innovations they may lead to long-lasting trade flows.

Theories of the product cycle which attempt to explain the displacement over time of comparative advantage from one trade partner to another move in similar directions (Vernon 1966). According to Vernon's theory, new products are generated in high-income countries where there is both the demand for those products and the required technical knowledge for their production. At first, production is limited to the country where the products originated. Demand in the other high-income countries is covered through imports. But when demand rises then production starts in those countries as well. Over time, trade flows in these products between their country of origin and the countries to which production subsequently spread diminish. In the end, the direction of trade may be reversed if the new producers, the imitators, have lower costs and can become relatively efficient in the production of the new products. Comparative advantage is displaced from the frontrunner to one of the laggards. As the products mature, they become standardized and their production process can be more easily transferred to other countries.

Incidentally, the theory of the product life cycle turned out to be useful in the explanation of Leontief's paradox. When a product is new, it is produced with maximum labor intensity because as the product matures its production requires lower labor intensity. Leontief's paradox can be explained if American exports consisted of "new" products and American imports consisted of mature products.

Another strand of research relates to the importance of imports and exports of technology in the shaping of the structure of trade for a country. The possibility of the future incorporation of theories on foreign investment, models on technology transfer, models of overcoming the technological gap and dynamic models of diffusion follows (Metcalfe 1988).

More recently, a number of studies (Krugman 1987; Grossman and Helpman 1990; Markusen 1989) have attempted to combine developments in the Theory of International Trade with endogenous technological growth and monopolistic competition in innovative intermediate inputs. In other words, they have tried to combine the Theory of International

Trade with the theories on endogenous economic growth (Romer 1986, 1990); Lucas 1988).

In more recent years, International Trade Theory has been confronted with yet one more difficulty – not the emergence of MNCs, for they have been around much longer, but their proliferation. By now, a sizeable proportion of international trade is carried out between the parent company and its affiliates, or among affiliates – or more generally between firms that are related. Such a development complicates things for theory, for it again poses the original question of how can trade be explained, what is it that gives birth to comparative advantage, and how and why is it displaced.

Neoclassical Theory of International Trade, based on a comparative advantage that derives from the relative abundance of resources, struggles to describe current developments. Attempts to render that theory more realistic, as through the "new" Theories of International Trade, although impressive, are nevertheless unsatisfactory because of the simplistic way they incorporate technology as an explanatory factor. Technological change is a much more complex process than can be portrayed through learning curves or through the creation of indirect inputs under monopolistic competition.

4.4.3 Economic Growth Theory

With most of the world in ruins after the Second World War, a revival of attention to growth was only natural and there was a genuine interest in understanding where growth comes from. Many possibilities were examined several of which did not consider the involvement of technological change. Among those that did, the first well-known attempt was the Harrod-Domar model (Domar 1946; Harrod 1939), but it was Robert Solow who won the Nobel Prize for his growth theory (Solow 1956). Growth is portrayed using a production function where output (Y) is a function of the volume of capital (K) and human labor (L), with technological change assumed unchanged.

$$Y = F(K,L)$$

Changes in the inputs K and L cause changes along the function. Technological change causes upward or downward shifts in the function. Neither Solow nor the swarm of studies performed in the wake of his original work (Denison 1962) were able to explain growth to a satisfactory level. The unexplained part came to be known as the Abramowitz residual; Solow called it "technical change" and others called it "a measure of our ignorance".

Solow's theory became known as "exogenous growth theory" with the variable standing for technology, A(t), added exogenously like a trend variable.

$$Y = A(t)\ F(K,L)$$

The call for new theories of economic growth was met (Romer 1986; Lucas 1988; Grossman and Helpman 1991), but with a time delay of three decades (Nelson 1994), and consisted of attempts to assimilate technological change into Neoclassical theory.

The new approaches underline the role of technological change as well as its constituents. In contrast with the traditional Neoclassical growth model where technological change is only a function of time, the new growth models take into account the endogenous determination of technological change, which means the endogenous determination of the sources of growth. The traditional Neoclassical model is characterized by a production function with constant returns – compatible with perfect competition – in which the factor "capital" can be accumulated but the same is not true for the other factor, "labor". This means that without a rising tendency for "labor" or a steady rate of technological change, growth simply evaporates due to diminishing returns of the only accumulated factor, "capital". Under such circumstances, growth can still be saved through external forces, but then the rate of growth is exogenous and independent of economic influences.

True, new growth theories attempt to assimilate many of the traits of the Schumpeterian tradition, but they remain loyal to the linear research model and ignore interactive feedbacks, organizational innovations and cross-influences between institutional changes, technological changes and investments. In short, what is missing is a better understanding of the nature of technology and the process of technological change, a determination of the explanatory factors of the capabilities of firms and organizations.

Lucas's (1988, 1993) argument on incremental learning is similar to Arthur's and leads to "path dependency". Romer's (1990) argument explains growth through the combination of private and public investments in R&D. Both these strands attempt to take stock of the fact that knowledge may be at least partly "appropriated" through intellectual property rights.

In addition,

> while the evolutionary literature focuses on a population of heterogeneous, "boundedly rational" agents that try to find themselves through trial and error (in an environment characterized by radical uncertainty), in the new growth theory this is reduced to one "representative", "perfectly rational" agent endowed with perfect information.
>
> (Fagerberg 2003: 31)

Additionally, the two approaches consider economic knowledge from different perspectives. Under the evolutionary approach, it is considered a distributed phenomenon residing in firms as shared routines reproduced through practice. For new growth theorists, knowledge is a "public good" that is potentially available, free for anyone to use, though with some limitations because of legal arrangements. "[T]he two streams of thought, while agreeing on the importance of innovation for long run economic development and some of the implications, actually look at the world through very different lenses" (Fagerberg 2003: 31).

Neoclassical theory expects countries with roughly similar rates of population growth and a tendency for saving to converge towards a common (moving) level of productivity. Inserting the hypothesis about exogenous technological change, common to all countries, still allows for the long-run development of productivity and convergence to take place, provided all countries benefit equally from the exogenous development of technology. In contrast, new growth theories imply the possibility of preserving differences among countries both at the levels as well as the rates of income rises.

Statistical investigations have led to the conclusion that convergence involves only a few countries, particularly those of the OECD, as they converge towards the technological champion, the US, and that this is a post-Second World War phenomenon. In the 1990s convergence was reversed as once more the US seemed to outdistance her followers. Convergence is not generalized and happens under specific circumstances. For the more advanced countries, convergence with the leader is explained through the reduction of the technological gap between them, and the rest of the countries attempting to lessen the gap are those with a higher level of education and higher rate of investment. The possibility of covering the gap exists but convergence does not occur automatically, and when it does, it is the result of specific policies intended to increase investments in education, in R&D and in capital equipment.

Finally, in Classical theory, one of the ways to achieve higher growth is to put resources to better uses in the Ricardian tradition (Lewis 1954; Fei and Ranis 1964) – for instance shifting resources from agriculture to manufacturing. With Neoclassical theory such arguments simply disappeared as the firm, free from any allusion to national growth, would produce steel or yogurt, whichever the market demanded. For evolutionary economics the differentiation of economic activities regains its importance as they differ as to their innovativeness, their links with other activities, their knowledge needs and a series of other characteristics that makes every activity unique as to the transfer and diffusion of technology, and hence as to the speed of growth and convergence.

4.5 Epilogue

Technological change has been an integral part of economic thought almost from the start but it was banished together with the whole Theory of Economic Growth during the heyday of Neoclassical theory. During its absence, mainstream theory came under pressure to explain current developments, and instead of replacing some of its basic hypotheses, chose the easier alternative of adopting various "theoretical patches" such as Vernon's product cycle or Romer's artificial increasing returns. There was no hesitation in even mustering solutions that derive from other bodies of knowledge. If increasing returns were not subsumed, how would the first player's advantage be explained other than through Game Theory? How would Google's or Apple's advantage rise with each new product or service innovation?

True, Neoclassical theory has taken many steps back. Nevertheless, despite this retreat, many of its propositions are still not convincing. Developments in convergence do not support theory, and the same is true for comparative advantage, growth and trade. In the meantime, technological change is knocking on our door and demanding urgently to be assimilated into the socio-economic system. This means the "de-sanctification" of some basic hypotheses and the assimilation of a constantly rising body of knowledge on the generation, use and diffusion of technology.

This bears direct implications at the level of the firm. Many difficulties arise from the fact that the firm is considered to be an agent applying the optimum technological practices handed over to it from a scientific and technological system that has developed exogenously – outside the socio-economic system. This representation of the firm totally ignores the concept of the firm as the main generator of technological change. Equally, it is impervious to the concept of knowledge – and time – and its differentiation from information, as well as of the human factor, the only factor capable of transferring knowledge. On the theoretical level, the only type of innovation it is in a position to recognize or examine is methods of production, but it has so far failed to develop the tools required to deal with either new products or organizational innovation.

Besides these problems, some of the theoretical hypotheses – such as perfect competition, immobility of factors of production as to movement abroad, and, in contrast, perfect mobility of factors of production within the country of production, immediate, and costless access to and diffusion of technology – are all unrealistic. In practice, innovation and technological change are elements of non-perfect competition.

Finally, we should realize that we are moving towards a world of increasing returns where "winner takes all", a world not compatible with comparative advantage. We are moving away from the "democratic" world of Ricardo, where even the weakest producer has something to offer, and towards the less democratic world of Smith, where the mightiest producer can produce everything. In addition, innovation and diffusion introduce us to a world of mechanisms of dynamic adjustments that are radically different from the distributive mechanisms under traditional theory. It is getting harder and harder to live with the concept of "equilibrium" and it is high time for inherited wisdom to assimilate the concepts, ideas and propositions presented above. Economic Theory does not need to be rewritten from the start; only parts of it do – urgently.

Bibliography

Abramowitz, M. (1956) "Resource and Output Trends in the United States since 1870", *American Economic Review*, 46, 5–23.

Arnott, R. and Stiglitz, J. (1988), "Randomization with Asymmetric Information", *RAND Journal of Economics*, 19(3), 344–362.

Arthur, W.B. (1989) "Competing Technologies, Increasing Returns and Lock-in by Historical Events", *Economic Journal*, 99(394), 116–131.

Best, M.H. (1990) *The New Competition: Institutions of Industrial Restructuring*, Cambridge: Polity Press.

Brewer, A. (1998) "Invention", in O.F. Hamouda, C. Lee and D. Mair (eds), *The Economics of John Rae*, London: Routledge, 129–143.

Cyert, R.M. and March, J.G. (1963) *A Behavioral Theory of the Firm*, Englewood Cliffs, NJ: Prentice Hall.

David, P. (1975) *Technological Choice, Innovation and Economic Growth*, Cambridge: Cambridge University Press.

David, P. (1985) "Clio and the Economics of QWERTY", *American Economic Review*, 75(2), 332–337.

DeBresson, C. (1987) "The Evolutionary Paradigm and the Economics of Technical Change", *Journal of Economic Issues*, 21(2), 751-62.

Denison, E.F. (1962), *The Sources of Economic Growth in the United States*, New York: Committee for Economic Development.

Dixit, A. and Norman, V. (1980) *Theory of International Trade*, Cambridge: Cambridge University Press.

Domar, D. (1946) "Capital Expansion, Rate of Growth and Employment", *Econometrica*, 14, 137–147.

Fagerberg, J. (2003) "Schumpeter and the Revival of Evolutionary Economics: An Appraisal of the Literature", *Journal of Evolutionary Economics*, 13, 125–159.

Fei, J.C.H. and Ranis, G. (1964) *Development of the Labor Surplus Economy: Theory and Policy*, Homewood, IL: Irwin.

Freeman, C. (1963) "The Plastics Industry: A Comparative Study of Research and Innovation", *National Institute Economic Review*, 26, 22–62.

Grossman, G. and Helpman, E. (1990a), "Comparative Advantage and Long-Run Growth", *American Economic Review*, 80(4), 796–815.

Grossman, G. and Helpman, E. (1990b) "Trade, Innovation and Growth", *American Economic Review, Papers and Proceedings*, 80(2), 86–91.

Grossman, G. and Helpman, E. (1991) *Innovation and Growth in the Global Economy*, Cambridge, MA: MIT Press.

Habakkuk, H.J. (1962) *American and British Technology in the 19th Century*, Cambridge: Cambridge University Press.

Hall, A. (1957) "Epilogue: The Rise of the West", in C. Singer (ed.), *A History of Technology*, vol. 3, Oxford: Oxford University Press.

Hall, B.H. and Rosenberg, N. (2006) "Introduction", in J. Fagerberg, D.C. Mowery, and P.R. Nelson (eds.), *The Oxford Handbook of Innovation*, Oxford: Oxford University Press, 1–13.

Harrod, R.F. (1939) "An Essay in Dynamic Theory", *Economic Journal*, 49(193), 14–33.

Hicks, J. (1932) *Theory of Wages*, London: Macmillan.

Krugman, P. (1979) "Increasing Returns, Monopolistic Competition and International Trade", *Journal of International Economics*, 9(4), 469–479.

Krugman, P. (1980) "Scale Economies, Product Differentiation and the Pattern of Trade", *American Economic Review*, 70, 950–959.

Krugman, P. (ed.) (1986) *Strategic Trade Policy and the New International Economics*, Cambridge, MA: MIT Press.

Krugman, P. (1987) "Is Free Trade Passe?", *Economic Perspectives*, 1(2), 131–144.

Krugman, P. and Obstfeld, M. (1988) *International Economics: Theory and Policy*, Boston: Scott Foresman/ Little Brown.

Kuznets, S. (1929) "Retardation in Economic Growth", *Journal of Economic and Business History*, 534–560.

Kuznets, S. (1930) *Secular Movements in Production and Prices*, Boston: Houghton Mifflin.

Landes, D. (1969) *The Unbound Prometheus*, Cambridge: Cambridge University Press.

Leontief, W.W. (1953a) "Domestic Production and Foreign Trade: The American Capital Position Re-examined", *Proceedings of the American Philosophical Society*, September, 232–249.

Leontief, W.W. (1953b) "Domestic Production and Foreign Trade: The American Capital Position Reexamined", *Proceedings of the American Philosophical Society*, 249–332.

Lewis, W.A. (1954) "Economic Development with Unlimited Supplies of Labor", *Manchester School of Economic and Social Studies*, 22(2), 139–191.

Lucas, R.E. (1988) "On the Mechanisms of Economic Development", *Journal of Monetary Economics*, 22, 3–42.

Lucas, R.E. (1993) "Making a Miracle", *Econometrica*, 61, 251–272.

Malthus, T.R. [1820] (1989) *Principles of Political Economy*, ed. J. Pullen, Cambridge: Cambridge University Press.

Markusen, J. (1989) "First Mover Advantages, Blockaded Entry and the Economics of Uneven Development", Department of Economics, University of Colorado, Boulder.

Markussen, J. and Melvin, J. (1988) *The Theory of International Trade*, New York: Harper & Row.

Markusen, J. and Venables, A. (1988) "Trade Policy with Increasing Returns and Imperfect Competition: Contradictory Results from Competing Assumptions", *Journal of International Economics*, 24, 299–316.

Marshall, A. [1890] (1948) *Principles of Economics*, 8th edition, London: Macmillan.

Marx, K. and Engels, F. [1848] (1951) *The Communist Manifesto*, as reprinted in *Selected Works*, Moscow: Foreign Languages Publishing House.

Metcalfe, J.S. (1988) "Trade, Technology and Evolutionary Changes", PREST, Manchester University, Discussion Paper.

Mill, J.S. [1848] (1923) *Principles of Political Economy*, London: Longmans and Co.

Needham, J. (1969) *The Grand Titration*, London: Allen & Unwin.

Nelson, R.R. (1994) "What Has Been the Matter in Neoclassical Growth Theory?", in G. Silverberg and L. Soete (eds.), *The Economics of Growth and Technical Change: Technologies, Nations, Agents*, Aldershot: Edward Elgar, 290–324.

Nelson, R.R and Winter, S.G. (1982) *An Evolutionary Theory of Economic Change*, Cambridge, MA: Harvard University Press.

OECD (1996) *Technology and Industrial Performance, Technology Diffusion, Productivity, Employment and Skills, International Competitiveness*, Paris: OECD.

Reinert, E.S. (1994) "Catching-up from Way Behind. A Third World Perspective on First World History", in J. Fagerberg, B. Verspagen and N. von Tunzelmann (eds.), *The Dynamics of Technology, Trade and Growth*, Aldershot: Edward Elgar, 168–197.

Ricardo, D. (1821) *Principles of Political Economy and Taxation*. London: John Murray.

Romer, P. (1986) "Increasing Returns and Long-Run Growth", *Journal of Political Economy*, 94(5), 1002–1037.

Romer, P. (1990) "Endogenous Technological Change", *Journal of Political Economy*, 98(5), S71–S102.

Rosenberg, N. (1982) "The Historiography of Technical Progress", in *Inside the Black Box: Technology and Economics*, New York: Cambridge University Press, 3–33.

Rosenberg, N. (1990) "The Historiography of Technical Progress", in *Inside the Black Box: Technology and Economics*, Cambridge: Cambridge University Press.

Rothschild, M. and Stiglitz, J. (1976) "Equilibrium in Competitive Insurance Markets: An Essay on the Economics of Imperfect Information", *Quarterly Journal of Economics*, 90(4), 629–649.

Schmookler, J. (1972) *Patents, Invention and Economic Change: Data and Selected Essays*, ed. Z. Griliches and L. Hurwicz, Cambridge, MA: Harvard University Press.

Schumpeter, J. (1934) *The Theory of Economic Development*, Cambridge, MA: Harvard University Press.

Schumpeter, J. (1943) *Capitalism, Socialism and Democracy*, New York: Harper.

Schumpeter, J. [1937] (1989) "Preface" to the Japanese edition of "*Theorie der Wirtschaftlichen Entwicklung*", reprinted in *Essays on Entrepreneurs, Innovations, Business Cycles and the Evolution of Capitalism*, ed. R.V. Clemence, New Brunswick, NJ: Transaction.

Simon, H.A. (1959) "Theories of Decision Making in Economics", *American Economic Review*, 49, 253–283.

Simon, H.A. (1965) *Administrative Behavior*, 2nd edition, New York: Free Press.

Smith, A. [1776] (1937) *An Inquiry into the Nature and Causes of the Wealth of Nations*, ed. E. Cannan, New York: The Modern Library.

Solow, R. (1956) "A Contribution to the Theory of Economic Growth", *Quarterly Journal of Economics*, 70, 65–94.

Stewart (Steuart), J. (1767) (1966) *An Inquiry into the Principles of Political Oeconomy*, vol. I, ed. A. Miller and T. Cadell, London: Oliver and Boyd.

Vernon, R. (1966) "International Investment and International Trade in the Product Cycle", *Quarterly Journal of Economics*, 80, 190–207.

Viner, J. (1937) *Studies in the Theory of International Trade*, New York: Harper.

Weber, M. [1905] (1930) *The Protestant Ethic and the Spirit of Capitalism*, London: Unwin Hyman.

5 Innovation and the economy

The impact of innovation at the level of the national economy and at the level of sectors

The repercussions of technological change on the real economy permeate the economic system. We may think of it as a bottom-up impact, which starts with the firm, diffuses through the sector in which the firm operates, and ends up reaching the overall economy at the national level. However, for purposes of a more comprehensible presentation we shall here follow a top-down approach, from the general to the specific.

Thus, in this chapter we first address the repercussions at the macro level, that is, at the national and international level. Then we move one level down and examine the effects of innovation at the middle level, i.e. the level of economic sectors.

The next logical step, still one level down, would be to address the repercussions of innovation on the firm, which is precisely the subject matter of the next chapter.

Through a historical narrative the text shows the impact of innovation on the world economy and national economies, explains the big cycles, periods of growth and recession, frontrunners and laggards up to de-industrialization, as well as presenting some of the most prominent theories on cycles.

We then descend one level and reach the level of the sectors of the economy, which grow, rise, fall and move under the overall economy and the cycle. The analysis discusses the way in which innovation impacts sectoral composition; technological opportunities, market structure and performance; appropriation conditions and performance; performance and evolution; and the evolution of the market structure. The product life cycle is also discussed.

Subsequently the text analyzes the mechanisms of innovation and their impact on sectors, their borders, their mergers, separations, and internationalization. Finally, we deal with the impact of innovation on the evolution of sectors: emergence of new sectors, reinvigoration of existing sectors, substitution, death of sectors and competition.

5.1 Technological change and long cycles

A short reconstruction of international economic history in the tradition of Freeman, Clark and Soete (1982) will prove helpful in understanding the role of innovation through time, and will allow us to discuss the changing importance of economic activities, the rise of technological frontrunners, and the role of laggards in technology.

5.1.1 From the Industrial Revolution to the New Economy

A century ago, in 1913, Pareto and Van Gelderen, respectively Italian and Dutch economists, each published the results of research carried out independently of each other on long-term economic data. They both came to the conclusion that there are long cycles of international economic activity which succeed each other and have an average duration of half a century each – from 40 to 60 years. Each cycle is characterized by a period of expansion followed by a period of contraction, which is not necessarily of the same duration. Later on, it was Kondratiev's work (1925) that contributed most to the dissemination of the existence of long cycles, which have since been named after him – meager retribution for his exile and execution.

The main conclusions that emerged from the long cycle studies are: (1) economic development is uneven through time; (2) technological progress is very uneven through time; (3) the evolution of economic activities and industrial sectors is very uneven through time: some new sectors are created while others perish; (4) each long cycle is led by different sectors; (5) inequalities are also reflected in the geographic dimension: there are alternations among frontrunner countries in terms of economic performance, but these alternations are not as frequent as the long cycles.

We start with the Industrial Revolution, usually placed by most researchers between 1760 and 1780 in Britain. The first Kondratiev cycle was clearly a British affair, and no other country was in a position to repeat its achievements, as none had its economic and political infrastructure. Mokyr asserts that Britain's edge during the Industrial Revolution was due to its "government's consistently and vigorously siding with the 'party' for innovation" (Mokyr 1990: 255–261). "This is where Douglass North's argument that Britain's advantage over other nations lay in its government and institutions carries the most force" (Mokyr 1992: 331). Mokyr gives us an additional reason for why the Industrial Revolution took place in Britain.

> The real miracle is not that these Malthusian societies grew so slowly, but that they were in the end, replaced by a society in which rapid growth became the norm. At the core stood something I have called the Industrial Enlightenment (Mokyr 2002) … an attempt to carry out Bacon's dream that useful knowledge would become a "rich storehouse for the Glory of the Creator and the relief of Man's estate" (Bacon 1996, p. 143).
>
> (Mokyr 2010: 18)

Britain was beyond any competition as it produced two-thirds of the world's coal, more than half the steel, half of cotton textiles, 40 per cent of metallic constructions and had more steam engines than the rest of the world (Ray 1980). It is said that even the uniforms of the Napoleonic army bore the label "made in England".

However, despite the fact that throughout this period Britain outlawed the export of machinery (Jeremy 1973: 26), it did not manage effectively to restrict the diffusion of its innovations during the contraction period of the first Kondratiev cycle. The reason for this is that British innovations at the time were bright but simple; they concerned methods of production, and were restricted to already existing products – in other words, characteristics that enhance the spread of these innovations to countries that already produced those products.

The first Kondratiev cycle, also known as the steam engine cycle, was followed by the second cycle, which was characterized by the applications of the first: locomotives, railways, the steel industry. During the expansion phase of this cycle Britain retained its hegemony, despite the fact that in certain sectors its leadership was being challenged (Ray 1980: 18). Britain's abrupt demise during the contraction phase of the second cycle has attracted considerable interest as different explanations have been advanced. For instance, it has been claimed that by that time the industries that had propelled England to hegemony had lost their momentum, that by now its innovations were being imitated, and that it was futile to stick to the gold standard in a world ruled by protectionism and import substitution. But none of these reasons could offer a plausible explanation for the fact that at the time Britain could no longer assimilate not only foreign innovations but also its own, which were in the end exploited by others – i.e. Bessemer, Siemens (1866), Thomas (1878). Others found the British decline very natural, in effect:

> that Britain's advantage as "the first Industrial Nation" was inherently ephemeral, and the much-discussed British decline in the second half was little more than an equilibrating process, in which the technological capabilities of the other Western nations roughly caught up with Britain.
>
> (Mokyr 2010: 23)

An explanation that seems plausible is that by that time, innovations had gradually started to depend on science. Britain, which had traditionally neglected technical education and the linking of scientific centers and industry, was not well prepared to meet that challenge.

Another plausible explanation is suggested by Mokyr:

> the resistance to technological change played an important role in the demise of Britain's leadership … It found its most effective routes through an anti-technological ideology which permeated the education system, and through labor relations, in which unions developed mechanisms to short-circuit new technologies … as resistance increased, no leadership from the British government was forthcoming to encourage and protect technological creativity. The laissez-faire policies that had served so well a century earlier were slowly becoming an instrument for the defense of the technological status quo.
>
> (Mokyr 1992: 336)

Whatever the reason for England's demise during the contraction phase of the second Kondratiev cycle, it entered the third cycle as an interested observer, ceding its place of hegemony to the US and Germany. Explanations abound for the continuing distancing of England from the new frontrunners: reasons given vary from the antiquated structures in organization, management and finance (McCloskey 1981) to the lack of adaptation of the British industrialist to the new prevailing conditions (Lazonick 1991). The fact that England did not attempt to emulate the serious efforts made by both Germany and to a lesser extent the US to bridge the gap between science (hence research) and industry by setting up research centers did not help either.

The main innovations of the third Kondratiev cycle, also known as the cycle of electrification and the internal combustion engine, differ substantially from those of the first two cycles. The emphasis now is on new products rather than methods of production: cars, telephones, gramophones, typewriters, photographic machines, airplanes, radios, etc. Third-cycle innovations are much more complex, are much better linked to science and do not offer themselves easily to imitation. This means that they were much more difficult to produce by others, and their diffusion lasted longer. For the innovating countries the rise in production was phenomenal, and created needs for raw materials from the periphery: copper, rubber, petroleum. This rise in international economic interdependence did not last long. The contraction phase of the third Kondratiev cycle, which characterizes the period between the two world wars, had catastrophic effects on countries in the periphery, which by now had come to depend on the frontrunner countries for their exports in raw materials.

After the Second World War, the US emerged as the uncontested economic and technological champion and rushed into the fourth Kondratiev cycle, characterized by electronics, the new sectors of industrial chemistry, the jet engine, nuclear power, etc. In Europe, the simultaneous presence of destroyed capital and a scientifically and technologically proficient population provided the impetus for rapid economic development. The massive economic support from the US was followed by an equally impressive wave of technology transfer through FDIs by US MNCs. Several European countries seem to have benefited from those circumstances and rose rapidly to the point of coming close to challenging US leadership in some sectors.

However, the most impressive event of the fourth Kondratiev cycle was the emergence of Japan among the technological frontrunners. Japan seems to have assimilated the achievements of two Kondratiev cycles during the time period of one; it did not only overtake France and Germany but also approached the position of the US and prepared the ground to later threaten leadership in a few sectors. Several observers of what at the time was becoming known as the "Japanese miracle" came to conclusions about the reason for this ranging from chance to the psychology of the Japanese people. Others have observed that Japan's preoccupation with technology transfer has played a significant role, i.e. its insistence on appropriating foreign technology but not through FDIs. More recently, it has been observed that Japan's attempts at learning and adapting Western technologies has been titanic. What seems to have eluded most observers is the discrete but clearly leading role played by the state. Japan was not alone. Few noticed that during most of the fourth Kondratiev cycle a group of countries, the Asian Tigers (South Korea, Taiwan, Singapore and Hong Kong), copied Japan's policies, following its path and catching up with or reducing their distance from the frontrunners.

The expansionary phase of the fourth Kondratiev cycle was characterized by impressive achievements. The world economy had never known such high rates of growth. International economic interdependence was not only reinstated after its disintegration during the war, but was considerably reinforced by the Bretton Woods Agreement which set the US dollar as the basis of the exchange system, and created its watchdog, the International Monetary Fund (IMF). Even the less developed countries seem to have benefited from these achievements, though of course much less than the developed ones. This was the era of FDIs and MNCs, a time when further continued growth seemed unavoidable.

The start of the contraction phase of the fourth Kondratiev cycle dates to the end of the 1960s, that is, when "stagflation" (the simultaneous presence of stagnation and inflation) started. Later the two petroleum crises (1973) and (1979) aggravated an already difficult situation. As in all previous cycles during their contraction phase, the drop in the percentage of profits led a decrease in investments, only this time, it was accompanied by substantial labor substitution. It was this that harmed employment more than the drop in investment. The rationalization of productive capacity was brought about in two ways: the replacement of older units by new larger ones, increasing capital intensity but without substantially increasing productive capacity; and the turning at the national level towards some specific industries while withdrawing from others. As a result, there emerged a new division of production and employment at the international level.

Additionally, as has always happened in previous cycles during their contraction phase, competition became more intense with tendencies towards protectionism resulting in oligopolistic industrial concentration. The main characteristic of the rise in competition was that it was accomplished, to a much greater extent than at any time during the previous cycles, through innovation, that is, either through new methods of production, new products or new organizational changes.

What we are describing here is now referred to as "de-industrialization". This does not mean a decrease in industrial products nor does it mean the shrinking of industrial production. It refers either to the drop in the percentage of employment in industry in relation to total employment, or to the drop in the percentage of industry's participation in total domestic product. There is little point in discussing whether de-industrialization can be avoided or not, for the simple reason that all countries will end up experiencing it, whether they have been successful in meeting their policy goals or not.

In the first instance, the movement to the tertiary sector (services) as a result of de-industrialization will result from the success of the secondary sector (industry) in raising its productivity, just as in the past the movement to the secondary sector was the result of the success of the primary (agricultural) sector in substantially increasing its productivity. The frontrunners, with the luxury of planning ahead and charting their way towards that state, are the lucky ones. The laggards, in contrast, will be dragged towards the tertiary sector. Their failure in sufficiently raising the needed productivity in industry is not just due to weak performance but also due to inferior performance relative to other countries in terms of production, international trade and viability.

The relevant question then to be faced by every country is not whether it will de-industrialize but how it will reach that state: as part of a successful economic march or by being dragged there by a relatively less than successful performance.

De-industrialization, as already mentioned, does not mean that industrial products are no longer produced but rather that that these products are produced with ever-increasing productivity. This means that there will be a tendency for industrial products gradually to concentrate in the countries with the highest productivity. The implication is that those countries end up having expanded production, hence relatively better employment and steady income rises. Exactly the opposite is expected to happen to the laggards in productivity increases. For the frontrunners, the technology champions, it is a question of whether the demand in job creation by their innovative sectors is larger than the job losses from the rest of their sectors. For the technology laggards, there are no innovative sectors to demand job creation and counterbalance job losses in the rest of the economy. So, as expected, during the 1970s, the technology champions, i.e. the US and Japan, performed considerably better than the laggards, such as Europe. Both the US and Europe expanded their output by about 30 percent; however, while in the US employment rose by almost 9 percent, in Europe it shrank by almost 11 percent. In fact, ever since, Europe has had a backlog of serious unemployment which keeps growing over the years and presently stands around the 20 million mark.

There is another aspect to the fact that all countries will reach the stage of de-industrialization. The countries that reach that stage through their successful attempts at raising productivity will end up creating knowledge-demanding, very-well-remunerated service jobs. This will not happen with the laggards, so there will be a qualitative difference in jobs in the two types of countries, which translates into vast differences in incomes.

At this time, it would be rather premature to attempt a description of the fifth Kondratiev cycle, which seems to have started in the early 1980s. This cycle is characterized by the emergence of nanotechnology, the increasing importance of information technology and telecommunications as they diffuse more and more widely to more numerous activities, and the tendency of biotechnology to yield rewards. This is the era of the Internet, the new bar code, e-trade, e-banking, cryptography, security technologies, Skype, digital photography, digital television, digital image transfer, the VCR, the CD, the DVD, mobile telecommunications, mobile Internet, newspapers, libraries, mobile everything with an incredible amount of applications and of course social media. In medicine alone, technological change during this cycle has been impressive: the repercussions from the discovery of DNA, the trend towards personalized medicine and the perpetual practice of preventive medicine.

This is the era of the demise of the Soviet block, which had been exhibiting economic fatigue, especially since the emergence of the microprocessor – and the concomitant loss of ability to copy foreign technology through reverse engineering – and after five or six years of low export earnings due to low oil prices. This is the era when the victorious capitalistic system demanded and imposed the conditions for its further application and integration by freeing markets and services while capital acquired further mobility. In the presence of its two main ingredients, technological change and unbridled financial freedom, globalization thrived, and millions of at first unskilled jobs and then gradually more sophisticated ones emigrated to the East, especially China and India. Economic activity further concentrated in North America, Europe, Japan, the Asian Tigers (South Korea, Taiwan, Singapore, Hong Kong) and the BRICS (Brazil, Russia, India, China and South Africa). The rest of the world seems to have watched economic development from afar, until the phenomenal growth of China led it to invest in raw materials especially in Africa, which seems to be showing some signs of life. Within each country, the tendency from the previous cycle to recognize the value of knowledge is further enhanced, thus increasing the distance in terms of income between those that have knowledge and those that do not.

For a while, what came to be known as the digital divide threatened to add an additional hurdle to developing countries in their efforts to catch up or simply grow. But cheaper informatics, developments in telecommunications and especially the mobile Internet are in the process of circumventing that threat. In the waning days of the twentieth century, a number of developments suggested that the cycle was rushing more abruptly than expected towards its contractionary phase, despite the continued rise in importance of the copyright industries and the serious disequilibrium in favor of demand for content versus supply since the introduction of digital television.

First, the effects of the New Economy were making their presence felt, with a mixed bag of outcomes from the gales of creative destruction. Second, there was a large imbalance between supply and demand in telecommunications in the US, as demand, quite normal by other standards, proved insufficient to meet the exorbitant supply (billions invested in laying cable) fed by high expectations. The rather slow advances in security technologies and the naïve – due to lack of familiarity with the new environment – business plans seriously limited the spread of e-trade.

Then, at first the US and then the rest of the world were faced with a colossal financial crisis brought about by unbridled deregulation based on the belief that in capitalism stakeholders act according to their interest. A direct consequence is that globalization seems to have stopped and has actually retreated a little. Will things remain that way? The answer depends on whether the forces that propelled globalization will be active or not, and for all intents and purposes they have since only partly been mended.

5.1.2 Why long cycles exist

Most students of long cycles, including *Kondratiev* and *Schumpeter*, thought in terms of the general investment climate, accumulation of capital and fluctuations in the general level of investment. In this light, the expansionary phase of a Kondratiev cycle corresponds to a period during which investments rise because of perceived new profitable opportunities. In the contraction phase such opportunities are absent.

Schumpeter did not simply subscribe to the existence of the Kondratiev cycles but additionally provided an explanation of his own for their existence. Each cycle is unique in terms of the diversity of technological innovations and the diversity of exogenous factors it comprises, i.e. wars, gold discoveries, failures of agricultural production. The most important idiosyncrasy of the system is innovation. It constitutes the main driving force of capitalistic development and source of entrepreneurial profit. The ability and initiative of entrepreneurs – who are not necessarily innovators – create new opportunities for profits, which attract "swarms" of imitators who rush to also invest so as to benefit from the new opportunities, thus generating the conditions for explosive growth. The competitive process generated by the rush of the swarm of imitators erodes the margins of innovative profits. Before the system reaches equilibrium the whole process starts again through the disequilibrating consequences of a new innovative wave.

Schumpeter expected the process he proposed to be a sufficient explanation for the cycles, although he acknowledged the interdependence of this process on other idiosyncrasies of the economic system used for the explanation of conventional business cycles. Of course, the extent to which Schumpeter's proposition offers a convincing explanation for Kondratiev's cycles depends on whether some of the innovations are so large and so discontinuous as to their consequences as to bring about lasting fluctuations, or if they all bunch together (Kuznets 1940). Small technical improvements of the sort we notice routinely in most sectors are unable to provide fluctuations of the type referred to by Schumpeter.

For Schumpeter discontinuity was likened to "a series of explosions" accounted for by three hypotheses: first, that innovations are not randomly distributed throughout the economic system but are concentrated in specific sectors and are discontinuous by their nature. Second, the diffusion process is by its nature very uneven as innovations do not remain isolated events but tend to bunch together in time, with firms following every successful innovation. Third, Schumpeter suggested that the previous two features would be sufficient to impart cyclicality on the system and impose a special adjustment process.

Schumpeter's first hypothesis has been proven correct, provided innovation is interpreted in its narrow sense of being the result of R&D (see chapter 2). It is also a fact that there are differences in the rate of growth between sectors, and that while some sectors develop rapidly others atrophy and even perish, thus creating disequilibria and hence reasons for fluctuation. On the whole, however, Schumpeter's interpretation of the cycle is not totally convincing, especially in explaining the downturn and the (inevitable) succession of cycles.

Of course it has been recognized that sectors that grew rapidly owed their growth to an intense flow of technological changes and to their rapid diffusion. However, it has been argued that causality could be thought to run in the opposite direction, or even that both phenomena could be due to some third factor such as demand or entrepreneurship. Schumpeter underlined the importance of autonomous innovation and entrepreneurship. *Schmookler* (1966) emphasized demand.

Schmookler maintained that he was impartial on demand and supply and accorded them equal importance. However, his positions (Schmookler 1966: 204–208) prove the opposite. He managed to prove the co-existence of investments and patented innovations, but failed to prove the existence of a stable time delay between the two long-term series and it is not at all clear which variable precedes and which follows. All the studies that have attempted to investigate Schmookler's position, in particular regarding to what extent fluctuations in investment are followed by fluctuations in innovations, thus explaining Kondratiev cycles, conclude that this theory does not correspond to historical reality (Freeman et al. 1982: chapters 5 and 6). The same is true for theories, which rely exclusively on supply. On the contrary, Schumpeter's theory seems to relate much better to the way events happen. At first there is an autonomous impulse on the side of supply, resulting from scientific and technological progress and which transformed by charismatic entrepreneurs is then gradually substituted by demand as industry takes hold. Once a radical innovation takes place, it is very often followed by a series of marginal innovations originating from demand (minor improvements, adaptations) thus giving semblance of upholding Schmookler's thesis, which does not differentiate between radical and marginal innovations. The fact that, as we have seen in chapter 4, there are two Schumpeter models in no way hinders his explanatory contribution to the Kondratiev cycles.

Out of a number of theses that appeared on the mechanics of the long cycles a few facts emerged: basic innovations have a tendency to bunch together, but not specifically during the recessionary phase; the time elapsed between invention and innovation is not necessarily shortened during deep recessions; basic innovations are fewer towards the end of the long cycle; firms tend to decrease both their R&D activities and their attempts to obtain patents during deep recessions.

Freeman's theory gives added emphasis to scientific discoveries, to the technical and social interrelationships of "families" of innovations and to the many second-order innovations that follow an original innovation during its diffusion. He refers to these accumulations as "new technological systems" because beyond the fact that they relate to the rapid growth of some industry or industries, their impact is wider and reaches still more industries and services.

For Freeman, there are several processes of accumulation of which the phase of the cycle is only one. They include developments in basic science, specific technological discoveries, wars and preparations for war, developments in marketing and organization within the system of production and distribution, and the "natural trajectory of technologies". This does not mean abandoning Schumpeter's concept of the interrelationship between innovation and long-term economic fluctuations. Instead, the nature of this relation is much more complex and less tidy than the accumulation of a large number of basic innovations during specific decades every half a century or so. Additionally, there are innovations of wide applicability for which the links between related innovations are especially strong and move on multiple levels, i.e. scientific, technological and economic. Thus, there are "constellations" of innovations that are fundamentally interrelated and are much more than a random statistical accumulations of innovations during a specific time period.

At first, the long-term implications of a basic innovation are not visible, and it may take many years before their presence is felt at the macro-economic level. In the meantime it diffuses, a process that Schumpeter had described as a process of accumulation by imitators who foresee potential profits. This accumulation may not necessarily happen soon after the innovation's emergence, although it may if conditions are ripe. It may be delayed for a decade or more until its profitability is apparent, or until facilitating or enabling or organizational innovations appear, or social change takes place. From the moment accumulation is generated, there emerge strong multiplicative effects, which generate additional needs for capital goods, raw materials, distribution infrastructure and, of course, labor. These in turn will induce a new wave of innovations. The combination of related and induced innovations imparts expansionary impetus on the whole economy.

So the body of innovations that are being used or are diffused during the expansionary phase of the Kondratiev cycle includes not only innovations that emerged during the immediate previous recessionary phase but also innovations that emerged earlier, some which belong chronologically to the recession, and some of which emerged during the expansionary phase of the same cycle. Recessionary periods may help bring about important changes in the political and social climate, which in turn may encourage the formation of more positive conditions, which will eventually help the emergence of innovations, large and small, during the expansionary phase.

The last feature that Freeman introduced in order to finalize his theory is the "natural trajectory" of technologies, which refer either to specific industries and products or have a wider applicability. The latter are general processes of technological change such as electrification, mechanization and automation, which typically last for more than a century and whose impact may gradually shift, for instance, from transportation to control applications. The main idea is that industries benefit from technological trajectories and take advantage of economies of scale that are closely linked to them. Large investments will allow a sizeable introduction of new technology, the result of innovation in methods of production and tools, and even more general innovation.

Perez (1983) uses the term "techno-economic paradigm" instead of "technological paradigm" to signal the fact that the changes involved are beyond the engineering trajectories for some specific methods of production and to underline the fact that what binds technologies together is not uniquely technology. From the moment that natural trajectories are established as the main impact on engineers, designers and managers, a technological regime persists for several decades. To Schumpeter's successive industrial revolutions there correspond successive "techno-economic paradigms".

To recapitulate, long cycles could be generated if innovations were either really very important and with very long duration, and/or if some of them were interdependent or interrelated for technological or social reasons, or if general economic conditions encouraged their simultaneous development. Freeman emphasizes the "new technological systems" and not the random coexistence of discrete basic innovations. The groups of innovations are linked together in a technological web in which new industries and services are represented by new groups of firms with their own culture and discrete technology while they address new types of consumer behavior. Schumpeter himself considered that the first Kondratiev cycle relied on a group of textile industry innovations and the wide use of steam in industry, that the second cycle was the cycle of the railway and steel, and that the third cycle was the cycle of electrification, the internal combustion engine and the chemical industry.

5.2 Innovation and economic sectors

5.2.1 Technological change and the performance and evolution of sectors

We already know that long cycles are due to the growth and/or the generation of sectors as a result of innovations and technologies that emerged during the same or previous cycles, and that each long cycle is driven by different sectors. We also know that the evolution of sectors is very uneven as some new sectors are generated while others perish.

In this section we explain how technological change affects sectors (Dosi 1988: 1159) – or what is happening in each cycle, that is, how their performance, and evolution are affected. Performance refers to the sector's rate of growth, its rate of productivity increases, the sector's weight in the economy, and its capability to introduce new products. Evolution of the sector refers to its structure, composition, level of concentration and/or competition, and the rate of entry and exit of firms in the sector. We shall see that the main actors behind performance and evolution are: innovational learning, diffusion of innovational knowledge, and the selection process among firms, i.e. competition. Let us consider some of the most important situations.

5.2.1.1 Innovation, sectoral composition and performance

Sectoral composition matters. Every innovation brings about an asymmetry among the firms within a sector, which leads one or more firms to improvement in its competitive position. The greater the asymmetry, the greater the possibility for the technological leaders (or at least the most efficient producers) to change the structure of the sector to their advantage. In so doing, they exclude the laggards and hence improve the sector's overall performance. If the asymmetry among firms is not pronounced, then the overall level of improvement of the performance of the sector will depend on wide learning and the diffusion process.

5.2.1.2 Innovation, technological opportunities, performance, market structure

Every technological paradigm involves technological opportunities. The more numerous the technological opportunities the greater the sector's improvement in performance is expected to be. Of course, the more numerous the technological opportunities, the more there will be innovative learning on the part of some firms, and conversely more selection pressure on the laggards. The more numerous the technological opportunities, the greater the probability that some firm will "learn a lot" – a lot more than its competitors. Given its impressive performance it will drag the whole sector to higher performance levels, and by so doing, will eliminate the laggards, thus assuring the preconditions for the creation of highly concentrated structures (Nelson and Winter 1982).

5.2.1.3 Innovation, technological opportunities, appropriation conditions, performance

It is important to remember that technological opportunity is a necessary but not a sufficient condition for its exploitation. However, the speed of the sector's performance improvement does not depend on technological opportunity but on its exploitation, which depends on the existing conditions surrounding its appropriation, as well as other market factors such as market size, demand elasticities as to price and quality improvements, and the level of concentration. The rates at which technological opportunities are exploited and diffused affect the rates of change of sector performance, i.e. rates of productivity increase, changes in product prices. It is not by chance that we meet situations with impressive productivity increases and even real price declines in sectors characterized by promising new paradigms and important technological opportunities, i.e. semiconductors, computers.

5.2.1.4 Cumulativeness of technological change, performance and evolution

Cumulativeness is the most important characteristic of learning on innovational capabilities. The more cumulative technological change is at the level of firms, the more success breeds success. Firms that reach high levels of innovativeness (competitiveness), at the same time increase their chances of conserving or even increasing their competitiveness (innovativeness). In such cases, technological diversity and diffusion will play only a limited role in the sector's dynamics. In contrast, the rates of innovative learning of the technological frontrunners will determine directly the rate of change of overall performance in such sectors, which are usually very concentrated.

The opposite is true in the reverse situation where cumulativeness is relatively low as in the case of sectors, e.g. textiles, which depend technologically on other sectors to provide them with machinery and equipment in which technology is embodied (Pavitt 1984). In these situations, innovation occurs in the sectors producing the capital goods and not in the user sectors. Changes in structure and composition of the user sector result from the diffusion of new generations of capital goods, while the selection process is usually weak. Performance in these sectors depends on innovation in the capital goods' supplier sectors and their diffusion conditions.

5.2.1.5 Innovation and evolution as to market structure

If innovation derives from sources outside the sector, that is, either scientific development or capital goods suppliers, then the role of R&D within the sector is limited. If, nevertheless, a firm manages at some point in time to innovate, this does not mean that it has increased its innovative capabilities to the point of being able of take advantage of important future opportunities. And of course, the opposite is true when technological change is cumulative at the level of the firm, where present efforts depend on its previous achievements. Here, the tendency towards oligopolistic structures is intense and it will become very difficult indeed for laggards to catch up with the frontrunners (Nelson and Winter 2002).

The conditions of appropriation and the relevant efficiency of the sources of appropriation, such as patents, head start time in innovation, acquired knowledge at the level of the firm, all change during the development trajectory of a technology, its life cycle. The same is true of technological opportunities and their sources, as well as the size and rate of growth of the market. These factors, in conjunction with the prevailing conditions of competition – barriers to entry, minimum size, price/quality demand elasticity – determine how the sector develops in terms of performance and structure.

5.2.1.6 The product life cycle

The product life cycle belongs to the body of research which studies the historical evolution of sectors by analyzing the simultaneous evolution of technology and the sector's structure, and focusing on whether the sector has a natural life cycle (Nelson and Winter 2002; Utterback and Abernathy 1975). Therefore, it deals with technological change, competition and changes in the sector's structure, and the type of market that ensues. We present it here, despite the criticism it has been met with, as a tribute to the popularity it has received and its wide application as background theory in a number of fields (international trade, marketing, management).

The main argument behind the product life cycle runs as follows: when a technology is new, there is uncertainty as to how it can be improved and as to market preferences. The presence of these two types of uncertainty renders more difficult the selection between ways of development. New innovative firms trying out new solutions keep entering the sector. Those who fail leave the sector. Gradually, with accumulated effort and knowledge there emerges the most efficient selection, while the products of the new technology attract a widening market. Slowly, there emerges a dominant design, which evolves into a critical characteristic as to success and survival in the sector. The firms whose products embody the dominant design survive and can even achieve very positive results, while the firms that chose not to embody the dominant design in their products will be forced either to follow suit by redesigning their products – which is a very difficult proposition – or will have to exit from the sector.

With steadiness in product design, efficiency of R&D in methods of production rises. In the meantime, the degree of difficulty of entrance to the sector rises as the level of technological capabilities of the firms remaining in the sector also rises. Entries in the sector diminish while exits remain at high levels. As a result the number of firms in the sector diminishes, sometimes impressively, despite the fact that production may be on the rise. Usually, a few large firms remain in the sector and control it.

Three different stages can be recognized. During the first one, the discovery or embryonic stage, the size of demand is small, the level of uncertainty is high, product design lacks sophistication, and non-specialized equipment is used in production. There are many entries to the sector and the especially intense competition that develops is based on the introduction of new products.

During the second stage, the middle or development stage, the rate of increase in production is high, product design steadies, product innovation diminishes, innovation in methods of production rises, the production process becomes more sophisticated, and capital equipment becomes more specialized and substitutes for labor. Entries into the sector diminish and there is a "shakeout" among the firms in the sector. The third or mature stage corresponds to a mature market. The rise in production slows down, entries to the sector diminish further, market shares are established and innovations are less important, while the techniques in production, management and marketing have all improved.

By far the most serious criticism against the product life cycle hypothesis is that the evolution process it describes corresponds to the evolution of only a few sectors – not a very few, but not many either: out of the sectors investigated about half comply. Even worse, there is no way of knowing in advance whether a sector is evolving according to the prescriptions of the product life cycle before investigating it. The number of sectors investigated keeps gradually, rising but the uncharted territory is still vast (Mowery and Nelson 1999).

Thus, we know that beside the automotive industry – on which the original analysis of the product life cycle rests – other sectors investigated and found to comply with the product life cycle prescriptions are: typewriters (Klepper 1996), car tires (Jovanovich and

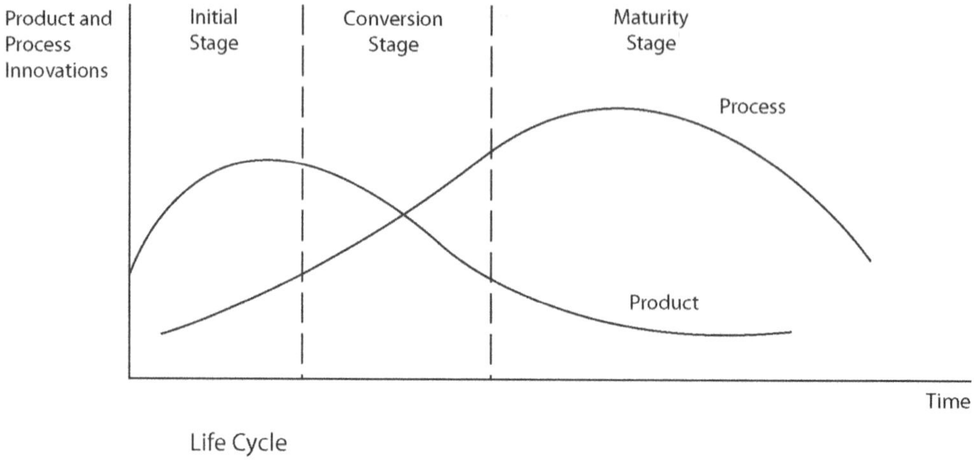

Figure 5.1 The product life cycle

MacDonald 1994), commercial aircraft (Phillips 1971), television sets, television picture tubes (Utterback and Suarez 1993) and penicillin (Klepper and Simmons 1996).

With regard to the sectors which have been investigated and found to comply with the prescriptions of the product life cycle, these have been found to do so to some extent but not necessarily in relation each and every of its details (Klepper 1997). Even Utterback (1996) conceded that the different phases of the model do not show up in the totality of industries. In addition, the crucial role allotted to dominant design in the sector's evolution has been challenged as being only one of several possible outcomes: as every technology is subject to changes, differentiations in its applications tend towards standardization, but not exclusively. Finally, there is the point concerning the "shakeouts", which turn out in some cases to be either milder than suggested or absent all together.

Thus, despite the conceptual attractiveness of the idea of the product life cycle, it offers an answer for only part of the whole. For the remaining sectors, reality is much more complex than a single alternative can explain.

The product life cycle is not in a position to explain or predict tendencies and types of innovative activity, either at the level of the firm or at the level of the country, for three reasons. First, technological opportunities, ease of appropriation and cumulativeness in innovative activities are not necessarily the same for each product life cycle. Second, the regularities of the product life cycle may be a mere reflection of changes of the ease of appropriation rather than the autonomous acceleration or deceleration of innovation. And lastly, the product life cycle is unable to predict innovative activity in a series of mature products whose innovative activity accelerated because a change occurred in their leadership or location of production. For instance, as soon as both the automotive industry and the color television sector moved to Japan, their level of innovation accelerated in new products as well as in methods of production. More recent studies shed additional light on some of the aspects of the evolutionary dynamics within the sectors.

At first, profits were high in the US television set producing sector, thus allowing entrance into the sector by firms with differentiated knowledge (Klepper and Simmons 2000a). Gradually, the erosion of profits due to the rise of competition and the constant technological change in semiconductors obliged the smaller firms to exit the sector. Only the larger US firms, most of which had already been large producers of radio sets before their entrance to the sector (in other words, besides size, they also disposed of prior knowledge in a related activity), were able to withstand the pressure. They managed to increase their market shares against their competitors but then they also lost out to their Japanese competitors. The latter, also disposed of size, had even more relevant knowledge, and were additionally supported by state development policies. They had also been radio set producers before entering the sector, but more importantly they had been frontrunners in solid-state radio, which turned out to be useful for further technological change. Of interest here is the fact that technological change, the result of innovational learning, turns out to be the source of advantage for firms with the greatest experience.

Similar results derive from the analyses of other sectors, which, like the television set producing sector, were driven towards oligopolistic structures, as was the case in the tire producing industry in the US (Klepper and Simmons 2000b). Here it is all too clear that technological change played a crucial role in the shaping of the level of survival of the firms. The results as to age, size, technological choices and level of survival confirm the fact that increasing returns that derive from technological change constitute, if not the unique, at least the strongest force in the shaping of the sector's structure.

With respect to firm entry into a sector, prior knowledge amounts to a "dowry" paid on the part of the parent firm to the firm in question, which enters the sector as a spinoff company. This entry is much more related to developments relevant to the parent firm than to the whole sector or the level of profitability of its established firms. Of course, we should notice that the parent firms are usually among the largest, the most successful and those that have accumulated the most knowledge (Klepper and Sleeper 2005: 33).

We do know that, historically, sectors with high technological intensity are subject to faster technological depreciation than others. In general, with regard to the life expectancy of firms, there is no real difference in the level of survival of firms that belong to sectors with high technological intensity and those that belong to others. However, this general result has two dimensions. During the early years, the survival rate is higher for firms in sectors with higher technological intensity and lower for the rest. In later years, this trend is reversed – a fact that could be ascribed to technological obsolescence (Agarwal 1996).

In the framework of an evolutionary process where firms are obliged to search for technological changes, the success of the solutions they choose is put to the market test. Some solutions will be met with great success, others with less, and still others will fail altogether. But the market's selection does not refer only to the solution in question but also to the firm that provides it. This evolutionary process differs from its biological counterpart because it does not depend on a random mechanism for change, but on learning. Firms learn from their environment, their competitors, their successes and their failures. The ways in which they learn are specific as to the body of knowledge characterizing each technology, each technological paradigm. As a result, the characteristics of the evolution of each sector are dictated by the ways of learning and how they affect the competitive process.

5.2.2 The mechanisms of innovation and their impact on sectors

There are several mechanisms through which technological change affects sectors. Technological change can merge sectors, separate them, and internationalize them. It can affect the way sectors evolve by creating whole new activities, it can reinvigorate sectors, it can substitute technologies and sectors relying on them, it can destroy sectors, and finally, within sectors, it can alter the competitive mechanism itself.

5.2.2.1 Innovation and the merging of sectors

When technological change strengthens the competitive influence of a technology shared by different activities and diminishes the importance of technology that addresses each specific activity, we have a "blurring" of borders between sectors, a merging that results in a new larger sector. The new success factors, now common to different activities – due to common technologies – supersede older ones specific to each activity.

As recently as two decades ago, we used to refer to the "grey" area between informatics and telecommunications. Now this area has vanished. Firms originally active in one activity were led to engage in another or others. For instance, IBM (International Business Machines), which originally started with typewriters and then moved on to computers, later attempted to enter the telecommunications sector by acquiring Rolm in 1984. Moving in the opposite direction, AT&T (American Telephone and Telegraph) moved from telecommunications to computers through its acquisition of NCR (National Cash Register) and part of Olivetti.

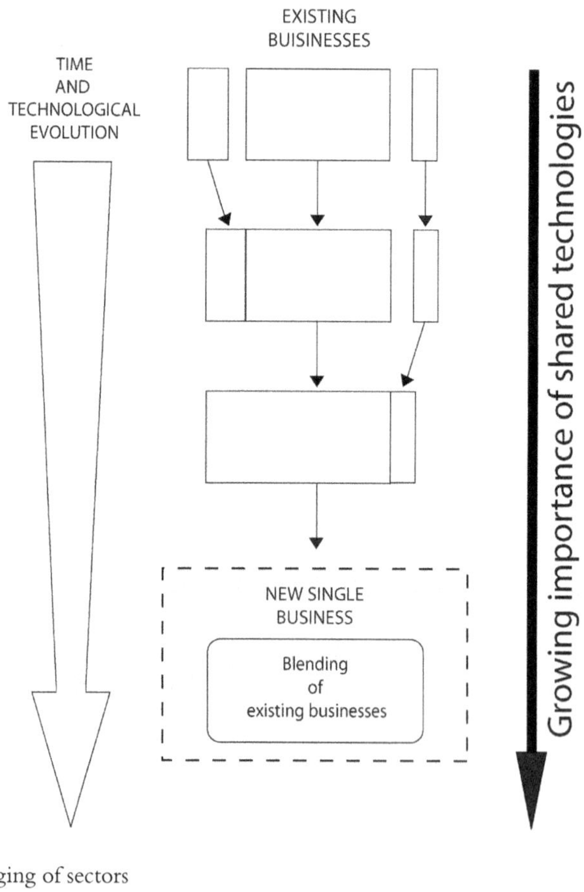

Figure 5.2 The merging of sectors

The blurring of the borders between sectors resulted in products and systems, that are difficult to categorize: word processors, fax machines, electronic mail. Ever since the Internet provided the official wedding certificate for the two sectors, we have had a torrent of such products. Skype, social media, games on the Internet.

Within the sector of machine tools, historically we had many specialized suppliers of specific products, each with its own characteristics in specialization and equipment – and its own market share. This sector was one of the few where smaller firms could survive and innovate – at least until some years ago. However, the emergence of numerically controlled machine tools – the blending of machine tools and computers – had strong repercussions on this sector. Large producers are able to develop new technologies and apply them to a wide range of products. Through robotics and flexible manufacturing, large firms are able to produce complete systems, which substantially diminish the need for many specialized models, thus worsening the small firms' survival prospects.

Instead of several activities we have one sector, an industry of industrial technology. A grouping of this kind is usually attended by a change in the cost structures of the industry. The percentage of R&D and design costs for each specific activity tends to diminish while the corresponding percentage of costs directed towards the common technologies rises.

5.2.2.2 Innovation and its impact on the borders (and composition) of sectors

If technological change is not shared by more than one sector but is localized in just one, then any sharing of activities between sectors is likely only to be partial, with movement in just one direction. An example is helpful here.

In the early 1960s, technological change generated the new product urea, a byproduct of oil refining, which was also to become one of the three main ingredients in chemical fertilizer. Simply put, oil companies were suddenly given the gift of being able to offer both a product in itself and one of the main production ingredients in a manufacturing activity different from their own. In contrast with the two-way merging of the computer and telecommunications industries, here we had a one-way-only industry movement involving the takeover of almost all fertilizer firms by rich oil firms.

5.2.2.3 Innovation and the separation of sectors

When technological change raises the percentage of costs of a specific activity while decreasing common costs, there is a tendency for the sector to fragment. In the defense industry, for instance, the level of efficiency and commercialization that defense products require sets them apart from other products that could belong to the same category. This differentiation tends to become more pronounced as a result of the divergence of the rate of technological change in the two categories. This differentiation may occur even within the same category, as for instance in the differences between the production of fighter planes and commercial aircraft and missiles. In addition, in consumer electronics goods, technological change has resulted in heightened separation, as for instance between appliances and hi-fi equipment – which dates back to at least four decades ago.

Technological change that requires the imposition of standards may bring about a separation of activities within a sector until it reunites the different standards or until one of the competitors manages to impose its own (de facto standard) – for instance PAL-SECAM in television or VHS in VCR.

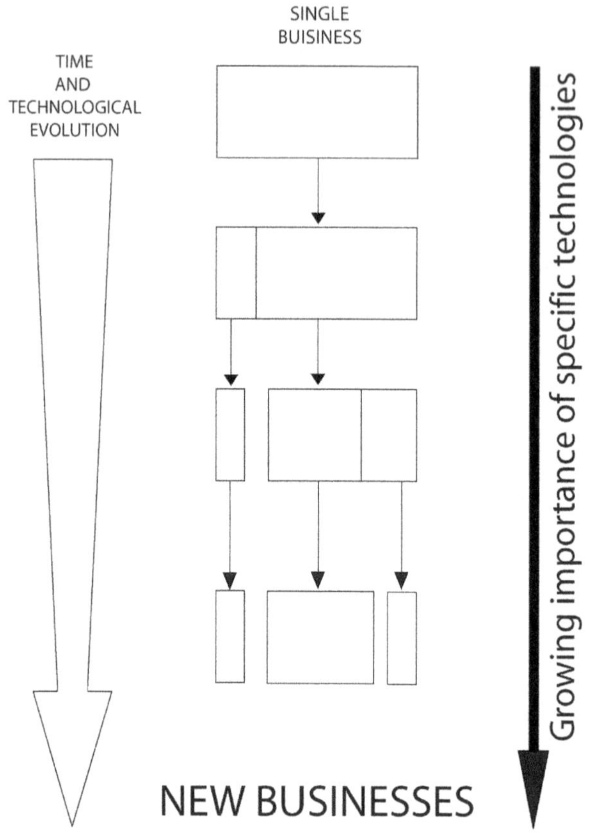

Figure 5.3 The separation of sectors

5.2.2.4 Innovation and internationalization

As the cost of technology influencing a sector rises substantially, the size of the market the sector addresses needs to expand to make room for the rising costs – i.e. distribute costs over larger production volume, hence lower per unit cost, which is a way to remain competitive. In this way, technological change and the rising impact of technology in many industries encourage internationalization. Among the products with the highest levels of internationalization are computers, cars and semiconductors.

5.2.2.5 Innovation and the evolution of sectors

INNOVATION AND THE EMERGENCE OF NEW SECTORS

Technological innovation is the most common cause of the emergence of new activities. Innovation is literally transformed into widespread economic activity and examples abound. From the steam engine to the telegraph and from the phonograph to the light bulb, important and marginal innovations create or help and improve new activities and in this book the role of new products has been often emphasized.

Of course, for technological innovation to lead to the emergence of new activities, this will only happen if the applications that ensue meet current real needs in terms of products or services. We distinguish innovations that are "pushed" by technology from those that are "pulled" by market demand. In the first instance, a market for the innovation has to be found; in the second instance, a technology has to be developed to meet the demand. It is usually said that photocopying belongs to the first category while the jet engine to the second. When the R&D department at Xerox came up with a machine that produced photocopies in 1957, the firm had no clue of what it could use it for. It is difficult for us to understand now as we cannot possibly imagine an office without a photocopying machine, but at the time a market for it had to be developed. In contrast, the jet engine was an improvement in efficiency on what the industry had to contend with at the time, i.e. a hoped for and welcome change.

This duality as to how activities emerge and develop carries weight in relation to the strategy of the firms responsible for their emergence. Some firms, those belonging to the technology-push group, base their strategy on a set of technological capabilities; others, those belonging to the market-demand pull, try to satisfy specific market needs or market categories using the most appropriate technology.

INNOVATION AND THE REINVIGORATION OF SECTORS

Technological change reaches high-intensity technology sectors as well as low-intensity ones, both new and old activities. It can reinvigorate sectors that seem to have reached saturation levels or that are already in the process of contraction, either due to product efficiency or price decrease.

In the first instance, that is the rise in efficiency; it may increase the customer base for those who are interested in additional functions, but at the same time it has the tendency to render generations of products obsolete faster, thus generating more rapid replacement: sales of reflex photographic cameras were stagnating in the mid 1980s. The introduction of autofocus raised sales by more than 20 percent.

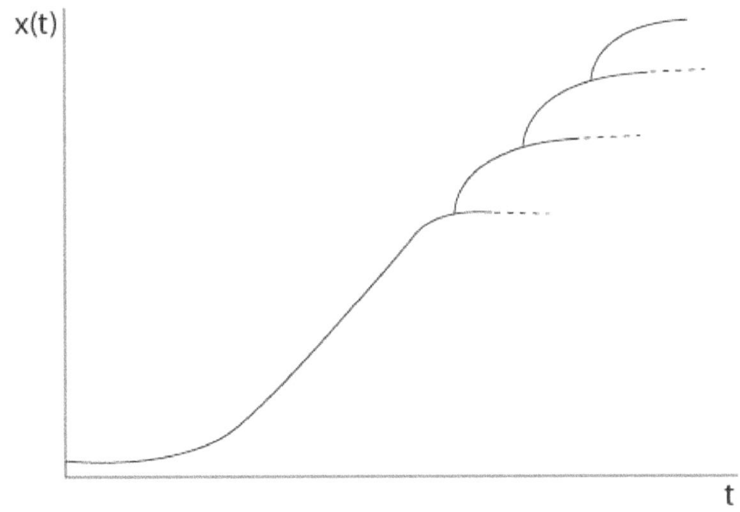

Figure 5.4 The revitalization of sectors: extending the product life cycle

In the second instance, the one associated with the price decrease, if demand is elastic, it will lead to the sector's growth. For example: an innovation in aeronautics which may lead to the construction of larger airplanes that are cheaper and more reliable may bring about lower ticket prices, which, in turn, may lead to a rise in the production of airplanes.

INNOVATION AND SUBSTITUTION

Substitution can be abrupt, causing sudden death for the product used until that time or for the firm(s) producing it or even the whole sector. Alternatively, the effect may not be so sudden and the old product may linger on for a while.

The slide rule, in its beautiful light or dark brown leather case, used to be the trademark of engineering students on university campuses, just as tees were the emblem of architecture students. The slide rule, a mechanical analog computer, was the most common scientific calculating device even during the 1950s and 1960s, as its use kept rising despite the introduction of digital computing devices. But in 1974 it became a sudden death victim at the hands of the scientific electronic calculator made available thanks to the microprocessor developed by Intel for Busicom, the Japanese calculator company.

Similarly, vacuum tubes were replaced by semiconductors. In watches, a crystal, quartz, vibrating at time's frequency, replaced the mechanical working mechanism. The new technology, launched by the Japanese firm Casio, shook down to its roots the world's biggest watch-making industry, the Swiss watch industry. Some of its firms like Vacheron Constantin, Patheque Philippe, even Rolex, plus a few others bet on brand name, while in order to survive the rest had to acquire and apply the new technology.

When substitution takes time, it is usually a matter of compatibility or sunken costs. Replacing the vinyl music record with the CD implies also giving up the record player for a CD player. Even worse, it implies replacing one's vinyl disc collection with CDs – a rather long-drawn-out proposition.

The diffusion of the jet engine was apparently slowed down by the fact that jet planes require longer takeoff and landing runways than those existing at the time, which had been built to serve propeller aircraft.

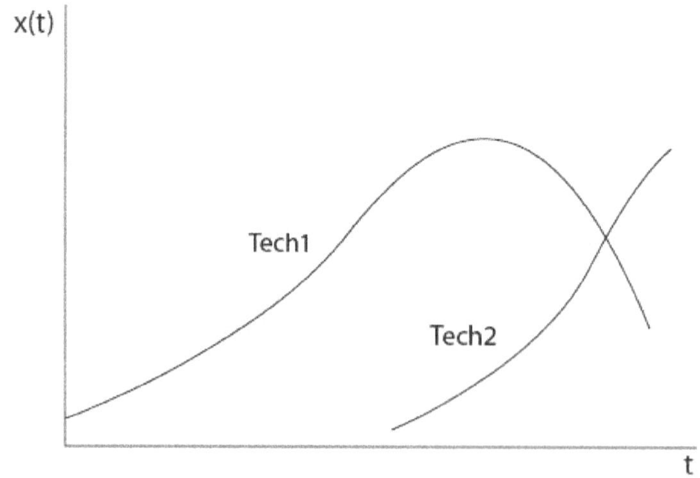

Figure 5.5 Substitution of technology

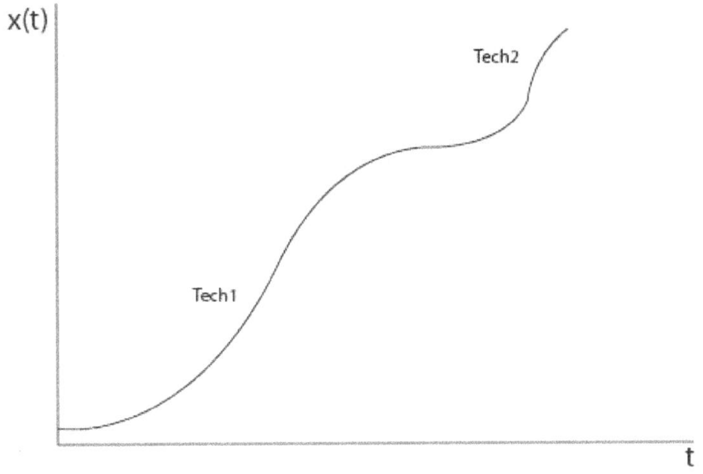

Figure 5.6 Change of technology

It has been repeatedly observed that when the threat of substitution looms, the existing firms in the sector fight back, either through lower prices or higher efficiency, thus prolonging the substitution process or rendering it partial. An often-cited case for counterattack by the old technology is the wooden ship, which came up with excellent designs while under threat of substitution by the steel ship.

Substitution may be total or partial. Some substitutions are by their nature only partial. The television–cinema relationship is like that and it is determined by the fact that from being substitutes initially they later came to complement each other, each supporting the other (see case study in chapter 6). Another example of a relationship, which is bound to remain one of co-existence and partial substitution, is the airplane–train relationship, where train travel has made inroads and reclaimed some of its older territory. Here, the travel distance of the train's advantage in terms of time and hassle has been increased during the last two decades both because of lengthy and irksome airport security checks and the introduction of faster trains (such as the TGV or *train à grande vitesse*).

INNOVATION AND THE DEATH OF SECTORS

Substitution of old technologies by new ones has sent to oblivion hundreds, even thousands, of technologies over the centuries. Incidentally, the same happens to associated activities, tools and skills. In this book, we have referred to several cases we have run into in our own time: the New Economy, for instance. New technologies replacing the old is after all what "the gales of creative destruction" are all about.

There is still a different case that needs some explanation: the case where the new technology's success itself becomes an impediment to the product or sector's long-term growth potential. This usually happens with technologies that stretch the life and increase the reliability of products in saturated or quasi-saturated markets. The problem here is that increased reliability does not generate additional demand. This is the case with car tires. The technology in their production has improved impressively and so has their reliability, but demand has been limited mostly to new car production and only limited replacement demand for the existing stock of cars.

INNOVATION AND COMPETITION

Even if technological change does not affect the borders of an activity, nor has an impact on its present growth or receptivity, it may still bring about drastic transformations of the basic factors for success in a specific activity, and above all change the relative importance of these factors. This is equivalent to a change in the rules of competition by decreasing the competitive advantages of some competitors – the strongest – and the disadvantages of others – the relatively weaker. The result is a new structure for the activity.

Here we do not refer to some innovation launched by one of the competitors to strengthen its position. Rather, the change is imposed on the whole of the activity and structurally alters the nature of competition. In this situation, each competitor has to adapt to the new conditions, and, if possible, take advantage of them in its own way.

Such technological changes usually derive from sources beyond the activity, most often upstream, such as from activities producing equipment, components or raw materials. They can impact upon the production system by minimizing or emphasizing the importance of production volume and the size of the production unit as important success factors in a given activity. For instance, in steel production, mini mills, the result of technological development allowing small volumes to be produced at a cost similar to that achieved by large units producing high production volumes, have radically altered the rules of competition. The cost advantage of large units established at specific locations to minimize transportation costs (close to mines or ports, etc.) is no longer so important in relation to smaller units located farther away from the sources of raw materials or ports. The same has happened in the aluminum industry, where technological change has considerably reduced power needs and economies of scale (moving directly from bauxite to aluminum and skipping the intermediate stage, alumina).

Technological evolution can have an effect upon other success factors such as the geographic location of production and hence the possibility of acquiring raw materials and energy at relatively low cost. In aluminum production, for instance, it has minimized the advantage of older and larger established firms, who had secured contracts for low-cost energy or proximity to hydro-electric stations.

In this light, technological change can affect the structure of competition by changing the location of activities upstream or downstream or by adding/removing barriers to entry in the industry and thus encouraging or discouraging entry. Increasing standardization in microcomputers, the localization of innovation in the microprocessor, the systematic use of subcontracting in the production of components and cloning of the most successful products have all contributed to the commoditization of microcomputers. In so doing they have reinforced the importance of factors such as distribution and after-sales services. As the firms can no longer differentiate themselves through technology, there is a tendency for increased competition through prices started by clone producers in their effort to bypass the brand name and the after-sales services offered by the large producers.

Simply put, the rules of competition are significantly affected by technological change.

Bibliography

Agarwal, R. (1996) "Technological Activity and Survival of Firms", *Economics Letters*, 52, 101–108.
Bacon, F. (1996) "The New Atlantis", in B. Vickers (ed.), *The Major Works*, Oxford: Oxford University Press.
Dosi, G. (1988) "Sources, Procedures and Microeconomic Effects of Innovation", *Journal of Economic Literature*, 26(3), 1120–1171.
Freeman, C., Clark, J. and Soete, L. (1982) *Unemployment and Technical Innovation, A Study of Long Waves and Economic Development*, London: Frances Pinter Publishers.
Jeremy, D.J. (1973) "British Textile Technology Transmission to the United States: the Philadelphia Region Experience, 1770–1820", *Business History Review*, 47, 24–52.
Jovanovich, B. and MacDonald, G.M. (1994) "The Life Cycle of a Competitive Industry", *Journal of Political Economy*, 102, 322–347.
Klepper, S. (1996) "Evolution, Market Concentration and Firm Survival", Mimeo.
Klepper, S. (1997) "Industry Life Cycles", *Industrial and Corporate Change*, 6(1), 145–181.
Klepper, S. and Simmons, K.L. (1996) "Innovation and Industry Shakeouts", paper presented at the Business History Conference, Ohio.
Klepper, S. and Simmons, K.L. (2000a) "Dominance by Birthright: Entry of Prior Radio Producers and Competitive Ramifications in the US Television Receiver Industry", *Strategic Management Journal*, 21, 997–1016.
Klepper, S. and Simmons, K.L. (2000b) "The Making of an Oligopoly: Firm Survival and Technological Change in the Evolution of the US Tire Industry", *Journal of Political Economy*, 108(4), 728–760.
Klepper, S. and Sleeper, S. (2005) "Entry by Spinoffs", Working Paper, Carnegie Mellon University, March.
Kondratiev, N. (1925) "The Major Economic Cycles", English translation in *Review of Economics and Statistics*, 18, November 1935.
Kuznets, S. (1940) "Schumpeter's Business Cycles", *American Economic Review*, 30(2), 257–271.
Lazonick, W. (1991) *Business Organization and the Myth of the Market Economy*, Cambridge: Cambridge University Press.
McCloskey, D.N. (1981) *Enterprise and Trade in Victorian Britain: Essays in Historical Economics*, London: Allen & Unwin.
Mokyr, J. (1990) *The Lever of Riches: Technological Creativity and Economic Progress*, New York: Oxford University Press.
Mokyr, J. (1992) "Technological Inertia in Economic History", *Journal of Economic History*, 52(2), 325–338.
Mokyr, J. (2002) *The Gifts of Athena: Historical Origins of the Knowledge Economy*, Princeton, NJ: Princeton University Press.
Mokyr, J. (2010) "The Contribution of Economic History to the Study of Innovation and Technical Change", in B.H. Hall and N. Rosenberg (eds.), *Handbook of the Economics of Innovation*, Amsterdam: Elsevier–North Holland, 11–50.
Mowery, D.C. and Nelson, R.R. (eds.) (1999), *Sources of Industrial Leadership: Studies of Seven Industries*, Cambridge: Cambridge University Press.
Nelson, R.R. and Winter, S. (1982) *An Evolutionary Theory of Economic Change*, Cambridge, MA: Belknap Press.
Nelson, R.R. and Winter, S. (2002) "Evolutionary Theorizing in Economics", *Journal of Economic Perspectives*, 16(2), 23–46.
Pavitt, K. (1984) "Patterns of Technical Change: Towards a Taxonomy and a Theory", *Research Policy*, 13(6), 343–373.
Perez, C. (1983) "Structural Change and the Assimilation of New Technologies in the Economic and Social System", *Futures*, 15(4), 357–375.
Phillips, A. (1971) *Technology and Market Structure*, Lexington, MA: Lexington Books.
Ray, G. (1980) "Innovation in the Long Cycle", *Lloyds Bank Review*, 135, 14–28.

Schmookler, J. (1966) *Invention and Economic Growth*, Cambridge, MA: Harvard University Press.
Utterback, J.M. (1996) *Mastering the Dynamics of Innovation*, Boston: Harvard Business School Press.
Utterback, J.M. and Abernathy, W.J. (1975) "A dynamic model of process and product innovation", *OMEGA*, 3(6), 639–656.
Utterback, J.M. and Suarez, F.F. (1993) "Innovation, Competition and Industry Structure", *Research Policy*, 22, 1–21.

6 Innovation and the firm

The previous chapter was devoted to the impact of technological change on the overall economy, and to the impact at the level of sectors. In the present chapter we move down one more level to analyze innovation's impact at the level of firms.

Over a period of less than two centuries, innovation has evolved though we have been rather slow in realizing both this fact as well as the changes that this has brought about for the firm. Innovation has generated new opportunities (as well as some threats), and in order to adapt and take advantage of them the firm has had to change dramatically in all its aspects: size, functions, organization, location, use of resources, aims, values, and the way it sees itself and the environment in which it has to survive and with which it must cooperate.

Overall, technological change has imposed on the firm five broad direct imperatives plus an indirect one – rules to be complied with just for survival. In connection with innovation, still more directives are to be observed by all firms. For high-tech firms the demands are even more stringent as the threats multiply, and opportunities get harder and harder to discern and require added boldness, deep pockets and knowledge. They rely on different economic principles than the rest of firms, have narrower windows of opportunity, face greater competition and faster technological change, and are more liable to fall victim to disruptive technologies.

The firm evolved from the family-owned small firm, producing a single product in one location to the decentralized multi-departmental, multi-product, multinational firm, carrying out its own research but also sharing knowledge with other firms and striving to turn into a learning organization.

The firm used to attempt to optimize mass production of specific goods and the challenge for its management was to forecast demand in slowly rising markets. Internally, respect for hierarchy and loyalty were valued most. Externally, the firm's relations with its environment could be described as "business is war". Under the pressure of competitiveness the firm "rationalized" by keeping its core activities and abandoning activities considered more distant from its centre. These were either sold to other firms through mergers and acquisitions or were "outsourced", that is, performed by outsiders to the firm more efficiently. Rationalization and outsourcing resulted in trimming the size of the firm and in flattening its structure.

Gradually, within the time span of five or six decades, the firm moved from a producing entity to a learning one alongside production and learned to face more ephemeral opportunities in fast-growing markets while large profits awaited whoever dared make the first move (innovation). Firms tend to congregate in ephemeral groupings and in "galaxies" of cooperation. Competition is accompanied by cooperation, a state of "co-opetition".

The chapter closes with a case study illustrating the evolution of a whole sector under the impact of technological change: the US motion picture industry with its evolving structure, studios, products, finance, links, value chain and prospects.

6.1 Innovation and the firm

During the second half of the nineteenth century, important changes took place with respect to the organization of the firm. Small, family-owned enterprises with a single activity, engaged in production or trade in a single location, were gradually replaced by larger enterprises, in which management became as important as the main producing activity of the business. The geographical spread of the firm's activities, the increasing volume of trade, as well as the process of forward and backward integration through which extracting units of raw materials were acquired, led to the specialization of management either geographically or functionally. As a result, the "centralized" firm, functionally departmentalized, emerged.

Since the 1920s, the development of existing production lines in firms, the search for new markets and new sources of raw materials, and the opening of new markets through a wide range of new products has led to the development of decentralized multi-departmental firms.

Innovation contributed to bringing about organizational changes. Progress in transportation and cooling, which allowed the geographical development of firms and markets, contributed to the generation of more complex managerial functions, and showed the necessity for new types of organization.

What is implied here is not that innovation is the unique and direct cause for the organizational changes that have taken place within the firm for more than a century. Instead, the implication is that technological change is one of the factors that shape firm strategies, which in their turn lead to organizational changes in firms. To it put simply, the firm attempts to benefit from the opportunities offered by all sorts of changes – i.e. demographic, income, preferences, plus those offered by the new technology by adopting a new strategy – but in order to do so it needs changes at all levels of its existence.

The evolution from the family-owned small firm to the decentralized multi-departmental firm relied on a gradually rising specialization of managerial roles leading to growing coordination problems. Coordination schemes like the "general office" or "central office" and business planning were introduced in the interwar years, but became widely accepted and were applied mostly in the 1950s and 1960s.

6.2 The R&D process and the firm: structure, organization, functioning, strategy

6.2.1 The migration of innovative activity from the independent workshop to the R&D department within the firm

The organizational changes described so far established the differentiation between the sales and production functions within the firm. In the meantime, the firm acquired another function, R&D, which was destined later to become a formidable weapon in its competition with rivals.

From the time of the first German research laboratories in the 1870s, followed shortly afterwards by their US counterparts, the location of research has gradually moved from independent workshops to organized laboratories within large firms. These workshops were used for simple tasks by firms that did not carry out their own research. In addition, these workshops never took the initiative to develop new products on their own, nor did they ever present a wide spectrum of research services even in areas with limited risk. At their peak, they employed about a quarter of all research personnel in the US.

What then was the cause of the movement from the independent workshop to the R&D department within the firm, placing it alongside the marketing and the production departments (Teece 1988)? If specialization were the answer, then the independent workshop would be the best choice. Indeed, in developed countries and in areas such as defense, substantial research is carried out by independent workshops on a contract basis. At the same time, in most of the firms engaged in sectors with very rapid technological change, research needs are covered within the firm's research facilities.

Two types of factor are at play. The first has to do with economies of scale and specialization. Innovation is an activity, which requires specific knowledge and capabilities, use of expensive tools and equipment, and people able to work collectively. The second has to do with the appropriability and safeguarding of the rights of the firm as to the results of its research. If research is occasional, or if there are economies of scale that the firm does not possess, then the use of the specialized external workshop is the natural choice. In that case, however, the risk involved in the contractual relation with the external workshop is especially high. The higher the rate of uncertainty, the greater the difficulties encountered by contractual arrangements, for it is very difficult to foresee what the end result of the research should be and to define it in contractual form before it has even started.

There is a still further danger for the firm inherent in having research performed repeatedly on its behalf by the same external workshop, no matter how well previous contracts might have been executed. This danger arises from the fact that the independent workshop may be familiar with a specific technology or technologies, and tends to see solutions to problems only in those technological terms. The result may be that the firm may find itself trapped or, as we usually say, end up "locked in" to a technology that may not turn out to be the best available choice (David 1985).

The development of organized R&D within the contemporary firm brought about a dramatic change in the capacity of industry to create a flow of new and differentiated products, and to use changes in this capacity as a competitive weapon among firms and among countries. In a way, it is an attempt to internalize and control prospective benefits from technological change rather than to end up as one of its victims. It is an effort on the part of the firm to minimize future uncertainty and to gain control in an environment "under negotiation".

6.2.2 The changing nature of R&D and its conceptualization

While the process of innovation was migrating to its new location, it was also changing in nature and so was our way of conceptualizing and realizing it.

As a result of the admiration, respect and awe inspired by the scientific and technological prowess of the war years, technology-push or supply innovation remained unchallenged as *the* source of innovation until the end of the 1960s and early 1970s. Then a number of studies, many of which were not comparable among themselves due to differences in the methodology used, came to the conclusion that demand (or need) is the most important factor behind innovation activity (Rothwell et al. 1974; Utterback 1974).

The "technology-push" argument suggests that innovation is led by science, which in its turn leads technology and its application. The "demand-pull" argument suggests the opposite, that is that demand is the crucial factor and that markets, users and applications are the main drivers of innovation. Underlying both these positions is a linear model of the innovative process with science at one end of the spectrum and users and markets at the other.

In the first instance, the process is simple, linear and in series. The emphasis is on R&D, while the market is a receptacle which accepts their results. In the second case, the process is also seen to be simple, linear and in series except that the emphasis now is on marketing. The market is the source of ideas that direct R&D, which simply reacts (Chiesa 2001: 6).

The simple "linear" model, whether in the form of "push" or "pull", and referred to as the traditional model, remained unchallenged until the early 1970s, when a third model emerged, mostly provoked by the riposte of critical studies in favor of supply innovation (Mowery and Rosenberg 1979). The most important criticism raised was that the ways definitions are given in the first group of studies tend to bring about the conclusion that demand is all-important. In addition, there was confusion between the innumerable unsatisfied human wishes and market demand. Besides, the recording of successes (at a specific point in time) ignores the possibility that some would-be innovations could have succeeded if the market were to be given more time. Moreover, the failure of a product may not necessarily be due to technology or the market. Finally, the products with the smallest possibility of producing immediate profit making – that is, products which do not belong to the "demand-pull" – are the first to be abandoned by firms in dire times. For all these reasons, therefore, there is a strong bias in favor of demand-pull in a success–failure type of analysis.

The most crucial element arising from these groups of studies is the importance of the dimension of time, as the recording of events should not capture a single isolated moment but instead the phenomenon as it develops over time. Otherwise, the research method risks affecting the result (Chidamber and Kon 1994).

In the short run, marginal innovations, which constitute the majority of successful innovations, move towards existing markets or markets whose short-term needs are well known. In such situations it is clear that demand is a condition for success. Another point emerging from the above-mentioned studies is that successful innovations originating from the "market-pull" tend to be marginal, with smaller long-term value than innovations originating from "supply-push".

Further analysis of the results and the way these studies were conducted lead to the conclusion, as already noted in chapter 4, that innovation is a very complex procedure which cannot be explained by a single factor, as success requires the simultaneous presence of both. Rather than being competitive, the two forces are complementary. The majority of commercially successful innovations may originate from "demand-pull" but they still rest on the existing knowledge base.

Studies conducted on synthetic materials and on chemical and electronics industries show that the question of demand versus supply does not arise, but that instead there is complementarity through time as one precedes the other (Walsh 1984). In the first stages of development of an industry "technology-push" tends to be more important while "demand-pull" gains importance as higher levels of maturity are approached. The analogy would be like examining a movie's 35 mm reel frame by frame. In the first part of the movie, "technology-push" will turn out to be the main actor. In later frames it is "demand pull" that gets the credit. Demand starts gaining importance once users start using the

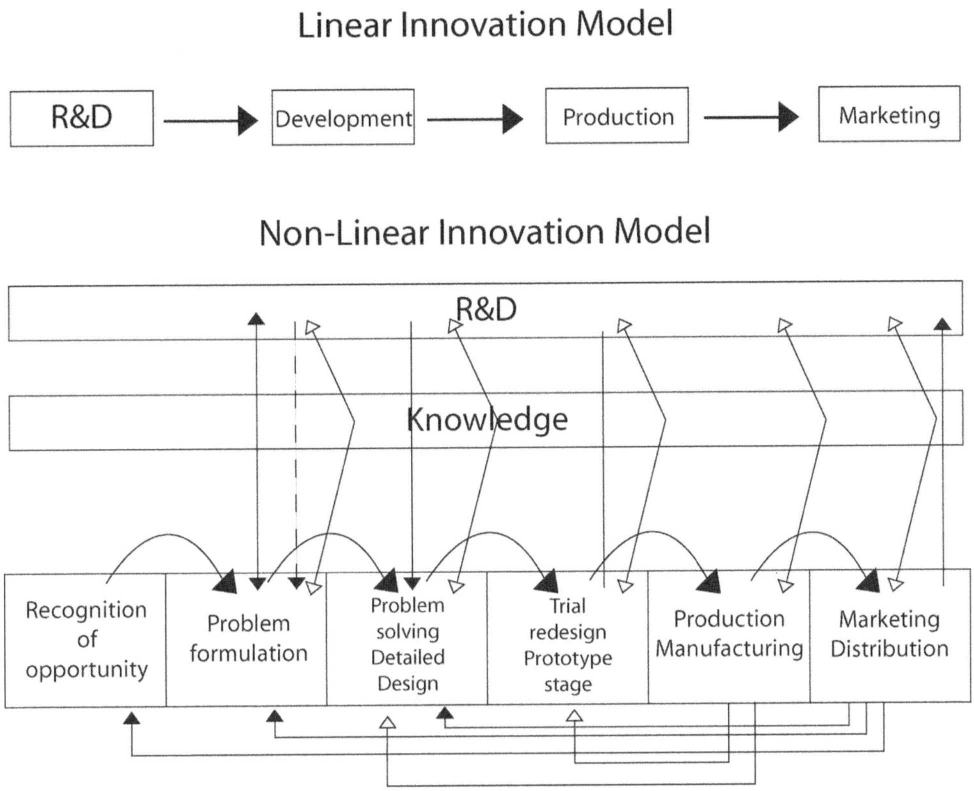

Figure 6.1 The linear and non-linear R&D process

innovation, as they realize that it could be improved, made more functional, acquire additional functions, become more reliable, more user friendly, cheaper, or applicable to more tasks.

Box 6.1 "Respect for the new, déjà vu for the old"

As a doctoral candidate at a well-known US university, this writer's main thesis task was to construct and solve an econometric model consisting of a few tens of non-linear equations. Once the long theoretical process of generating the equations was over, there remained the task of finding a way to solve the model. Of course, at the time, there was no program to deal with the problem at hand. The only way available was for our author to write his own program, and in order to do so one had to learn Fortran IV. Which he did. He wrote his program, and then in order to get closer to machine language, perforated each command on a Hasp card. Then came the day of revelation. Carrying about six thousand perforated cards in elongated cartons, he entered the temple and deposited his cards next to the reading machine. In the distance he could distinguish the altar, an IBM 360.

A day later he came back to the temple and collected the neat piles of hundreds of yards of printed-paper that were the answers to his problem. Amazed, elated and thankful he left the temple, and out of respect never turned his back to the altar.

Several decades have passed and our writer isn't much more than an average computer user. Yet, he has become demanding, even arrogant: machines are too slow, not sufficiently user friendly, programs do not cover all his needs, most graphics leave something to be desired. Things have changed since the time when the technology was emerging and our writer had no way of knowing what it was all about, how it was used or what could be expected from it in the future. But now he knows, not much, but enough to be able to articulate what his demands are. Demand starts playing a role as an emerging technology matures.

This third model was an interactive combinatorial model, still a linear model but one that recognizes that innovation comes from the combination of market needs and technological opportunities. The process is still in series but there are interactive loops. The roles are distributed between R&D on the one hand and marketing on the other. But the emphasis is on the point where the two meet, that is, the point at which technological opportunities are combined with user satisfaction.

Subsequently, two more generations of models have emerged (Rothwell 1992: 221). In the fourth-generation model there is a displacement of the innovative process and activities; rather than moving from one function to the next in series, they may now move in parallel. For instance, R&D is performed simultaneously with prototype development or production. The emphasis here is on the combination of R&D and production. In addition, there is a close relationship with suppliers and lead users as there are also horizontal forms of cooperation. Finally, the fifth generation of models is characterized by total adoption of parallel development. Links to suppliers and users are particularly strong throughout the innovative process. Horizontal links take many forms while the emphasis is on organizational flexibility and the speed of development.

As we shall see, the characteristics of the way innovative activities are carried out, more specifically the process of R&D, will have important implications as to the behavior, functioning, strategy and structure of the firm.

6.2.3 The characteristics of the R&D process and their impact on the firm

The major characteristics of the R&D process are: non-specificity, hystereses, uncertainty, and high costs (Kay 1988).

Non-specificity refers to two different levels: the product level and the firm level. An important part of R&D is non-specific as to product – which means that R&D may help create more than one product, i.e. it creates technological synergies or economies of scope. Low specificity as to product is desirable to the firm, as it may be useful in research on other products and hence may allow the firm to distribute R&D costs among several products.

In contrast, low specificity as to the firm creates external economies and ownership problems for the firm, which is totally undesirable since competitors would stand to benefit from the firm's efforts in research and so effectively be funded by it.

Hystereses are time delays and these are common in research. By themselves they do not constitute problems, although management always prefers tasks, including innovations, to have been delivered yesterday. Time delays may lead to or help the generation of problems such as danger of loss of private knowledge, and the rise in costs and uncertainty. Time delays may be helpful in allowing competitors of the firm to catch up with it.

Uncertainty is a pervasive problem, which may show up as an issue concerning future decisions, as technological uncertainty or as uncertainty concerning market success. High *costs* are an important characteristic, though it varies among sectors. Finally, economies of scale play the role of barriers to entry.

The repercussions of the characteristics of R&D change as the research program passes from one stage to the next: basic research, applied research, development, and finally introduction and launch of the innovation. As the research program unfolds, three of the characteristics, namely, non-specificities, hysteresis and uncertainties, tend to diminish, while the opposite happens with costs, which, low at first, tend to keep rising. All four are important, but their importance may vary from one technology to another, from one firm to another and over time, and may differ between stages of the same research program.

The earlier the research stage, the less room there is for "demand-pull" theories. But they may find their raison d'être in the later stages of research when the task is almost accomplished, faces diminished uncertainty and is more concrete. Of course, the opposite is true with respect to "supply-push" theories.

Briefly, the important direct consequences of the characteristics of the process of R&D are the following. First, uncertainty has a direct impact on the financing of R&D which cannot be covered by market sources – but only from internal (to the firm) funds – as the firm cannot divulge information on the research it is about to perform, or risk having its ideas, plans and objectives taken over by some other firm. This puts the new or small innovative firm, in particular, in an impossible situation. (See the section on finance in chapter 7.)

Second, in the early stages of research, non-specificities, hysteresis, and uncertainties are the main recipients of the research budget. This is especially so in basic research where uncertainties and hysteresis may discourage investors. This is one of the reasons why state intervention becomes a necessity in the light of market failure, and is directed towards universities rather than firms.

Third, concerning firm strategy, high costs, uncertainty and hysteresis may discourage small specialized firms from applying offensive or defensive strategies. Non-specificities and advantages in being second may encourage firms to follow a defensive rather than an offensive strategy.

Fourth, being first to innovate does not guarantee success for the firm, because all four characteristics of the R&D process unite against the (first) innovator. The firm, which is first in innovating, may have to face problems of uncertainty, hysteresis and costs, while non-specificities may hasten the diffusion of its newly generated knowledge to would-be competitors.

Fifth, during the last decades, the cost of R&D has risen drastically, a fact that has encouraged the formation of joint ventures. At first, this tendency was restricted to the first stages of research, the ones most removed from the expected final outcome, the products. EU programs called for "pre-competitive" collaboration. In more realistic terms, it is almost two decades since NEC and IBM "collaborated" to such an extent as to produce wholly new products whose only differences are the brand name they bear and which pre-agreed half of the world they trade in.

The new function of R&D, besides altering the structure of the firm by adding a whole new department, has also created important coordination problems in relation to the rest of the functions within the firm as well as changes in the external environment of the firm – changes that are important in the design of strategy in R&D itself. A development of this kind was to be expected since besides other difficulties of compatibility, expectations and goals, researchers employed in different departments within the firm tend to consider existing and future products from totally different points of view.

For instance, the marketing department staff's constant concern centers on the firm and its competitors' prices and market shares. They design sale strategies for existing goods and collect information that may prove useful for future production. In other words, they observe both long-term and short-term trends in demand. The longer-term trends seem to derive from changes in the characteristics of consumers, such as per capita income, or from social and cultural changes. Other changes may be due to short-term moves in demand by the firm's competitors in their attempts to alter the quality or range of their products, or changes brought about through the power of advertising.

The technologists employed in the firm's R&D department may live under the same roof with the staff of the marketing department but they certainly do not share their concern for demand. They belong to the scientific community at large, and it is through their interaction with that community (scientific journals, conferences, texts, announcements, etc.) that they keep up with the state of the art in their field, which is usually how they generate ideas for new or improved products. This happens independently of the efforts of the firm's management to organize and direct their research. Of course, they listen to what the marketing department and management may want as a new product, but they know they will be able to deliver (if all goes well) only what technology will allow to be feasible in the future.

Coordination then among the marketing department, the R&D department and the production department is not an easy task for management, as each of these departments has its own conception of the firm's future products. The outcome will be a compromise between demand and what is feasible. The precise point of compromise will depend on the influence of a number of factors, such as the size of firms, the geographical location, the type of market, the range of technical capabilities, even the personality of the people involved. The answer may lie in some "strategic management" or some business plan of the firm. Suppose, for instance, that two firms different in terms of some (not necessarily all) of their attributes, such as market, past experience and capabilities, set themselves the task of achieving the same goal: the result is bound to be different. A car producer and a producer of agricultural machinery both trying to produce a tractor will approach the problem from totally different angles.

As for the achievements of the R&D process, the rest of the departments within the firm tend to rate them very poorly. In general, until recently at least, managers did not consider R&D to be a reflection of their work. Non-specificities strengthen the tendency of managers at the departmental level not to invest in R&D if externalities lead benefits to other centers within the firm. Problems involving the long term, non-specificities and high uncertainty are best met at the higher echelons of the firm where strategic decisions are taken. Heads of the R&D departments are lucky to be on the executive board and are generally very hard put to convince the board to invest in their projects.

Overall, the institutionalization of R&D has brought about important implications as to the way we consider technological change. First, we can no longer consider that innovations are exogenous to the economic system. Since flows of funds are budgeted and given to specific projects, the direction and intensity of R&D is decided from the start on the basis of economic criteria.

Second, since R&D creates new knowledge and new uncertainty, its existence is incompatible with the hypotheses of perfect information and identical technology, which are embodied in the Neoclassical theory of the firm. Third, the firm has a whole spectrum of strategies to choose from.

In relation to the structure of the firm (Coombs et al. 1989: 34–39), there is a difference between U-type firms – pyramid shaped with specialized functional departments such as production, sales, finance – and M-type firms, which have departments based on products or geographical departments. It has been suggested that the M-type firm is an organizational innovation to combat the loss of control that is brought about as the U-type firm grows.

In the last part of the nineteenth century and the beginnings of the twentieth, large U-type firms developed in a few industries in the US. These were the industries in which the combination of technological factors and market size encouraged the forward and backward integration of firms in raw materials production or in intermediate goods production such as parts, and in the distribution of final goods. In other words, industries in large-scale production destined for large markets (food and drinks products), even including mechanical products and complex but standardized products (electrical products and chemicals).

The technological and market factors that helped these developments materialize were electrification changes in manufacturing technology, changes in communication systems and changes in income, which enlarged the size of markets. In industries where these factors did not co-occur, either only a few U-type firms developed or none at all. Gradually, the large U-type firms extended their activities so as to maximize the exploitation of their resources. However, in their efforts for gigantism, they started facing the problems associated with the U type, so they started shedding their highly hierarchic character and adopted M-type characteristics (Chandler 1962).

The spreading of R&D departments in firms coincided with the shift towards the M type. The common feature of these two developments is the effort on the part of the firm partly to control its environment instead of simply reacting to it. The role of the R&D department is to generate technology and, in so doing, expand the firm's technological capabilities and thus serve its strategies for growth. However, the R&D department is not expected to generate every possible technology, nor is it expected to solve all functional problems that may arise. In reality, the use of technology will be limited by a few factors including the following.

- First, both scientific and technological knowledge generate limitations. Very simply, some things cannot be invented.
- Second, for any firm, its managerial and technological capabilities limit the range of technological ingredients that it could integrate in its growth strategies.
- Third, some technologies offer different economic or growth incentives, irrespective of the technological difficulties entailed.
- Fourth, technological possibilities do not constitute an infinite whole.

6.2.4 Innovative activity and firm strategy

The choice of strategy stretches along the entire spectrum from offensive innovation strategy to traditional innovation strategy, with innumerable points in between (Freeman 1982: 169–186). We shall refer here to a few of the most characteristic strategies. Of course, we should keep in mind that firms do not always behave according to the strategy that their capabilities dictate, and that their behavior may change over time and switch to a different strategy.

6.2.4.1 Offensive innovation strategy

The firm applying this strategy aims at both technological and market leadership by being first in bringing new products or services to market or being first in applying a new method of production. This can only happen if it relies either on an intimate relation with the international scientific and economic system, on a very strong and independent R&D department of its own, on the capability to muster very rapidly and commercialize new capabilities, or on some combination of all these advantages.

Under this strategy, the R&D department is called forth to play an especially important role. The firm is expected to be very active in terms of innovative activity, which means being able to afford substantial amounts of funds, turning size into a very important characteristic. The firm is also expected to allot sufficient importance to the safeguarding of its intellectual property. As it aims to be first, even internationally, it expects to gain high monopolistic profits, which should cover the incurred high research costs, including failures.

Part of research involves a searching process in the realm of older and more recent knowledge. Simply watching developments on the national or international scene in basic research through scientific journals is clearly insufficient. What is required is direct access to the results of research, and this can be achieved in a number of ways. Even if the offensive innovating firm does not perform focused basic research there will always be a need for it to communicate with those firms that do. This could be achieved either through collaboration with such firms, by performing applied research, by hiring new specialized staff or by hiring consultants. And this bears direct implications as to the firm's behavior in relation to the human factor it employs and its links with the scientific and technological community.

Several dexterities may be needed, since for the offensive innovator the most crucial tasks are those connected with experimental development, i.e. design engineering and applied research. The firm will also need good scientists, technologists and technicians, as well as specialized personnel for the marketing of the new products.

In addition, the generation and processing of information should be an important function within the firm rather than be considered a deadweight, as in the case of traditional firms. The firm could be described as one of "learning intensity" as it employs highly specialized human resources and keeps adding to knowledge.

The firm, besides being expected to invest heavily in R&D, should be prepared to educate both its own human resources as well as its customers or prospective customers. Examples of this type abound: the Marconi school for telegraph operators; BASF consulting agricultural stations; the technical services of ICI in polyethylene and plastics; services for computer users by IBM and ICL.

Few firms are able and willing to play the role of the offensive innovator, or to keep doing so forever. Does that mean that the successful offensive innovator could or should rest on his laurels? The answer is no, for several reasons, not least among them the fact that there are many reasons for an incumbent to lose out to a challenger. (In fact we shall see later in this book that this may happen anyway, even if the firm does everything correctly – the case of disruptive technologies.) An ex-president of Matsushita – the giant Japanese electronics firm and parent firm of Panasonic among others – used to say that the good innovating firm is the one that is first to destroy its own innovation.

Many very well-known firms possess the characteristics of the offensive innovator, among them GE, 3M, IBM and Microsoft.

6.2.4.2 Defensive innovation strategy

Defensive strategy does not mean absence of R&D. On the contrary, it may involve a research effort as intensive as that of an offensive innovation strategy. The difference between the two lies in the nature and the timing of innovation. The defensive innovator does not strive to be first, but also does not wish to fall far behind technological developments. It may not want to carry the risk associated with being first. It may also expect to benefit from the mistakes of the offensive innovator and from the fact that the road to the market has been prepared.

The defensive innovator may have remained defensive because it lacks the ability to achieve some complex innovation because of lack of links with basic research. It may also turn out to be an offensive innovator resting on its laurels, or an unsuccessful offensive innovator. Internationally, most industrial R&D is defensive or imitational and is preoccupied with small improvements, changes in already existing products and production processes and technical services. Defensive innovation characterizes oligopolistic markets and product differentiation.

At the risk of being left behind to perish, the defensive innovator must be close enough to the offensive innovator and move fast enough to catch up with it when the time is appropriate for action. It should be in a position to design new products as well as the offensive innovator, and if possible it should endow these products with features that would make them cheaper and differentiable.

Education of the human element in the form of employed personnel and consumers may be as important as in the case the offensive innovator. However, advertisement and sales organization, traditionally important factors in oligopolies, carry more weight than for offensive innovators.

Knowledge intensity and use of scientific personnel is the same in both cases. Scientific and technological information services are crucial as they are needed to accelerate decision making because survival and growth depend on time. R&D is based on speed and efficiency in design and development.

Defensive strategy characterizes firms that cannot risk following an offensive strategy. Among others, it fits firms in the less developed countries, which cannot benefit from the local market or a scientific environment.

6.2.4.3 Imitative innovation strategy

The imitative innovator does not strive to surpass nor even to catch up with the frontrunners. It is content to follow at some distance, often at a great distance. The imitative innovator needs to have some advantages so as to be able to compete with the real innovators: such as captive markets or low cost advantages. Captive markets can be achieved in several ways including state inducements, concessions or tariffs. Low-cost advantage may derive from low salaries and through economizing on R&D by performing only the barest essentials in research and purchasing knowledge from abroad.

The frontrunners of innovation try to hold on to their position through constant innovation. But if they relent in their efforts they risk being caught by the imitators, especially in mature technologies and products.

Imitators aiming at lower costs usually attempt to achieve them through improvements in methods of production which means that within R&D design should be one of the strong points for the imitator. It should also be aware of improvements in production techniques, which implies good scientific and technical information. This will allow it to choose the products to be imitated and the sources from which know-how should be sought.

From the above, it must have become clear that the prospective imitative innovator in developing countries is not in an enviable situation unless national policy is specifically designed to favor technological progress. Some developing countries, such as China and India, are in the privileged position of being able to trade market share for knowledge – as China has done in its recent dealings in telecommunications.

Imitative innovation strategy has served well in cases where meeting consumer demand fully and promptly was not of high priority as in the example of consumer goods in the former Soviet Union. R&D consisted only of applying reverse engineering to an appliance, a refrigerator purchased in the West for example, and imitating it or at most adapting it to local realities. When the Chinese decided to make something better than the bicycle available to the average citizen, they bought Zundapp of East Germany lock, stock and barrel, shipped it to China and started producing the noisy light motorcycle en masse.

6.2.4.4 Dependent innovation strategy

The firm applying such a strategy accepts a role of dependence from other stronger firms. The firm with a dependent innovation strategy does not try to introduce technological changes in its products, unless a client requests it to do so or the firm it is dependent asks it to. The firm is open to suggestions and technical advice. The typical case here is a sub-contractor, who does not engage in R&D, and has no interest in design. The firm actually constitutes a department of the larger and stronger firm on which it depends. Dependence is real and such firms constitute the most competitive part within a sector. Typical examples of such firms are the sub-contractors of large assemblers such as car makers. When Toyota decided to locate part of its production in the US, some 400 of its sub-contractors also moved there.

6.2.4.5 Traditional innovation strategy

The product of the traditional innovation firm is not liable to technological change. Neither the market nor the competition asks for it. It may, however, embody changes in design as fashion rather than technological change. Restaurants and hotels are examples of firms with a traditional innovation strategy. They seem to be able to survive even in developed countries and under fast-changing technological environments. They are the "peasants" of technology.

6.2.4.6 Opportunistic or niche strategy

Small firms of particular inventiveness can address specific market needs or a particular part of the market (niche) that other firms may have overlooked or that larger firms may have bypassed due to small expected market demand. Some belong to the "spin-off" category, where a group of technologists working in the R&D department of a large firm on developing a new product is suddenly fired because management has reassessed the targeted market and found it too small to bother with. If the research conducted so far has been substantial then the group of technologists may decide to set up a small firm – sometimes with assistance from their ex-employer – to make the remaining effort to bring the product to market. In this case, there is little requirement on R&D since it has mostly already been conducted within the large firm – except for the last stages, usually plus design.

Box 6.2 The Scandinavian alarm

In the 1950s, Scandinavia, moving against the fashion at the time, decided that older people should be integrated in society rather than put in old people's homes. A number of social services and some contraptions were generated to make life possible for older people living alone, even at some distance from their closest neighbors. One such innovation, worn around the wrist like a watch, at the touch of a button transformed the house telephone into a transmitter-receiver, thus allowing a call for help, even by an incapacitated person at some distance from the telephone. The alarm, offered free of charge by each relevant community since 1970, reached saturation point in a few years' time. For example, in Finland all 462 communities had distributed it to all of their constituents who needed it by 1982. Produced by Nokia, the alarm had a minor problem: the button was stiff. Rather than fix it, the producer decided to ignore the small, 25,000-unit per year niche market, which a small firm was only too glad to inherit.

Niche strategy is particularly suitable for the small specialized firm with limited resources that cannot therefore undertake R&D on a sizeable scale. It addresses a small market with specific needs, often in high technologies, deriving high value added and hence able to afford high salaries. At the country level, it fits well with countries that already have high incomes, and are among the frontrunners in few very specific fields of technology.

6.3 The dictates of technological change and the firm

At the level of the paradigm, technology imposes a number of dictates, including: electrification, mechanization, miniaturization, automation, digitization or becoming mobile. Up to now, we have seen how technological change influences the economic system at the macroscopic level, where long cycles are generated at the national and international levels. We have also examined what happens under the surface, at the level of sectors, when new activities are born, perish, merge, separate, are enhanced or discouraged, mature, emerge, are led to internationalization, etc. These changes, involve, among other things, implications for each sector as to structure, composition, level of concentration, size of firms, mergers and acquisitions, entries and exits from the sector.

Beyond these consequences, technological change has had an impact at the level of the firm: on its structure, organization, even the firm's values. Not directly but indirectly it has offered the firm opportunities, which in order to be taken advantage of demand new strategies, that is, changes in the firm's functioning, production and distribution. Last, we have seen how the firm's own method, extent, and type of R&D activities imposes additional limitations on its behavior.

At the firm level, we can distinguish five plus one technological imperatives – that is, five deriving directly from technology and one that results indirectly as a byproduct. These are: the imperative for flexibility in structure and organization; the need for collaboration among firms; the imperative to subscribe to the realities of globalization, which is itself supported and enhanced by technological change; the imperative to obey the rules of competitiveness; the necessity of recognizing the value of human capital. These imperatives are intended to allow the firm to survive, that is, to better compete, and competition for the firm means, above all, innovation – the sixth, or indirect, imperative.

6.3.1 The imperative for flexibility

During the 1980s, as a result of pressures and opportunities brought about by new technologies and the instability of the international macroeconomic environment, some new organizational types developed in firms. The "Fordist" model, which had characterized industrial production for decades, from the late 1960s started facing difficulties in the US, where its effectiveness was under scrutiny. Ironically, at the same time, Europe was still envious of the American method of organization and production (Servan-Schreiber 1968).

Priority that had until then had been given to economies of scale through the mass production of undifferentiated products implied a high level of inflexibility, something that was in total contrast with the rapid changes in demand and markets at the time. This inflexibility was not simply related to the large volume of investment or the organization of the long production line, but to the entire way of looking at the situation. Its grip on managers' minds was so intense that at the height of its glory the "Fordist" model was considered to be irreversible.

The inertia of mass production is not limited to the volume of standardized products produced, but also to their quality – such as the percentage of defective output – the lack of variety of products and the lack of flexibility with respect to changes in demand. Of course, all of these are symptoms of the long period of time that elapsed under the "Fordist" model between the diagnosis of market needs and their satisfaction. The Ford Mustang, when it first reached the market in 1956, satisfied only a fraction of revealed demand, and customers had to sign up with dealers and get a car many months or a year later. The Edsel Ford is usually mentioned as the most well known "lemon" of the automotive industry, but its production did not stop until more than 200,000 unsold units had piled up. Until the mid-1980s the situation did not much improve (Dertouzos et al. 1988). In Europe, 30 percent of cars by prestigious producers were considered defective and went directly from the production line to a vast area reserved for cars to be completely rebuilt.

Many suggestions have been made as to the reasons for the realization that the "Fordist" model had exhausted its usefulness. By far the most important among them was the infiltration of Japanese products in the West, effectively an infiltration by products of a more appropriate paradigm, and superior production and organization. "Toyotism" emerged as the Japanese answer to the weaknesses of "Fordism".

The main points that differentiated "Toyotism" from "Fordism" model were: (1) the adoption of the use of networking in sub-contracting, of "just in time" in delivery, a reversal of the tendency towards horizontal and vertical integration; (2) the reorganization of work at the level of the unit of production and at the production level; and (3) limitation of the degree of deconcentration of the R&D process, design in mechanical production and marketing. The combination of these three changes led to new levels of flexibility, while the advantages of standardization still exist, and new standards of quality are set – for instance, zero defective output.

Of course, in the West networking was a new phenomenon but the same could not be said for Japan, where it has always been the traditional way of doing business. The system, banned, but almost impossible to enforce, by the victors after the Second World War, amounts to six large *keiretsu*, the heirs of pre Second World War *zaibatsu*, each consisting of about 150 firms and accounting for a large share of national output. Each *keiretsu* consists of a variety of firms covering practically every field in the economy. A typical *keiretsu* includes a bank(s), an insurance company, and companies in steel production, shipbuilding, automobiles, cement, construction, chemicals, textiles, computers, appliances and optics, plus distribution and commercial firms. Trade takes place among the firms within the *keiretsu* and only in cases where needs cannot be met is there recourse to members of another *keiretsu*.

With interlocking directorships on their executive boards, and with the presence of at least one bank in the decision making process, this system promotes both long-term stability and the capacity for long-term planning, thus substantially enabling innovation. In addition, this system confers benefits on the firms involved, as they economize on transaction costs, which they would have otherwise been forced to carry, while at the same time avoiding the loss of control that could ensue.

One of the main implications that derive from this new management example concerns R&D for new products. In the "Fordist" model, R&D means in-depth but narrow technical specialization, differentiation of functions and local functional responsibility. The result: fragmentation, discontinuity, minimized resistance. R&D functions as a separate research institution involved in the development of products but with minimum contact with production and distribution.

In contrast, in the Japanese model, there is unification and overlapping of R&D with all the stages of product development, production and distribution. Ideas for innovation come from customers, users and personnel, in addition to those actually employed in the R&D department. The aim from the start is to select the most promising innovations and to introduce them in such a way as to take into account the dictates of design for the production stage and the minimization of time to the market. Of course, for efficient overlap to take place there needs to be efficient knowledge transfer, interactiveness and discourse between individual researchers and their teams.

6.3.2 The imperative for cooperation

The significant growth observed in cooperation in R&D among firms derives from the rising costs of conducting research and from the increasing tendency for fields that used to be separate to merge. There is an increasing need for firms to control, monitor and overview separate fields that touch upon its core interests.

Since science and technology began to be organized, there have been official networks among researchers and among research institutions that were either part of universities or state laboratories, or industrial cooperative laboratories and firms. In the last three or four decades, these networks have multiplied and have acquired a more official character.

From the organizational perspective, these new networks could be considered to be a cooperative exploitation of the until then unexpected "spillovers" generated in R&D and external economies. This is especially in the case of cross-fertilization between scientific fields and between scientific and technological advances. In the most eloquent example, informatics kept rising in importance as it diffused widely to research instruments, adding to its pervasiveness throughout the rest of the economy.

Thus, beyond the tendency for more and more R&D, there emerged a need for "technological capability", something that can be met less and less easily by the firm on its own in view of the multiplication of emerging technological fields and the limitations this implies. The more numerous the fields that touch upon the main body of the firm's interests become, the more difficult it will be to achieve efficient critical mass in these fields' R&D. Such pressures can be met in part at the firm level by increasing its own efforts in R&D, and partly at the national or international level through the creation of joint ventures meant exclusively for the conduct of R&D in collaboration with other firms. Very often what is required is the acquisition of knowledge through external sources, know-how and dexterities from other organizations and firms. As already noted in this book, the transfer of such knowledge is far from a simple process due to the role and properties of tacit knowledge which usually accompanies innovation.

In contrast with information, knowledge often requires much more complicated transfer mechanisms. It requires specific learning processes and it can be shared, transmitted, or transferred only through relations based on networks.

In addition to the official networks that have existed alongside organized science and technology, there have always been unofficial networks among researchers and among research institutions of differing organizational types (universities, state and collaborative laboratories and firms). Networks also link researchers in applied research in state laboratories with industrial firms. These networks are much more eclectic and closed than networks in basic science because, in this case, the knowledge that is exchanged – if it is transferable and whether it constitutes know-how or tacit, non-codified knowledge –is proprietary and constitutes competitive advantage.

The need to acquire knowledge from sources external to the firm in basic research is most often directed to universities. The nature and extent of such cooperation depends on the activities and size of the firm. As for the goals of such cooperation, the role that the university can perform varies.

Large firms in high technologies have at their disposal wide and well-managed networks in universities, at the national and international levels. They perform a substantial amount of R&D. Universities are considered an extension of their research and a radar for the firm's monitoring of whatever is new to emerges. In this relationship between equals, firms usually have more means at their disposal, but universities can be more meticulous in their research and can look deeper for solutions. A division of labor between the university and the firm develops.

Inter-firm agreements and alliances also play the role of networks. One aspect of such agreements and alliances is that they may serve as tools of strategic competition. In some cases there may even be collusion to protect basic technologies, or oligopolistic structures may arise. Whether through networks or otherwise, what emerges is that the concept of the firm as a unit operating solely on the basis of its own decisions and capabilities is no longer accurate. Each firm is linked to one or (most often) more firms, giving the impression of a galaxy. Some firms may have links with firms that are part of other galaxies. These links underscore the mobility in alliances, the flexibility in agreements, the variability in formations and the plurality of ways to coordinate.

The opportunities created through agreements differ, and in some cases are more important, than those arising from the internalization of tasks. In other words, they are better than those arising through vertical and horizontal integration or through foreign direct investment. The new types of agreements offer the firms flexibility of action and the possibility of securing the acquisition of basic scientific and technological assets from a third party. This is particularly important in an uncertain economic environment and in activities where technological progress is especially rapid. The sharing of research costs is an obvious benefit but even more important are the links created between firms or institutions that can be so beneficial to the emergence of contemporary technological developments (Imai and Itai 1984).

Aiming at production, exchange or commercialization of technology, there are many types of links devised between firms and they cover a wide spectrum. They may refer to the entire process of producing, beginning with R&D and ending with the launch of the product, or they may refer to only part of the process, or even to one specific aspect of it such as R&D or marketing. The shapes of these links also vary substantially among themselves.

We have already analyzed the cooperation networks between universities and industry. They fall almost entirely within the realm of cooperation in R&D in the domain of

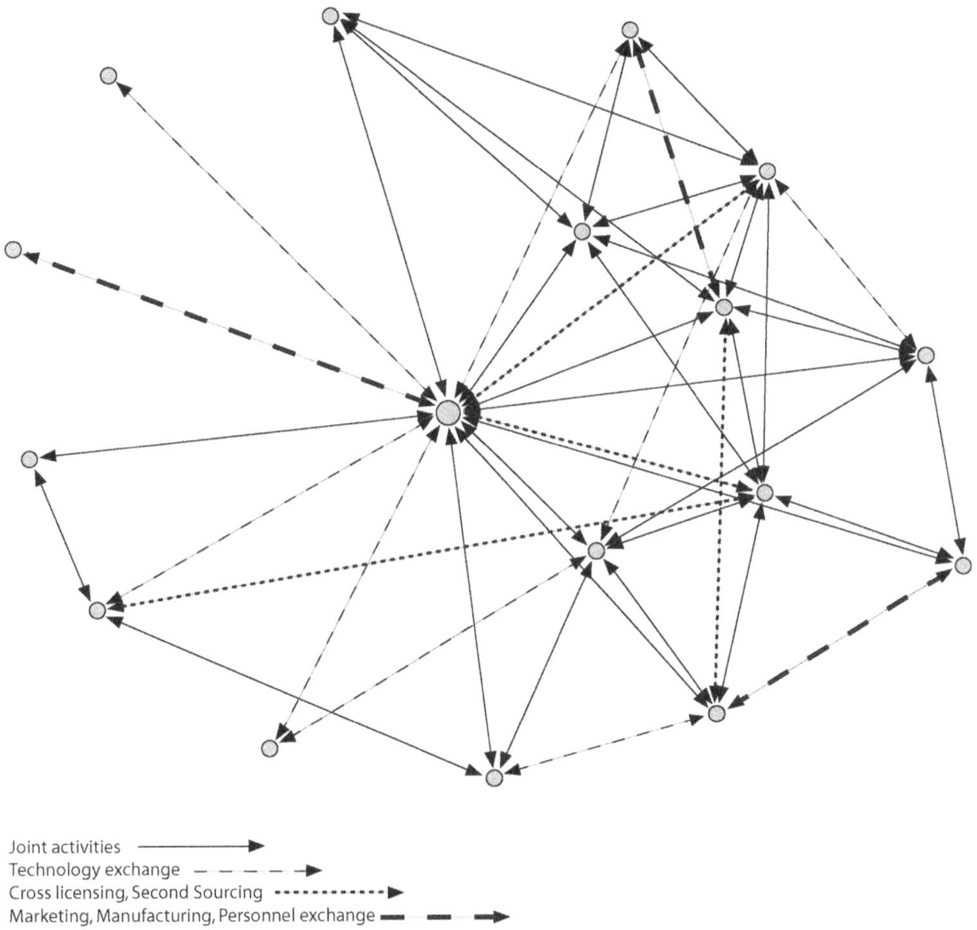

Figure 6.2 Constellation of firms and the links between them

pre-competitive research – meaning that we are still far from the final product and the market. Within this same area we have new research firms, private joint ventures financed by one or more larger firms participating in the firm's capital. Research is conducted at their own premises, which are different from those of the participants. It is generic, hence pre-competitive, but with a definite commercial orientation, naturally in the direction of the participants' interests. Examples abound, but we may just as well mention the Microelectronics and Computer Technology Corporation, with 21 participants in the US, and the European Computer Research Centre in the EU, with the participation of ICL, Bull and Siemens.

Moving now to the competitive range in technology research cooperation, there are participatory agreements between large firms which take shape in the form of venture capital firms intended to detect innovative processes in small firms and to monitor the development of new technologies that emerge. Typically, processes of managerial decision and R&D remain under the jurisdiction of the venture capital firm and not the large firms that control them.

Further, there are research cooperation agreements which aim at providing solutions to specific problems. Small numbers of firms participate in such schemes and they usually possess complementary structures in R&D. These agreements usually last for a limited period and aim at very specific results.

A further type of agreements has to do with the flexibility that firms would like to achieve in the process of benefiting from the results of their own research, that is, from technologies that have already proved their technological feasibility. Such agreements may take the form of sharing know-how between two competitors, or the exchange of patents or know-how that may have come about as a byproduct of research on complementary and non-competitive products.

We can say that firms conclude cooperation agreements to have fast access to new technologies or markets, to benefit from economies of scale in research and production, to exploit know-how produced by a third party, or to reduce the risk of certain activities.

Analyzing agreements we can only conclude that the basic target for firms involved in such arrangements is to acquire the possibility of following new strategies in innovation through cooperation with others, provided that such a step does not jeopardize their entity. This is evident in the revealed preference in such agreements for non-participation in the firm's capital, i.e. a preference for the most flexible cooperation that does not alter the specific entities of the parties involved. What is most looked for in such agreements is access to tacit knowledge, effectively proprietary know-how. Chesnais analyzed 1500 agreements between firms (Chesnais 1988); only about 30 percent did not involve some kind of technology transfer or sharing.

6.3.2.1 The user–producer relationship

Apparently, a large proportion of the resources devoted by firms to R&D involves cooperation with some other firm, and a large proportion (80 percent) of this cooperation takes place unofficially (Hakansson 1989). It most commonly takes the form of agreements between the producers of innovations and their users.

In relation to the innovation process, the producers of innovation have important reasons to generate and maintain close relations with users, and even to monitor aspects of their activities (Freeman and Lundvall 1988). In the case of mechanical equipment and industrial processes, the knowledge emanating from "learning by using" cannot be transformed into new products unless producers are in direct contact with users.

Problems in production and technological interdependencies constitute potential markets for the producer of innovation. Moreover, they have good reasons for the monitoring of the capability and learning potential of users because they need to understand the capability of firms to adopt new products. In turn, users usually need information concerning the new products: specific specialized information characteristic of the users' values and how they are related to their needs. And of course information as to the reliability of producers is always needed.

In the case of development of complex systems and specialized equipment and processes, cooperation between user and producer is expected throughout the development process. As for the producer, it has plenty of inducements to understand how the user learns so as to be able to use this information either in improving the current innovation or coming up with another.

6.3.3 The imperative to enlist in the globalized international environment

Globalization is a dynamic and multidimensional process. The word "globalization" is used more and more often to describe the internationalization of markets for goods and services, the economic and financial system, competition, technology, firms and industries. Globalized international exchanges in goods and services mean the opening of new and expanded markets. Globalization of financial markets signals very large increases in investment portfolios and large movements of short-term funds, with investors and borrowers interacting in ever more unified markets. The globalization of technology derives from the speed at which innovations are diffused, with international networks formed among research centers and as a result of the ensuing industrial standards. For firms, globalization of competition is characterized by the fact that new strategic factors have to be taken into account. It is no longer limited to foreign direct investments and changes in the location of production, but also involves joint ventures, cooperation agreements, strategic alliances and calls for the firms to reexamine their overall strategies.

In a system where national resources move internationally, the markets and production of different countries become more interdependent. A rising proportion of value and wealth is produced and distributed worldwide through a system of interconnected private networks.

Globalization is not a new phenomenon but a new stage in the process of internationalization of world production. None of its constituent dimensions is new, but their proliferation and the rise in their intensity in the last three decades has created a new economic system. In fact, the order of importance of the constituent dimensions of globalization changed as international trade ceded its relative dominant position to direct foreign investment, the role of MNCs was reinforced through their intra-firm (parent firm with affiliates) trade, and a high concentration on the supply side emerged at world level as fewer and fewer firms captured the totality of world production.

One of the most interesting developments in the process of globalization is the rising interdependence of its constituents. Most often, direct investments abroad generate exports for the countries of origin. These exports are usually accompanied by transfers of technology, know-how and capital movements (shares, international loans, repatriation of earnings, interest payments, royalties, etc.). There may even exist complementarities among the constituent forces of globalization.

Since the 1980s two basic factors have accelerated the changes to the existing structures of internationalization. The first has to do with the new technologies and their role in supporting the process of globalization and in leading it to even greater levels. The second is the liberalization of the international financial sector and of some services, and their globalization.

The greater mobility of factors of production and accelerated capital movements have prompted important changes in traditional trends of specialization in production both at country and firm levels. At both levels the effort has been directed at strengthening competitive advantage through specialization in differentiated products with increasing technological content. For instance, for the OECD countries, the percentage of their exports in high-technology products rose over a single decade (1990s) from 18 percent to 25 percent. Of course, the other side of the coin is that both countries and firms give up the production of some of their products and acquire these from others.

New technologies are at the center of the globalization process both as a supportive factor but also as pressure for greater globalization. Technological change in telecommunications and information technologies has brought about important modifications in location theory, and their merging has allowed MNCs to spread their resources and activities on a truly global scale. The usual mode of work for these firms is to have their own internal network linking banks and firms providing industrial services, and to unite production and distribution on a world scale. They adopt networks, forge alliances and develop other types of links accelerating other types of cooperation with suppliers, clients, even competitors.

As a result of all these changes, which are the result of technological progress but also of changes internal to the technology itself – i.e. the shortening of the product life cycle – there were strong pressures on firms to reevaluate their strategies, their way of functioning, their structure, even the basic values relating to their goals. The "business is war" philosophy, which reigned unchallenged until the 1960s, developed a new tendency for cooperation and then from the 1980s started being replaced by the spirit of "co-opetition", that is, cooperation and competition (Nalebuff and Brandenburger 1996). Cooperation, agreements and alliances coexist with competition, mergers and acquisitions.

As a result of globalization, technological change is not limited in its influence to the location of production, exchange and distribution of products and services, but also influences the location of the conduct of R&D (see chapters 1 and 7).

Finally, possibly the most important of globalization's achievements is the fact that it has imposed the concept of "global competitiveness", whereby firms are obliged to face their competitors in all markets. Fewer and fewer activities, even in services, can find shelter in local markets.

6.3.4 The imperative for competitiveness

In response to the call for "global competitiveness" the firm has no other choice but to attempt to become as competitive as possible. This means doing its best at competing and raising its market share to increase its profits and grow. There are two separate but not exclusive ways of achieving this: through costs and prices; and through organization, capability in the use of technology and product quality.

Successful competition requires product quality, improved technology in methods of production and a higher level of organization of production, swiftness in delivery and quality in after sales support services. What is needed is: successful management of production flows, raw materials and inventories; successful assimilation mechanisms for market programming, for official R&D, for design, for mechanical and industrial manufacturing; the capability to blend knowledge produced within the firm with knowledge acquired from other sources; a better understanding of market developments expressed in production strategies; the ability to create successful links; and finally the capacity to take steps in order to enhance human capabilities.

Equally desirable for the competing firm is the ability swiftly to exploit its innovations, to be able to bring to market different versions of its innovations and to spread them to several markets (Nevens et al. 1990: 3, 155). Moreover, in a system of production based on information, the competitiveness of firms increasingly depends on economies of scope and variety rather than on economies of scale, and the relationship with clients becomes one of the basic aspects of the search for competitiveness.

At the macro-economic level, the concept of competitiveness becomes more furtive. Methods describing competitiveness at this level are no longer confined to costs and prices but have to be complemented by analyses of long-term changes in market shares, profits, dividends, investments, changes in tendencies in international trade, etc. With no consensus on this issue, a debate followed which reached a number of interesting results. It turned out that competitiveness based on lower costs, which had been thought until then adequately to explain competitiveness, has some role to play but only a lesser one compared to non-cost factors. These include technological change, economies of scale, new markets, but also quality, services, financing, and the better adaptation of products to users' needs. Later studies came to the conclusion that the main explanatory variables for the differences in competitiveness among countries are technological competitiveness and the readiness to create new productive capabilities.

It also emerged that competitiveness at the national level is not directly linked to competitiveness at the firm or sectoral level but instead to the business environment: this means the way in which firms compete with each other but especially policy factors that shape the institutional framework in which this competition takes place. (Thompson 2004: 197–218). The structural characteristics of a national economy or a local sector have significant influence over the performance of a specific firm. Competitiveness at the level of a national economy is an overall characterization of efficiency in mustering its resources and in so doing changing its technological and social characteristics. The displacement of activities towards higher productivity generates better-paid jobs. Such efficiency in mustering national resources is referred to as "structural competitiveness" and in many ways affects the competitiveness of firms.

Research on structural competitiveness has covered a wide range of topics. From the mid 1990s, a new direction was inaugurated: the focus turned towards economic freedom (Islam 1996), property rights (Leblang 1996) and corruption (Mauro 1995), as well as other institutional barriers to competition. Another strand of research has investigated the size and role of the state, its efficiency and role in regulation (Hall and Jones 1999). The truth of the matter is that we are still in need of a theory to explain structural competitiveness. Two firms, in every way similar to each other, except that one operates in the West and the other in a third-world country, will definitely exhibit very different levels of performance.

6.3.5 The imperative to recognize the value of human capital

Considering our discussion so far, it would be a tautology to say that recognition of the value of human capital is a prerequisite for the firm in its quest for competitiveness. Besides, a number of factors led the firm to realize the value of human capital.

First, there is the fact that man is the only "means" capable of transferring tacit, accumulated knowledge. Second, in view of the rising participation of technology and hence of R&D, there is a chronic shortage of skilled personnel to serve research needs. Firms have been lavish both in recruiting and in retaining personnel. High salaries are complemented with many other perks: excellent food, vacations, even yoga lessons. The Head of Technology at eBay is reportedly paid twice as much as the Chief Executive. In the US, several high schools are offered equipment or are financially supported by large high-tech firms eager to secure sufficiently trained personnel. While at school, students who after graduation are expected to work for the "parent" firms mix work-practice with school. Abroad, large high-tech multinationals such as Cisco Systems, Sun and Oracle have agreed with universities in South East Asia, and India in particular, on a jointly designed curriculum in return for hiring their graduates. At the national level several countries have seen their immigration policies evolve in the direction of a strong preference for skilled human resources (US, Germany).

The more developed economies intensify their efforts to acquire knowledge and move towards the knowledge economy, the more human resources become important to economic progress. Productivity is increasingly determined by knowledge and the skills of the human factor involved. Human knowledge has already displaced equipment as the main explanatory factor of productivity. The firm is no longer solely a producing unit but a learning unit as well, and its values are also changing. The human factor is no longer a liability but an asset that should be best directed and managed. It is no accident that the last three decades have seen the emergence of human resources management, the fastest developing field in management, with a mission to use the combination of tangible and intangible investment most efficiently.

With the rise in automation, hence the need for less labor, the requirement for multiple dexterities on the part of personnel has increased along with the constant demand for its higher specialization. A workforce consisting of just a few highly specialized persons with multiple dexterities, running a highly automated producing unit (overseeing, programming, redirecting, fixing a number of diverse pieces of equipment) differs significantly from the traditional concept of the workforce we are used to. It goes without saying that the employer–employee relationship should also be different, something that approaches partnership.

While the firm's needs in terms of specialized personnel rose rapidly, its needs for non-skilled labor have dwindled. Of course, higher skills and dexterities have higher demands as to learning both within the firm as well as with respect to the educational system.

6.3.6 The imperative to innovate, innovate, innovate

This is the era in which "the emerging economy … is offering unprecedented opportunities, an ever-expanding choice of terrific deals, good products, excellent long-term investments, and great jobs for people with the right talents and skills. Never before in human history have so many had access to so much so easily" (Reich 2002: 5).

Both supply and demand are driven by technological change. On the supply side, technological change in the form of new methods of production, new services or new products makes all these new terrific deals available. On the demand side, technological change in communications, transportation, information technologies and the Internet have made it easier to search, locate, "and get better deals from anywhere … allow[ing] us to switch instantly to even better ones" (Reich 2002: 5). As a result, producers are less secure and more vulnerable to competitors. And competition is fierce. Producers have no other choice but to innovate and keep innovating. Winners are those able to come up fast with better products and lower prices, usually through trustworthy brands. But the concept of winning is ephemeral as there is no end to the story. Winners cannot rest on their laurels for fear others may catch up with them or that they may fall behind. The result is an explosion of innovation, better or cheaper goods, with productivity on the rise while inflation remains at low levels. Innovation is more frequent in activities where technologies give users the most choices and the greatest ease to migrate to other alternatives. Besides, these are the most rapidly growing activities.

At the macro-economic level, the economy is moving from a stable system based on large-scale production to a system of fast and continuous innovation. The imperative to innovate is the sixth imperative dictated by technological change, only this time dictated indirectly through the need to face competition.

In a broader sense, the firm is not simply trying to compete against rivals in terms of innovation but competing for funds against the entire business community. In order for the firm to keep its shareholders happy and possibly attract some more, it needs to deliver growth in profits, beyond the market's expectations. The best way to go about that is to invest in innovation (Christensen and Raynor 2003: 3–5).

6.4 The firm faces up to innovation, evolves, adapts

6.4.1 Additional requirements for all firms as to innovation-related hurdles

The five plus one imperatives just analyzed should be considered to be a minimum but not necessarily sufficient requirement for survival for all firms in the face of oncoming technological change. In addition, the firm should become aware of a number of innovation-related problems that may arise, and be prepared to act, whenever possible. One such situation concerns the possibility of *loss of leadership*, a paradox – since profit was the innovator's incentive – whereby the first to innovate is not the one to reap most of the benefits. This situation may arise in a number of ways.

Intuitively, one such situation is to lose to a fast second, a defensive innovator, an *imitator turned competitor*. Another possibility is to lose to a fast second follower who may be an *imitator but with complementary assets*. This is the case so vividly presented by Teece (1986).

Box 6.3 EMI

In the early 1970s, the British firm EMI (Electrical Musical Industries), mostly known as a music industry firm, was involved with music records, motion pictures and advanced electronics. EMI had a history in important innovations: high-resolution TVs in the 1930s, airborne radar during the Second World War, the UK's first solid-state computers in the early 1950s. In the 1970s, EMI came up with the CAT (computerized axial tomography) scanner. At first, it was successful but within eight years of its introduction in the US, EMI had not only lost leadership but pulled out of the CAT scanner business altogether, only to be rescued by Thorn Electrical Industries.

EMI had failed to capture the profits generated by the innovation it had pioneered and successfully commercialized. Its consolation prize was to keep receiving royalties. And of course there was the prestige associated with Godfrey Houndsfield,

EMI's senior research engineer, sharing the Nobel Prize in 1979 for the invention of the scanner.

The technical sophistication of the EMI scanner required continued support and servicing and a high level of training for hospital staff. EMI possessed no such capabilities. GE and Technicare both possessed such complementary capabilities. Besides, intellectual property law did not protect the scanner against copying through reverse engineering once it was on the market. Again, both GE and Technicare were technologically capable of achieving this. Finally, they were both knowledgeable in the market for medical equipment and they were both known for their reliability, quality and service. Based on the EMI scanner, they developed their own innovation with improved features, and rushed to the market. Subsequently, GE, which also won the race with Technicare, acquired EMI's CAT scanner business from Thorn, apparently for a pittance.

The case of EMI illustrates how a firm with outstanding technological capabilities, as witnessed by a tradition of impressive innovations and an excellent product already successfully commercialized, can fail to profit from its own innovation while its imitators succeed in doing so.

Still another situation: where a very promising innovation draws a *swarm of imitators* encouraged by the small size of the original innovator – therefore expected to be open to intimidation – and/or, also possibly by the ease of reproducing the technology. Here, the options for the innovator are to sell out to a larger firm, to disappear or to continue to cling to existence while fighting long, expensive legal battles.

Box 6.4 Tivo

Two years after its foundation in 1997, Tivo, a small startup financed by a venture capital firm, launched its innovation, which turned out to be the granddaddy of all DVRs (digital video recorders). There was much acclaim for Tivo, which has so far been awarded four Emmy Awards for its product which "lets you watch anything you want to watch when you want to watch it", along with rumors that it would forever change the public's viewing habits.

It could store up to 40 hours of programming, it offered the user the ability to fast-forward through commercials and to pause during a live program. It also included a unique feature that could tape shows based on the user's expressed preferences – that is, based on what the user had previously recorded.

The original idea became so popular that it began to be used as a verb (I xeroxed a page, googled a review, tivoed *Friends*). In fact, DVR adoption attests to the greatness of the original idea. As of February 2011, US households with a DVR represented 42.2 percent of all households. Yet, surprisingly, only approximately one-tenth of these households were Tivo customers. In fact, the company struggled from the beginning and has registered substantial losses over the years.

Tivo has felt the pressure of imitators from the start. In order to beat Replay TV, which introduced its own product at the end of 2000, Tivo jumped the gun and went to market with its product at the end of March 1999, several months ahead of schedule. It thus experienced what is referred to as "first mover disadvantage", which consists of bringing to market a product that is too expensive or too complex for the average consumer's taste, rather than waiting until the creases have been ironed out. Large firms can survive their gun jumping (Apple for instance with its proto-PDA, the Newton); smaller ones rarely do. The first Tivos, when introduced in the US, were sold at about $500 plus a monthly fee of $9.95, and were thought of as being complex to set up.

By November 2003, Replay TV was no longer a threat due to bankruptcy as a result of a lawsuit, but Tivo was facing competition from Microsoft and Sony. Hackers were already producing how-to guides on a do-it-yourself basis with memory space a multiple of Tivo's size. Several imitators were fearful of Tivo's power to change commercial television broadcasting forever. It was not at all popular with the TV networks which saw it as a threat to their lifeblood – advertising. Now the US has several television providers, Comcast, FIOS, AT&T's T-verse, Direct TV and Dish Network, all of them offering their own version of the DVR to record television.

"Tivo came early to the DVR business, but soon found its product being replicated in one way or another by many in the cable industry, as well as by entrants from the Internet … One response to this was to keep on improving the product and keep the customers happy" (Sherr 2013). Another way was to fight back at imitators through litigation. On the technical improvement path, Tivo extended programming hours to 450. It made recording possible from any available source, including the Internet, and you could watch recordings anywhere (tablet, mobile phone, etc.). By now it could be described as "allowing you to watch anything, from any source, you want to watch, when you want to watch it, and where you want to watch it".

As for litigation, Tivo sued several companies, which it considered to be in breach of its IPRs, and has reached settlement with many of them. Google-owned "Motorola and Cisco Systems agreed to pay $490 million, while Verizon said it would pay at least $250 million. Dish Network and its former unit EchoStar agreed to a $500 million settlement in 2011. In total, the company says it has secured about $1.6 billion in damages and consideration through its litigation [so far]" (Mullin 2013). This is far from impressive retribution for the stunting of its growth over so many years, especially if its own litigation costs are taken into account.

Even if a firm takes all the correct decisions and manages to avoid all possible pitfalls, or is lucky enough not to attract a swarm of imitators, there is still a further dimension of competition to be considered, a chance that *it may end up losing by falling victim to disruptive innovation*. How is that possible?

The leading, well-established firms within a sector, have no reason to change the status quo, either through sudden changes, or changes in the rules of the game. Consequently, they limit themselves to mostly marginal innovations in the form of continuously improved products, services and new methods of production. They move on established improvement trajectories of already existing products, along dimensions that have been historically appreciated by users. In other words, they are applying a *sustaining technological change*: airplanes with longer travel range, faster computers, longer-lasting batteries, higher-definition television.

Every so often, the leading well-established firm falls victim to an attack, usually by a smaller firm, which may not even belong to the sector, and loses its leadership due to a "*disruptive*" innovation.

Here, we are not referring to small, slow, badly managed incumbents, but rather to large, well-managed firms which at some time have been acclaimed as phenomenal successes and were the envy of every potential shareholder, firms like IBM, Xerox and Sears & Roebuck. However,

> good management was the most powerful reason they failed to stay atop their industries. Precisely because these firms listened to their customers, invested aggressively in new technologies that would provide their customers more and better products of the sort they wanted, and because they studied market trends and systematically allocated investment capital to innovations that promised the best returns, they lost their position of leadership.
>
> (Christensen 1997: xii)

The virtues encapsulated in profit maximization are the perfect tools for management as long as there is no threat from a disruptive innovator. When the threat becomes apparent, the incumbent firm realizes that it should have been prepared and should have abandoned its management principles much earlier, but it is usually too late. Continuing on the same course is fatal.

For disruption to occur, there must first be something to disrupt. Disruption is always relative to something else: large computer mainframes were disrupted by minicomputers, which in turn were disrupted by the personal computer (PC).

There are two types of disruptions: low-end and new-market disruptions (Christensen and Raynor 2003: 43-49). *Low-end disruptions* refer to innovations akin to already existing products only with lower performance criteria, at least in the short or medium term. They are usually cheaper, simpler, smaller and sometimes more convenient in their use. As incumbents develop their technology in their efforts to provide better products than their competitors and earn higher prices and margins, they offer some customers more than they need or are willing to pay for. Continuing on this path enlarges the "overshot" market, which in the long-term may develop to such an extent as to become a threat to the remaining high-end market itself. Discount stores (Wal-Mart), Japanese steel companies, Toys 'R' Us, Hyundai and Kia, and Amazon.com in its original incarnation as a bookstore, are examples of low-end disruptors.

New-market disruptions compete against "non-consumption" as their products are more affordable and easier to use and make them accessible to many more customers and in more convenient settings. Kodak, Bell Telephone, Sony, Intel and Wireless telephony are examples of new-market disruptions.

Most disruptive innovations are hybrid in the sense that they belong to both types at once. Some are closer to new-market, some are closer to low-end. Products which are first time ever innovations are at the new-market end of the spectrum and all the rest, depending on the situation, are located closer to this point or further away and closer to the low-end market position.

Disruptive innovations, the most vivid realization of Schumpeter's "gales of creative destruction" are a permanent aspect of competition within the evolving environment in which the firm is expected to survive and hopefully prosper – for a while. For today's disruptors are tomorrow's disruptees. IBM, the dominant firm in mainframes missed the rise of minicomputers where DEC (Digital Equipment Corporation) and Hewlett-Packard defined the market. But they in turn also missed the emergence of the PC (desktop personal computer) created by Apple, IBM's PC division, Commodore and Tandy. And all of them missed the rise of the portable computer market.

A study analysing over a thousand firms over a period of almost four decades, from the early 1960s to the late 1990s, concludes that only 16 percent of firms were able merely to survive over that period (Foster and Kaplan 2001). From the original Dow Jones Industrial Average list of 1896, only one firm survives to this moment: GE (*Economist*, May 11, 2013: 64). Like humans, firms are ephemeral, only they tend to have shorter lives – with some exceptions.

From the discussion above, it transpires that sustaining and disruptive innovation are not discrete subcategories of innovation (unlike a new product or method of production or open innovation) which is why they were not introduced among the different kinds of innovation in chapter 2. They do not characterize an innovation itself, but rather its impact on markets in terms of dynamism, structure, market type, and potential market dominance.

6.5 High-tech firms: further requirements

Up to this point, we have attempted to address all firms about the basic implications of technological change, to which they would be wise to adapt, plus a few of the salient innovation-related obstacles which they should bear in mind in the hope that they need only know about them but need not deal with them, as some of them can prove fatal. This is the least expected of firms interested in surviving rather than conquering the world. But if this is enough for most firms, high-tech firms need more. After all, they operate in a different, more dangerous sort of world. They function under much more difficult circumstances in an environment of heightened competition, rapid and constant technological change, and are therefore more likely to succumb to disruptive technologies. They enjoy narrower windows of opportunity, and work on the basis of totally different economic principles than the rest of firms.

In decreasing returns, the last additional unit of input adds successively less to total output, until products and firms that grow run into limitations. An expanding farm producing a certain product (saffron) may soon discover that new additions to its land may suit less its specific type of production. Similar limitations will eventually appear in factories, offices, etc. Information, like capital and labor, is subject to decreasing returns. By contrast, knowledge obeys the laws of increasing returns (chapter 4), and so do high-tech firms which rely on knowledge (because, as we may recall, technology is knowledge; chapter 2). At any given point in time, some firms rely heavily on knowledge/technology and are referred to as high-tech firms; others rely on it less and still others even less so. Over time, they all tend to increase their reliance on knowledge, only at differing speeds. As knowledge gradually becomes an integral and growing part of an economic activity, so does the latter's allegiance to increasing returns.

One of the consequences of increasing returns is the tendency of the frontrunner to increase his lead ahead of the rest of the firms. As soon as the frontrunner moves a little ahead of the pack, the market tilts in his favor. After winning the first round, his chances of winning the next are reinforced, and so on. We have a "first mover's advantage" which may take the form of "winner takes all" or "winner takes most" depending on the situation. Dominating the whole market or most of it implies that an inferior innovation or technology could very possibly be "adopted" in a market to the detriment of a better or cheaper one that is slower to reach the market. There could be a "lock-in" to an inferior technology and path dependence, meaning that history counts – it makes a difference how it got there. Inherent in the situation is the fact that potential profits can be phenomenal, which is why competition is so fierce. And all this emphasizes the fact that reaching the market first is quintessential: competition is based on innovation, not price.

We already know that the more an activity relies on knowledge, the more we should expect it to operate under conditions of oligopoly. With phenomenal potential profits we should also expect competition to be fierce. And this is precisely what is happening in the world as exemplified by developments in information technologies and telecommunications, which have become one by now as what used to be called the "grey area" between them ceased to exist more than two decades ago. In this sector things change so swiftly that IBM seems to have belonged to the Stone Age, and Microsoft and Nokia appear to belong to the distant past. Of the four large protagonists at present, Apple, Google, Amazon and Facebook, all are basically young firms, with Apple, the oldest, founded in 1976. As the web meets the phone and software may reside in the cloud, the lines between their territories are getting blurred and so are their strategies. A seemingly unrelated development or the appearance or coming of age of a new gadget may result in the crossing of so far apparently separate paths.

Google and Apple seemed to enjoy a more than peaceful coexistence, one that is so peaceful in fact that Google's chief executive sat on Apple's board until 2009. Google had acquired Android for $50 million back in 2005, together with several other companies, an event that went almost unnoticed by most executives in the company. It seemed at first, that the fate of Android would be no different than that of most firms acquired by giants: soon neglected and forgotten. The arrival of the iPhone in 2007 and the light it shed on the potentialities for smartphones alerted Google to the possibility of its services being left out of competitors' future smartphones. Android, which provides an operating system for mobiles, offered the perfect answer. By being given away Android soon attracted hundreds of thousands of applications and democratized the smartphone, while Apple, with its iOS operating system, still offers the "Rolls Royce" of smartphones. Google seems to have won the consumer's migration from PC to phone (Kowitt 2011).

This battle is far from being the only battle fought in that domain, as the protagonists plus a swarm of smaller firms are constantly at each other's throats. Apple is fighting Amazon over music through iTunes, while in video they both try to wrest ground from Netflix, which in turn is moving towards video streaming. Amazon is busy fighting off all the rest plus outsiders as they all envy its e-commerce trove. Put simply, it is not exactly competition but a free-for-all.

In contrast with the rest of firms, in the world of high tech the situation the manager faces is a little like gambling: not in a single static game but in a whole casino where a different game is being played at every table. At one table they play multimedia, at another cloud computing, at a third electronic banking, at the next digital cinema, and so on. The manager sits at a table.

> How much to play? (he asks the croupier).
> Three billion [dollars] (he replies).

Who'll be playing?
We won't know until they show up.
What are the rules?
Those'll emerge as the game unfolds.
What are my odds of winning?
We can't say. Do you still want to play?

(Arthur 1999: 157)

This extract from Brian Arthur describes vividly how difficult it is to predict what the next game will be, what the rules will be, and what it takes to win. With the hindsight of a little less than two decades since Arthur's description, the present writer would add that the manager does not sit at one table but simultaneously plays on different tables. Winning may require playing and winning at more than one table simultaneously.

The game of high technology requires technical dexterity, will, courage and deep pockets. As in gambling, bluffing should not be ruled out, and indeed many firms do so in an attempt to scare away competitors, and announce the coming of a future innovation, which sometimes never arrives. The big winners are those who in the mist of technological uncertainty, rapid changes and a sea of seemingly unrelated developments are able to visualize the rough shape of the next games to emerge, and guess the contours of the rules of play. In such a situation the firm is much less production oriented, it is mission oriented so hierarchy is no longer needed and the firm's structure is flat. In a world whose future is unknown, the manager cannot optimize. All s/he can do is adapt, preventively, like innovating in the mist.

So the high-tech firm will eventually have to change in practically every aspect to regenerate itself: every time a new challenge with a different team, like a new motion picture, with a new title, scenario, director, producer, and actors. The high-tech firm needs a different mode of behavior, style and culture than other firms. It needs different management techniques and strategies, and as we shall see in the next chapter on state Innovation Policy, new codes of state intervention or different state regulatory roles are needed.

6.6 Technological change and the evolution of an industry

Box 6.5 The US motion picture industry

It is only appropriate to close this chapter on the effects of technological change by presenting a case study involving a whole industry, the US motion picture industry. The aim is to analyze the transformation brought about by technological change in this industry as a whole, not just firms and products. So we consider development, organization, structure, actors, production, distribution, product characteristics, and the importance and prospects of the industry in relation to the national economy.

Technological change has been one of the strongest forces shaping the US motion picture industry, slowly at first, then at a constantly accelerating rate, and it promises to do so much more actively in the years to come. The challenges were many and were met differently in the US in comparison with Europe and the rest of the world. The result has been the hegemony of the former over the rest. In the US, the motion picture industry together with software, music, magazines, papers and publications constitute what have come to be known as "copyright industries", among the largest contributor industries to GDP and employment, one of the fastest developing sectors of the economy, and a major contributor to exports (Siwek 2013; see chapter 1). In 2010, the motion picture and television industry supported 2.1 million jobs, spent $143 billion in wages in the US and earned $13.5 billion in exports. Its competitiveness is underlined by its perennial balance of payments' surplus (Motion Picture Association of America 2012).

In the early phase of the US movie industry, technological change was mostly limited to methods of production. Adoption of new technologies was rather slow and took place at varying rates. It took much less time to adopt sound than it took to adopt color. Although the Europeans had a head start, the US film industry had the advantage of the English language, and from the start the indirect support of the state (Segrave 1997). From the late 1920s, the US film industry started a process that gradually came to differentiate it from the European one. The "star system" was borrowed from Broadway, and marketing developed into an integral part of distribution. The interwar period turned out to be "the golden age" of the American cinema industry. However, from the late 1940s, two developments changed the industry's course. The first was the result of antitrust legislation ordering the breaking apart of the hold cinema companies had on the chain from production to theater seat. The second was the gradual coming of age of television, a technological change of the previous decade, which had by then diffused sufficiently as to lay serious claims to a respectable share of consumers' leisure time. Cinema theater seat revenue dwindled and by the 1960s cinema companies had been acquired by conglomerates.

The golden era of Hollywood was gradually replaced by a "New Hollywood" (Scott 2002, 2005) in which the studios survived the transition but in the process reconfigured the basis of their oligopolistic power. New Hollywood rested on two major developments. First, film marketing and distribution underwent a process of vertical integration and globalization, as the studios were regrouped under larger international conglomerates. Ever since, the studios have increasingly operated as gatekeepers rather than film factories, with control over access to finances required to produce, market, and distribute films (Compaine and Gomery 2000). At present, the six studios which account for more than 90 percent of total industry revenues

> are now divisions of a broader circuit of copyright capital, which in turn is dominated by a so-called media oligopoly. This broader oligopoly comprises five firms (Bertelsmann, News Corporation, Time Warner, Viacom, and Walt Disney), which are supplemented by other firms (e.g. General Electric, parent of NBC Universal and Sony Corp.) through co-financing deals, strategic alliances and interlocking directorates.
>
> (Bagdikian 2004)

This grouping has all the features of an oligopoly with limited competition at the product level and cooperation among the leading firms. Together, these firms control media and entertainment (films, cable television, music, radio, newspapers, magazines, books) in the US, and increasingly so in the rest of the world.

The second development upon which New Hollywood rested was that film production underwent vertical disintegration leading to the rise of a "transaction-intensive" entertainment complex in Los Angeles (Coe 2001). With television and the studios now operating under the same conglomerate roof, no longer as competitors but as complements, a complex was generated. In this complex, the studios provide content, television is the main bread earner as it provides the main source of income for cinema besides box office revenues, and in some cases even becomes a shareholder or part owner of the studios through vertical integration. The prime mover of the whole complex is none other than advertising, which feeds television directly and so cinema indirectly. Besides cinema studios and television stations, the complex extends to scriptwriters, producers, directors, actors, technicians, advertisers, scouts, agents, insurance agents and financiers (Brown 1997). All these activities have developed and evolved into what is Hollywood today. The complex extends throughout the world via agents, representatives and theaters.

From the mid to late 1950s, the studios started operating as corporate units, following a business plan with a slate of eight to ten films each. While the fight with television continued, wider screens and better sound were used and the business plan, at first sporadically then systematically, included the precursor (epic films) of what the blockbuster was to become in the 1970s. It turned into the locomotive of the industry, absorbing the greater part of investment in the business plan, while the rest of the films produced were intended to cover marginal costs and thus allow for the full employment of the productive capacity of studios. This caused an explosion of costs and the concentration of receipts. Entry barriers rose and became very stringent, and since then only very few producers have been in a position to meet the demanding budgets.

In distribution, block booking is the norm, by which theater operators – usually chains of theaters – cannot choose individual films but have to accept whole packages of several movies. The package usually consists of one blockbuster and a number of lesser movies. Over the years, fewer and fewer theater operators both in the US and in Europe have remained independent of the studios. In contrast with the US, where legislation on competition forbids it, in Europe, all seven American studios are represented by a single organization, UIP.

Gradually the tools of the US film industry were improved. Marketing, consisting of the star system and advertising, was extended to R&D before production and prior to release. The concept of the blockbuster was further refined as the quest for a "scientific formula" for success continued. Factors of success or failure were analyzed: best dates for launching a movie on a global scale; choice of actors and scripts; what was it that made *Titanic* such a hit not only with its 19-to-24-year-old main target group but also many other age groups, including many repeat viewers even in the over-50 group?

At the level of financing, American film industry firms managed to achieve internalization. Part of finance of course continued to be supplied by the traditional banking sector. However, to a great extent financing is internal to the firms, and so is the risk that shareholders bear (Media Business School 1992).

> The distributor – whether a Hollywood studio or a powerful integrated foreign company – has invested (or coinvested rather than putting up 100% of a budget) in production costs, and has borne the majority of the prints and the advertising costs. In turn, the distributor has been the principal beneficiary of success in the marketplace.
>
> (Finney 2010: 9)

During the 1980s and 1990s, home entertainment gained in importance and the complex extended its activities to the production of video films and the development of music applications beyond soundtracks, then to DVD movies. In the chain of production value, applications multiplied with video on demand, pay TV, pay per view and video rental. As a result, theater receipts started representing only a small fraction of total complex revenue, giving the erroneous impression that cinema has ceased to be an important component of the complex. In fact, cinema was the foundation upon which the whole edifice was built and still stands. Whatever addition has been made to the value chain, it has never preceded and always followed cinema, and in any case, success of a film at the theater level remains a quasi-prerequisite for its successful transportation to the rest of the value chain.

The emergence of digital television in the US and the establishment of high definition from November 1998 multiplied the number of channels and thus generated a huge disequilibrium between supply and demand for content. The result was that the role of content producers was upgraded, which led to a number of spectacular mergers and acquisitions between content producers and cable owners, as in the case of Walt Disney with MCI, without however threatening the way the film industry complex works.

A potential threat to the complex that surfaced at the end of the twentieth and in the early twenty-first centuries came from a technological development in the area of television by the startup firm Tivo. For a moment this threatened to revolutionize the way we view television (Alsop 1999). Above all, it threatened advertising, the advertisement establishment on Park Avenue, and essentially set a time bomb in the foundations of the complex of the US film, television and related industries by threatening its source of financing. Luckily for the complex, the bomb was not so much diffused but turned into a dud, smothered by rights infringement and litigation – as we saw earlier in this chapter (see Box 6.4).

The threat of competition from abroad is rather unlikely for the foreseeable future. The film industry has developed independently in Europe, resulting in a totally different industry structure, in which a Hollywood-type complex never developed, and a distinctly different product treated as a piece of art addressed to cinephiles. It is much less important in terms of income and employment, exhibits much slower growth, and displays a large and rising negative balance of payments vis-à-vis the US. India and China are rising powers with impressive volumes of films produced (Nigeria too), but cater mostly to local audiences.

Production of an American studio is comparable to a European country's movie production, but European films have much lower budgets and enjoy much less R&D and script supply than American films (IDATE 1997). If, in addition, account is taken of the advantage of size of prospective market due to the English language, the securing of product demand through block booking, and far superior American financing it becomes clear why the American film industry has overwhelmed its European counterpart and has established its own more commercial product as the de facto example of good practice even within Europe (Finney 1993). The American product has conquered European theaters up to 80 percent on average (in some countries exceeding 90 percent), limiting national industry films to a tiny and decreasing share, and supporting the view that the European public is being Americanized (Gyory and Glass 1992).

While there is no foreseeable foreign threat, there may be instances of imitation as in the cases of *The Fifth Element*, a French film produced with the Hollywood blockbuster formula featuring Bruce Willis, or *The Full Monty*, a British film distributed by the Americans. For now, catering to large, fast-emerging markets may require only a few extra minutes of footage featuring Chinese actors (as was shot for the Chinese version of *Iron Man 3*) but in future something more may be required.

Initial developments in digital cinema seemed innocuous to the Hollywood complex. Digital production has been around since George Lucas's *Star Wars*; at the time he was obliged to film his output on 35 mm film as there was no way to transfer it, transmit it or project it digitally. However, the economic challenge to project digitally was substantial. Considering the fact that each copy of a 35 mm film costs \$20,000–\$30,000, and that the "psychological" requirements of a blockbuster for impact maximization dictate simultaneous opening projection in 1500 to 3000 theaters around the world, the amount that could be saved would be in the order of \$30–\$90 million per film! On the other hand, besides the technical difficulties that lay ahead, the cost to equip a theater for digital cinema was \$100,000–150,000.

Gradually difficulties were overcome. Problems of compression and decompression in transmission via satellite were solved and so was encryption in charging. Protocols briefly threatened to become an obstacle. At present, transmission is mostly through mobile storage, laptop or even DVDs. Within a decade, by 2012, two-thirds of the 130,000 theaters around the world had switched to digital cinema, while about 35 percent had adopted 3D (Motion Picture Association of America, Theatrical Market Statistics 2012).

Technological change may have benefited the incumbents by lowering the costs of distribution, but digital production left the door open to independents eager to rush into the industry. Combining such development with the Internet may send the film industry as we know it into a spin – perhaps in the way the music industry has evolved. Since 2004, DVD sales, then the motion picture industry's largest window outlet, have been on the wane as consumers shifted to lower-cost rentals and subscriptions to services like Netflix. Consolidation is very much present: Sony acquired MGM in 2004, Paramount acquired Dreamworks in 2005, Disney acquired Pixar in 2006, and so on. The number of films produced each year by the studios is declining while those produced by independents is exploding – since 2012, statistics have no longer been collected on films by independents with a budget of less than \$1 million. The existing model is under increasing pressure to change. The arrival of direct links between the producer and the consumer, in particular via the Internet, is now fundamentally changing the economic characteristics, and the architecture of the industry as well. The role of production is being enhanced at the expense of distribution's prior dominant position.

This is a major turning point for independent film. *Olive*, the first film shot entirely on a Nokia N8 Smartphone was screened at a Los Angeles cinema back in 2011. In the past, short-film makers were lucky to have their work seen by a few hundred festival goers. Now, there is Sundance, Young Irish Film Makers, even India's Dirt is Good competition, and several more. Atom and Current TV, to name but a few, have joined Youtube. There is also web TV which allows film makers to broadcast entire series, and *Persona*, the world's first smartphone drama which allows viewers to download daily "appisodes" intended to be viewed on smartphones and iPads.

Thanks to rapidly evolving digital technology, film making and distribution are undergoing a major revolution. However, "the potential transformation is unlikely to have a significant impact on all sectors of the film business" (Finney 2010: 17). The studio product will most likely be produced and distributed for at least the next decade in the same way as in the past, with the aim of reaching the 19-to-24-year-old market. It is the potential tightening of the film value chain that is of interest. As lines get sharper and sharper, distribution, the gatekeeper which essentially selects and determines which films get made, marketed and directed, is making it increasingly difficult to get movies into theaters, even for talents like Steven Spielberg and George Lucas (*Hollywood Reporter* 2013). As the blockbuster has been increasingly correlated with success, there is a rising tendency to abandon films with lower budgets, irrespective of quality. As a result, the number of films produced is expected to shrink and is already doing so, leading Lucas and Spielberg to predict implosion of the film industry.

As real "oligopolistic firms, the studios are characterized by collusion, strategic interdependence, maximization of revenues and the preservation of industry structure" (Currah 2006: 440). At stake is the complex system of release windows they have developed, "which is carefully balanced to maximize the revenues from the distribution of each product. That is, films are made available to consumers in different formats at different prices over time through a series of well defined and exclusive space-times: theatrical exhibition, home video, pay-per-view, pay-television, network television, syndicated television" (Scott 2005).

As oligopolistic firms, the studios also have "a tendency to neglect or marginalize emerging markets, especially those that are seen to threaten the status quo" (Currah 2006: 440). For them, the advent of the Internet was not a reason for celebration but an expected collision. They prefer to protect the structure of the industry rather than exploit the disruptive power of a new technology. The emergence of P2P (peer-to-peer) file sharing has attracted the legal wrath of powerful copyright owners such as the record labels and Hollywood studios because it has enabled "unacceptable and disruptive" forms of consumer behavior. In return, the studios have used their power in the industry to impose a model that aims at mending the open and decentralized nature of the Internet and reshaping it into a "walled garden" with total control over content. In other words, a model "with a centralized server-client architecture, which delivers films on a rental basis in the existing pay-per-view window directly to individual computers, where the films reside for limited periods before self-destructing (under the orders of DRM software)" (Currah 2006: 452). This would limit the Internet to being a pay-per-view window.

Among the studios some will fare better than others. While little can be done to better predict whether a film will be a success or not, to stop the proliferation of independents or losing viewers to Internet-related solutions, there appear to be some hybrid solutions leading to the transformation of existing firms and their ways of operating in the hope of staying alive. Selling stakes in expensive productions to third parties is a good example of such a policy. It reduces the risk attached to both failure (and success), but with less exposure on the part of the firm. Emphasizing her television side is another option for the firm as it helps spread the financial lumpiness of the blockbuster – which may explain why nowadays so many well-known movie film actors are met more and more often on television series. Another option for the firm is to release films for home viewing, on demand, simultaneously with the film's release in the theaters, or even to produce programs for the digital firms – the likes of Netflix. Of course, this last option is not open to the major studios, which are owned by large conglomerates whose interests in cable companies and broadcasting networks would be at odds with such a development. But the field is open to those who dare, those ready to bend the rules of the game, and those unattached, such as Lionsgate (*Economist* 2014: 53).

Beyond the studio enclave, the opening of the market leaves the role of the distributor ready to be taken over by a swarm of old and even new players, either on the producer or the exhibitor's side. For instance, large retail brands looking for marketing or sponsorship opportunities, or technology providers could become involved in generating and delivering content. Even pay TV and video operators, and end users like cinema owners and chains may be attracted. The same is true of the large aggregators such as Apple or Google. Whatever changes materialize, a new value chain is emerging in film production.

Bibliography

Alsop, S. (1999) "Goodbye to TV as we Know It", *Fortune*, August 2, 173.

Arthur, W.B. (1999) "Increasing Returns and the New World of Business", *Harvard Business Review*, July-August, 100–109.

Bagdikian, B.H. (2004) *The New Media Monopoly*, Boston: Beacon Press.

Brown, C. (1997) *The New Economics of Audio-visual Production, Film and TV Financing, Production and Distribution in the Digital Age*, London: FT Media & Telecoms.

Chandler, A.D. (1962) *Strategy and Structure*, Cambridge, MA: MIT Press.

Chesnais, F. (1988) "Technical Co-operation Agreements between Firms", *STI Review*, 4, Paris: OECD.

Chidamber, S.R. and Kon, H.B. (1994) "A Research Retrospective of Innovation Inception and Success: The Technology-push Demand-pull Question", *International Journal of Technology Management*, 9, 1, 94–112.

Chiesa, V. (2001) *R&D Strategy and Organization, Managing Technical Change in Dynamic Contexts*, London: Imperial College Press.

Christensen, C.M. (1997) *The Innovator's Dilemma: When New Technologies Cause Great Firms to Fail*, Boston: Harvard Business School Press.

Christensen, C.M. and Raynor, M.E. (2003) *The Innovator's Solution: Creating and Sustaining Successful Growth*, Boston: Harvard Business School Press.

Coe, N.M. (2001) "A Hybrid Agglomeration? The Development of a Satellite-Marshallian Industrial District in Vancouver's Film Industry", *Urban Studies*, 38, 1753–1775.

Compaine, B. and Gomery, D. (2000) *Who Owns the Media? Competition and Concentration in the Mass Media*, New York: Lawrence Erlbaum Associates.

Currah, A. (2006) "Hollywood versus the Internet: The Media and Entertainment Industries in a Digital and Networked Economy", *Journal of Economic Geography*, 6, 439–468.

David, P.A. (1985) "Clio and the Economics of QWERTY", *American Economic Review*, 75, 2, 332–337.

Dertouzos, M.L., Lester, R.K. and Solow, R. (eds.) (1988) *Made in America: Report of the MIT Commission on US Industrial Productivity*, Cambridge, MA: MIT Press.

Economist, The (2013) "Microsoft Blues", May 11, 64.

Economist, The (2014) "Fighting the System", January 25, 53.

Finney, A. (1993) *A Dose of Reality: The State of European Cinema*, London: Screen International, European Film Academy.

Finney, A. (2010) "Value Chain Restructuring in the Global Film Industry", the 4th Annual Conference on "Cultural Production in a Global Context: The Worldwide Film Industries", Grenoble Ecole de Management, Grenoble, France, June.

Foster, R. and Kaplan, S. (2001) *Creative Destruction*, New York: Currency/Doubleday.

Freeman, C. (1982) *The Economics of Industrial Innovation*, London: Pinter.

Freeman, C. and Lunvall, B.A. (1988) *Small Countries Facing the Technological Revolution*, London: Pinter.

Gyory, M. (1997) *La Distribution des Films Européens en Europe, Accès au Marché et Choix Libre du Public*, Federation of European Film Directors.

Gyory, M. and Glass, G. (1992) *Statistics of the Film Industry in Europe*, European Centre for Research and Information on Film and Television.

Hakansson, H. (1989) *Corporate Technological Behaviour: Cooperation and Networks*, London: Routledge.

Hall, R.E. and Jones, C.I. (1999) "Why Do Some Countries Produce so Much More Output per Worker than Others?", *Quarterly Journal of Economics*, 114(1), 83–116.

Hollywood Reporter (2013), "Steven Spielberg Predicts 'Implosion' of Film Industry", December 8.

IDATE (1997) *Analyse Sectorielle, Marché Mondial de l'Audio-visuel et du Cinema, «Industries et Marchés»*, IDATE.

Imai, K.J. and Itami, H. (1984) "Mutual Infiltration of Organization and Market: Japan's Firm and Market in Comparison with the US", *International Journal of Industrial Organization*, 1, 2.

Islam, S. (1996) "Economic Freedom, Per Capita Income and Economic Growth", *Applied Economic Letters*, 3(9), 595–597.

Kay, N. (1988) "The R&D Function: Corporate Strategy and Structure", in G. Dosi, C. Freeman, R. Nelson, G. Silverberg and L. Soete (eds.), *Technical Change and Economic Theory*, London: Pinter, 282–294.

Kowitt, B. (2011) "One Hundred Million Android Fans Can't Be Wrong", *Fortune*, July 4.

Leblang, D.A. (1996) "Property Rights, Democracy and Economic Growth", *Political Research Quarterly*, 49(1), 5–26.

Mauro, P. (1995) "Corruption and Growth", *Quarterly Journal of Economics*, 110, 681–712.

Media Business School (1992) *The Competitive Position of the European and US Film Industries*, London: London Economics.

Motion Picture Association of America (2012a) "The Economic Contribution of the Motion Picture &Television Industry to the United States", <http://www.americanbar.org/content/dam/aba/events/entertainment_sports/2012/10/forum_on_the_entertainmentsportsindustries2012annualmeeting/imgntd/mpaa_economics_fact_sheet.authcheckdam.pdf>.

Motion Picture Association of America (2012b) *Theatrical Market Statistics*, MPAA.

Mowery, D. and Rosenberg, N. (1979) "The Influence of Market Demand Upon Innovation: A Critical Review of Some Recent Empirical Studies", *Research Policy*, 8, 102–153.

Mullin, J. (2013) "Motorola and Cisco Pay Tivo $490 M to Dodge East Texas Patent Trial", *arstechnica*, June 7.

Nalebuff, B.J. and Brandenburger, A.M. (1996) *Co-opetition*, London: Harper Collins Business.

Nevens, T.M., Summe, G.L. and Uttal, B. (1990) "Commercializing Technology: What the Best Companies Do", *Harvard Business Review*, 68(4), 154–163.

Reich, R.B. (2002) *The Future of Success*, New York: Vintage.

Rothwell, R. (1992) "Successful Industrial Innovation: Critical Factors for the 1990s", *R&D Management*, 22(3), 221.

Rothwell, R. et al. (1974) "Sappho Updated – Project Sappho Phase II", *Research Policy*, 3, 258–291.

Scott, A.J. (2002) "A New Map of Hollywood: The Production and Distribution of Motion Pictures", *Regional Studies*, 36, 957–975.

Scott, A.J. (2005) *On Hollywood: The Place, The Industry*, Princeton, NJ: Princeton University Press.

Segrave, K. (1997) *American Films Abroad: Hollywood's Domination of the World's Movie Screens*, Jefferson, NC: McFarland & Co.

Servan-Schreiber, J.J. (1968) *The American Challenge*, London: Hamish Hamilton.

Sherr, I. (2013) "Tivo's Recipe: Persistence and Plenty of Lawyers", *Wall Street Journal*, August 27.

Siwek, S.E. and Economists Incorporated, (2013) *Copyright Industries in the US Economy: The 2013 Report*, prepared for the International Intellectual Property Alliance (IIPA), November 2013, available at <www.iipa.com>.

Teece, D.J. (1986) "Profiting from Technological Innovation: Implications for Integration, Collaboration, Licensing and Public Policy", *Research Policy*, 15, 285–305.

Teece, D.J. (1988) "Technological Change and the Nature of the Firm", in G. Dosi, C. Freeman, R. Nelson, G. Silverberg and L. Soete (eds.), *Technical Change and Economic Theory*, London: Pinter, 256–281.

Thompson, E.R. (2004) "National Competitiveness: A Question of Cost Conditions or Institutional Circumstances?", *British Journal of Management*, 15, 197–218.

Utterback, J.M. (1974) "Innovation in Industry and the Diffusion of Technology", *Science*, 183, 620–626.

Walsh, V. (1984) "Invention and Innovation in the Chemical Industry: Demand-pull or Discovery-push?", *Research Policy*, 13, 211–234.

Part IV

Innovation Policy

7 Innovation Policy

In the distant past, recourse to Innovation Policy, which at the time was thought of as a simple policy action, was a sporadic isolated event, with very narrow and specific goals. Gradually, the goals of our evolving society changed, while at the same time more came to be expected from Innovation Policy. A contemporary definition of Innovation Policy amounts to what could government do to use innovation and progress in scientific and technical knowledge for the purpose of realizing specific national objectives and taking the necessary steps to ensure the meeting of these objectives.

Now, generating growth, raising income and creating employment are the universal objectives but not all countries are able or willing to do what it takes to implement Innovation Policy to its fullest extent. Most developing countries do not possess the prerequisites and some of the developed ones, in favor of "small government" or due to ideology are willing to use Innovation Policy as sparingly as possible and only up to the point of eliminating "market failures", that is, in cases where the effort expended on innovation is less than the situation necessitates.

For those countries that wish to use Innovation Policy beyond the elimination of market failures and to further address the realities of our imperfect world, the recipe is straightforward. It consists of possessing (building, ameliorating and maintaining) a National System of Innovation (NSI) consisting of the whole research community, i.e. universities, research institutions and organizations, plus the hopefully enterprising business firms, finance and the state. When accompanied by appropriate policies and regulations this system becomes an ecosystem conducive to the generation, diffusion and commercialization of innovation. Furthermore, it is rendered hospitable when supplemented with intellectual property rights, and appropriate policies on skilled immigration, steady and low taxation, an efficient civil service, and a low level of corruption, while balance is sought between variety and selection. The productive sector of the ecosystem is expected to be internationally competitive and able both to withstand external threats such as globalization and to take advantage of the opportunities generated. Diffusion of the benefits from innovation is expected, and steps are taken to guarantee the system's perpetuation.

Each of the four sections of the ecosystem labeled "the building blocks of the ecosystem" is followed by a rationale for policy action and by policy recommendations. Three topics that stand out as in need of additional explanation are treated individually at greater length: finance, which is part of the NSI; intellectual property rights (IPRs), which help render the ecosystem more hospitable; and the balance between variety and selection. They appear in shaded boxes. They are also followed by rationales for policy and policy recommendations, as is the section on globalization.

Other important aspects of the design and implementation of Innovation Policy are also discussed: specificity, flexibility, the nature of policy, degree of intervention, public funding and governance.

Reference is made to salient characteristics of Innovation Policy in various situations: England and Germany in the nineteenth century, present-day Switzerland, the EU, Israel, China and the US.

The main current challenges to Innovation Policy are then discussed and analyzed: the lukewarm stance with regard to "real" innovation expressed either as lack of interest or risk aversion or lack of true entrepreneurship. For some among the very large group of countries aspiring to development but clinging to rentier values the problem of overcoming risk aversion may be met by the encouragement of the creation of venture capital firms, as in Israel, or following the Asian model (South Korea).

At the world level, we note (briefly) the lack of interest of policy planners, with the exception of those in the US, in the regeneration and perpetuation of the system.

Emerging challenges for Innovation Policy are then discussed such as the waning of benefits of innovation in terms of employment, income and the rising level of inequality. The chapter ends with action points and brief proposals for future Innovation Policy and some concluding remarks: the spread of the Internet; the need to support startups but not all startups – only those that promise to become gazelles; the need to encourage the formation of venture capital firms through lower capital gains taxes and by allowing limited participation of pension funds; the formation of technology rating agencies; the articulation of SBIR-type programs; and turning to the state, the "entrepreneur of last resort".

We also discuss issues we should worry about: the precariousness of the best Innovation Policy – the American one; the need for greater reliance on diffusion; the need to abandon the "Washington Consensus" and alternatives for developing countries; the social impact of innovation; income inequality arising from innovation; the need for intervention; the role of standards; the need for antitrust laws based on dynamic competition.

Policy recommendations are made in the "Policy recommendation" sections throughout and at the end of the chapter. At the very end of the chapter there is some discussion of the passage from science to technology to Innovation Policy.

7.1 What is Innovation Policy and what are its objectives?

Simply put, Innovation Policy is what government could do to use innovation and progress in scientific and technical knowledge for the purpose of realizing specific national objectives and taking the necessary steps to ensure the meeting of these objectives. This is a contemporary definition, as the goals of an evolving society change. At the same time society gradually learns to expect more from Innovation Policy.

Innovation Policy has been with us for quite some time, literally for centuries. In the first half of the sixteenth century Henry VIII took the initiative of producing iron cannons desperately needed to win the ongoing war with France (Lundvall and Borras 2006). The success of this initiative rested, significantly, on using the skills of expert forgers who had moved to England from the Continent. In the 1880s, Denmark changed the mode of agricultural production and adopted new dairy processing in order to replace the collapsed Danish exports of rye with butter. The transformation to a system of cooperative ownership supportive of rapid diffusion of dairy techniques was a social rather than a technical innovation.

What do Henry VIII's cannons and Denmark's adoption of dairy processing technology have in common? They are both examples illustrating the fact that Innovation Policy concerns a wide set of issues that have been important throughout history and are still important today. Further, they are successful examples of Innovation Policy. But that is not also so, as illustrated in the case of futile British attempts to prohibit the diffusion of textile machinery beyond its borders (1781) (Jeremy 1973: 26), or the unsuccessful initiative on the part of the French state to attract clock masters (Boorstin 1985: 67–72).

At first, recourse to Innovation Policy, which at the time was thought of as a simple policy action, was a sporadic isolated event, with very narrow and specific goals. At the time, of course, Innovation Policy was not institutionalized, but it emerged occasionally in specific instances when the situation seemed to call for it, as the intermittent interest of the relevant authorities saw fit. It was seen in a different light than we tend to see it now and was aimed at very specific goals, and inevitably it was not as sophisticated as it is now. In the sixteenth century, Henry VIII's initiative was not referred to as "a mission-oriented policy", as we would call it now, or as "technology transfer through immigration". As for the Danish attempt to adopt dairy processing technology, it was not identified at the time as "a diffusion-oriented policy".

Judging by the incentives used we may infer that Innovation Policy at the time was rudimentary, but incentives did exist and so did checks to regulate or limit the use of innovations. There were also prizes. The Portuguese, then the Spaniards, the British, the French and the Dutch, all announced prizes for the innovation of instruments that would help their seafarers locate their position at sea (Boorstin 1985: 46–53). Over the years, the patent system gradually developed and so did laws governing the use or limiting the transfer of technology.

A convenient way to understand the changing character of Innovation Policy over time, is to consider it as an evolving series of refinements, whose succession gives the impression that Innovation Policy is chasing a moving target. At present, as is usually the case, Innovation Policy is still at some distance behind its target.

7.1.1 From Science Policy to Technology Policy to Innovation Policy[1]

While all these refinements were taking place there was a clear transition from science to technology to innovation policy as there was a shift in emphasis in the entire set of policies considered. However, this evolution does not mean that when we pass from one stage to another, one type of policy substitutes the preceding ones, which are abandoned, but that new forms of policy are added that may overlap and coexist with the previous types (Lundvall and Borras 2006). In fact, some of the main issues that arise over time defy the chronological sequence. They may appear under two or even all three headings: Science, Technology and/or Innovation Policy.

7.1.1.1 Science Policy

Up to the start of the Second World War, in the US as well as in other industrialized countries, the responsibility for science and technology was largely in the hands of the private sector, through industrial research laboratories, university–industry linkages and corporate foundations, such as the Carnegie, Ford and Rockefeller foundations.

Science Policy after the Second World War is usually traced back to Vannevar Bush's report (Bush 1945), prepared by specific request of President Roosevelt, which in a sense set the agenda for US postwar Science and Technology Policy. In the last days of the Second World War, Vannevar Bush, science adviser to the President and the most important technocrat and "engineer" of the wartime scientific effort, wrote a report that served as a blueprint for how R&D should be carried out in time of peace. Among other things, it was also the blueprint for what would become America's enormously successful information technology industry. In "Science, the Endless Frontier", Bush laid out a vision for government-funded science and engineering that would unite academia, industry and the armed forces. It did so by keeping them apart. Universities were expected to research basic science and then industry would develop these findings to the point where they could get to market. The idea of R&D as two distinct activities was born. Firms soon organized themselves along similar lines. The result of this approach was a stunning success with AT&T's Bell Labs, IBM's research laboratories and Xerox's PARC.

Science Policy was seen to be contributing to national security, health and economic growth. The scientific successes during the war years led to euphoria fueled by the belief that science, applied science and technological development could produce solutions to almost all problems. The development of radar, penicillin and the atomic bomb and the establishment of operations research are all examples of successful "big science" projects in the allied war effort. These successes contributed to the respectability of Planning and Science Policy. They also contributed to the hegemony for a number of years of "science-push" ideas in Science and Technology Policy. Government-sponsored war research developed very rapidly with huge increases in investment in R&D. The notion of permanent policy-making bodies for R&D after the war gained support.

By 1950, the governments of many advanced countries had adopted policies aimed at fostering economic growth. Above all, at the time it was believed that growth was the result of a high volume of investment, particularly in innovation. In the 1950s and early 1960s expenditure on science and technology was accepted as an unfailing recipe for technical progress and growth.

Science Policy was expected to allocate sufficient resources to activities within science, and to ensure that such resources are used efficiently and contribute to social welfare. Therefore, the organizations and institutions to be focused upon were universities, research institutes, technological institutes and laboratories, and the task of Science Policy was their internal regulation. A starting interest in how they link up to the entire environment, especially to industry and government, could be discerned at the time, but strengthening of this linkage became more important and a stated goal at a later stage, in Technology and Innovation Policy.

By the late 1960s, skepticism followed, as "science-push" policies were being questioned by decision makers in both government and industry (see chapter 6) while "science-push" theories were being modified by ideas about "demand-pull" in the study of innovation. By the early 1970s, the "demand-pull school had scored a victory on points" (Freeman 1979: 206), at least in the interpretations of the policy makers (Mowery and Rosenberg 1979).

Thus, the early upsurge in spending on R&D, followed by cutbacks justified with reference to the results of research on innovation, are mere reflections of the changing fortunes of "science-push" and "demand-pull" theories.

7.1.1.2 Technology Policy

During the 1970s and early 1980s the world economic environment deteriorated, forcing a shift in policy emphasis. Technology replaced science as a more effective base from which to support national industry and economic performance.

> At the same time, the understanding of the innovation process started to change dramatically. New interactive models, significantly different to the previous linear approaches, emphasized the central role played by diverse activities out of the sphere of basic research. Attention shifted then to issues such as the feedback effects between the phases of the innovation process which are related to the market and those related to technology and to the diverse interactions between science, technology and other activities concerning innovation that occur at firm or at industry level ... Attention thus shifted to a whole set of measures aimed not only at the generation and diffusion of knowledge, but also at stimulating the economic, institutional and social factors that influence the absorption and generation of technological knowledge.
>
> (Lopez-Martinez 2006: 78)

In terms of objectives, Technology Policy is not much different from Science Policy but it represents a shift to a higher level of focus on economic objectives, i.e. focus on technologies and sectors. The era of Technology Policy is one in which especially science-based technologies were seen as being at the core of economic growth, and Technology Policy means different things for different countries.

> In big high-income countries the focus will be on establishing a capacity in producing the most recent science-based technologies, as well as applying these innovations. In smaller countries it might be a question about being able to absorb and use these technologies as they come on the market. Catching-up countries may make efforts to enter into specific promising established industries using new technologies in the process of doing so.
>
> (Lundvall and Borras 2006: 608)

In several cases, Technology Policy government initiatives resulted in fostering a policy intended to produce national champions, as in the case of France, Germany and the UK. In others, such as in the case of Japan and the Asian Tigers later, catching up was the goal of the strategy followed. Overall, the West felt the brunt of the pressure brought by the contact with Japanese products, that is, products of a higher quality produced by a more efficient organization and production system in activities that were considered to be their preserve (semiconductors, automobiles) and, as perceived, the result of a coordinated national strategy. In the US this prompted a search for determining what was at fault (Dertouzos et al. 1988) for long-term strategic thinking about technology (Dodgson 1989), for focusing on specific technologies deemed strategic or critical (Branscomb 1993), and for outright pleas for government intervention (e.g., semiconductors agreement).

As in Science Policy, the elements of the innovation system in focus remained universities, technological institutes and R&D laboratories. However, the interest in universities became more particular and concerned more specifically engineering and transition sciences, and with respect to the universities' internal organization there was a shift of focus towards their external links to industry and other organizations.

7.1.1.3 Innovation Policy

Innovation Policy has two different versions, which coexist: the so-called "laissez-faire" version and the "systemic" version, and each of these versions leads to a list of rationales for policy action. The reason for this dichotomy is the theoretical rift between Neoclassical and heterodox economics referred to on many occasions throughout this book, which underlies each of these versions. We shall analyze each version followed by the rationales for policy action they each imply.

THE "LAISSEZ-FAIRE" OR "MARKET-FAILURE" APPROACH

The first version, referred to as the "laissez-faire" version, rests upon an application of standard Neoclassical economics on innovation. It is non-interventionist and holds that the emphasis should be on "framework conditions" and not on specific technologies or sectors. The advocates of this type of Innovation Policy fall into two categories: those espousing an extreme version where general education and basic research are considered to be the only legitimate public activities, and the protection of IPRs the only legitimate field of government regulation, and those in favor of a more moderate version that tolerates public initiatives aimed at fostering entrepreneurship and promoting a positive attitude to science and technology.

The laissez-faire version of Innovation Policy, both in its extreme or more moderate expression, rests upon the standard assumption made in economics that firms always know what is best for them and that they usually – in the absence of market failure – act accordingly. Therefore, the only case where intervention would be considered admissible is in the case of a market failure, since if markets can do the job there is no reason to intervene. Market failures could be the result of many causes but the one most often implicated in the case of Innovation Policy is the lack of incentive for investment in the production of knowledge. When knowledge is rival but non-excludable, IPRs can be guaranteed and enforced by government. Where knowledge is both non-rival and non-excludable, government should subsidize knowledge production intended to address public use, or take charge of producing it by itself.

The theoretical underpinnings of the laissez-faire version of policy derive from Nelson (1959) and Arrow (1962). Both argue that technological knowledge has distinctive features, which lead business firms in a market economy to perform less R&D than is socially optimal. In short, Arrow, who essentially equates the production of "technology" with "science", discovers elements of market failure and infers the necessity of public financing or production of research results in order to correct this market failure. This view prevailed for a long time, and in many ways remains the basis of Innovation Policy.

As for the distinctive features of market failure that Arrow sees, the first is "uncertainty", which means committing resources to a knowledge production process without knowing the results with any accuracy. The second feature has to do with "appropriability": if producers cannot appropriate the benefits of knowledge, then they have no incentive to produce it. The third feature of technological information is "indivisibility": the underlying knowledge must exist on a certain minimum scale before any production at all can take place, and this necessary minimum is independent of the rate of production. This means that there are scale economies in the use of indivisible capital goods. And this applies to technological information.

This "market failure" approach to knowledge production leads to a relatively simple set of policy proposals. In this setup, the basic policy task is to encourage discovery-oriented activities, and then to protect the use of the results. The problems of risk and indivisibility lead to straightforward under-provision of knowledge, and suggest that the public sector should either produce knowledge directly or provide subsidies to knowledge-producing institutions. The appropriability problem implies the existence of a major positive externality, and suggests either subsidization or the creation of property rights (via patents or other intellectual property protection). The basic problem with this approach is that it does not give any secure guide to how to identify areas of market failure, or the appropriate levels of public support that might follow from it (Metcalfe 1994).

This type of approach to Innovation Policy accords very well with what is sometimes called the "linear model" of innovation. This is the view that the process of innovation is essentially a process of discovery, in which new knowledge is transformed into new products via a fixed (linear) sequence of activities or stages.

MARKET FAILURE APPROACH: RATIONALES FOR POLICY ACTION

Whether implicitly or explicitly stated, the underlying rationale for government support of R&D is usually based on theories relating science and technology to economic growth. Besides assumptions as to this causal relationship, a number of reasons in the sphere of "market failure" are used to justify government sponsorship and regulation of technical change (Coombs et al. 1987: 205–209). They center on the failure of the market mechanism to provide adequate resources for the development of science and technology considered necessary for a contemporary society. They relate to the likelihood of a "social return" on R&D being significantly higher than "private returns" to investing firms.

- First, the scale of capital investment or R&D investment required is often such that individual firms cannot raise the necessary funds nor can they accept the high risk involved in the development of the new technology.
- Transportation, telecommunications and energy are activities of interest to industry and society at large. Yet in these activities a firm may not necessarily benefit from an investment aiming at innovation.
- The length of time between investment in basic research and the economic benefit resulting from it could be so long that a shift of resources could result in a movement away from basic research towards applied research.
- There are many areas of basic research, which do not lead directly to technological advances. The market mechanism would allocate resources to these areas even less efficiently.
- There are sectors composed of small units, and the market alone does not generate all the technical change that is economically or socially necessary or desirable.
- Health is an area where it is widely accepted that the market mechanism alone should not reign supreme, while defense is considered by definition to be a topic for government policy.
- In addition, governments provide funds to support industry in the face of international competition, either to support sectors, which for strategic reasons the government believes should be competitive, or to protect others that are not.
- Finally, in the area which concerns the control of unwanted effects – "externalities" – the market fails to direct technology.

Some innovations are partial in distributing their costs and benefits among different groups in society. Government therefore regulates technical change with a view to optimizing costs and benefits to society as a whole or intervening in conflicts between groups with different interests as to the positive or negative consequences of technical change.

Given these market failures recognized since the mid-1980s, as concern about the environment and sustainability (Courvisanos 2009a, 2009b; Ruttan 2001) joined society's preexisting anxieties in the 1990s, and as even more recently, health and safety at work, product safety and technological change have come to be considered as possible cause, effect or cure, the introduction of related regulation has followed.

THE "SYSTEMIC" OR "(NATIONAL) SYSTEMS OF INNOVATION" APPROACH.

The second type of Innovation Policy, the "systemic" version, encompasses the concept of an "innovation system". This concept underlines the fact that firms usually innovate, not in isolation, but through collaboration with other firms, and are interdependent with other organizations. First, we have a constellation of organizations, whether firm entities (such as customers, suppliers, competitors) or non-firm entities (such as universities, schools, research institutions, government ministries, etc). Then we have laws, rules and norms – institutions in general – which act upon the behavior of organizations and end up either as incentives or obstacles to innovation. These organizations and institutions are components of systems for the generation, use, diffusion and commercialization of knowledge. Thus, innovations emerge in "systems of innovation" (Edquist 2006). The firm is no longer the unit; instead the system is.

An innovation system is a conceptual and functional framework, which integrates theoretical perspectives and empirical insights. It makes use of empirical material and analytical models developed in innovation research, as well as institutional and evolutionary models. It recognizes the fact that competence is unequally distributed among firms, and that good practice in terms of developing, absorbing and using new technology is not instantly diffused among firms. In a way, it is a response to the view held under the laissez-faire approach that agents know what there is to know and to disregard competence building, which is at the heart of competition in the learning economy. Nor does it assume that markets are pure and governed by anonymous relationships between producers and users. Additionally, it holds that "failures" are not exhausted by the Neoclassical concept of failure but may extend to "systemic failures", that is, failures of institutions to coordinate, link or address systemic needs (Edquist 1997).

The concept of a system of innovation or national system of innovation (NSI), was developed independently and from different perspectives by Lundvall (1988) and Freeman (1988). Ever since, it has diffused rapidly and consists of several definitions to fit different nuances in terms of meaning, scope and use (McKelvey 1991).

According to Lundvall (1992) we can distinguish two definitions of the concept of national system of innovation: a narrow and a broad one. The narrow definition includes only organizations and institutions involved in searching and exploring. The broad one refers to all parts and aspects of the economic structure and the institutional setup affecting learning as well as searching and exploring. In this instance, the production system, the marketing system and the system of finance become subsystems in which learning takes place.

The most comprehensive definition is provided by Edquist, who states that national systems of innovation include "all important economic, social, political, organizational, institutional and other factors that influence the development, diffusion and use of innovations" (Edquist 1997: 14).

7.2 The building blocks of the innovation ecosystem

Now most countries expect to generate growth, income and employment through Innovation Policy. Due to the emphasis given to the importance of the innovative effort as a measure of innovation during the last five decades, its maximization has become a universal goal. But a high innovation effort by itself means little. (The Soviet Union at some point apparently reached 4 percent of GDP in R&D.)

Certainly, achieving a high innovation effort is commendable but the real tool of Innovation Policy in order to seek growth, income and employment is to possess a well-functioning NSI supported by a number of policies and regulations that bestow a number of desired attributes. Depending on the situation, this may amount to starting building from scratch, or ameliorating and maintaining whatever already exists of a well-functioning ecosystem conducive to the generation, diffusion and commercialization of innovation. This means ensuring that it is a hospitable ecosystem with balance between variety and selection; with a productive structure favoring innovation; capable of meeting external threats such as globalization and profiting from the opportunities it offers; and able to diffuse widely the benefits of innovation. The ecosystem must be allowed to evolve while insuring its perpetuation in the future.

This amounts to building, ameliorating and maintaining an NSI which when accompanied by appropriate policies and regulations becomes conducive to the generation, diffusion and commercialization of innovation. It further becomes a hospitable ecosystem when supplemented with IPR and appropriate policies on skilled immigration, steady and low taxation, an efficient civil service, and a low level of corruption, while balance is sought between variety and selection. The productive sector of the ecosystem should be internationally competitive and able both to withstand external threats such as globalization and to take advantage of the opportunities generated. Diffusion of the benefits from innovation is expected, and steps are taken to guarantee the system's perpetuation.

We shall examine each part of this proposition in turn.

7.2.1 The building, amelioration and maintenance of an ecosystem conducive to the generation, diffusion and commercialization of innovation

This is tantamount to building, ameliorating and maintaining an NSI with all components interlinked, including finance; well-functioning universities with effective research; good research organizations; a motivated, interested and entrepreneurial private producing sector; and a system of support of research beyond basic research. Every element in the structure must perform its task well with the state leading – supporting each element, coordinating and orchestrating the overall functioning of the mechanism set up. A simplified version of such a system is portrayed in Figure 7.1, showing research laboratories, institutes and research organizations, universities, firms, finance and the state.

It is only to be expected that there should be a multitude of NSI constructs with innovation and learning at their center – one for every economy – to accommodate the multitude of differences in country characteristics that determine innovation in its many dimensions (level, type, intensity, location, direction, etc.).

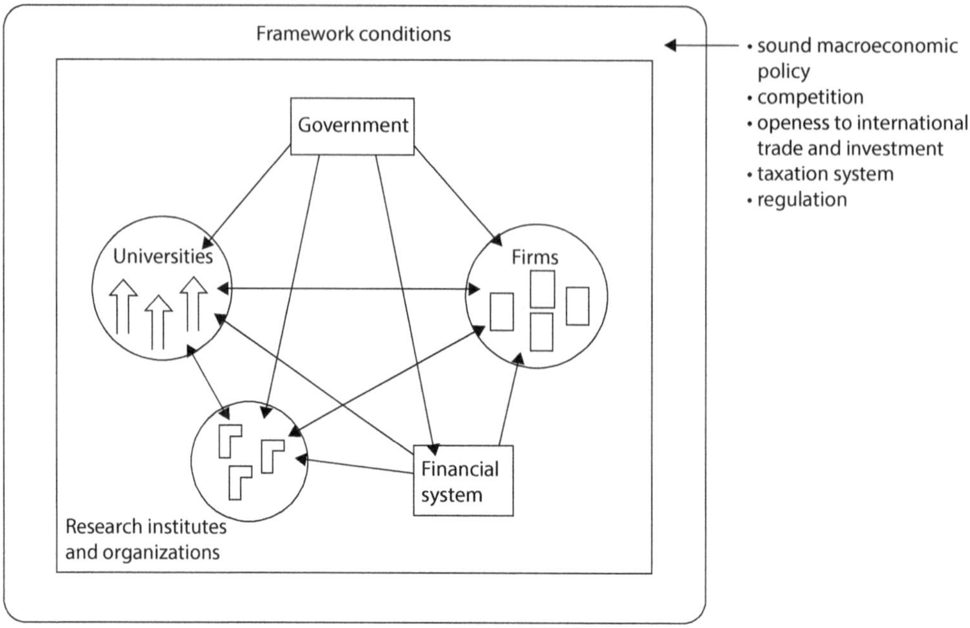

Figure 7.1 A National System of Innovation

Differences in characteristics among countries that have at times been advanced pertain to institutions, organizations, laws, the whole apparatus of their national innovation systems; the structure of their economies and of the specialization of their productive capacities; their level of development; the degree of government involvement; the ways in which their market players cooperate, learn, generate, diffuse and absorb knowledge; their history and accumulated experience; their social conventions; the ways they have developed in facing social problems.

Some elements of systems of innovation are consciously designed by actors such as government policy makers, while other elements evolve spontaneously over long time periods. History matters as the natural resource base of some countries (Sweden, Canada, Australia) and their economic histories strongly influence the present anatomies of their NSIs and determine to a large degree the direction of their innovation.

Thus, one size does not fit all. As it is highly unlikely that two economies will share exactly the same characteristics, it follows that they should not be forced to share exactly the same NSI in all its details.

Of all the elements of the NSI construct the one that needs further understanding as to its role, functioning and impact is finance, which we shall treat separately.

7.2.1.1 National Systems of Innovation: the rationales for policy action

Arrow's market-failure analysis, based on the simplified market models of Neoclassical economics, where the forces of demand and supply determine the optimum use of resources, simply prescribes a policy of public financing or production of research results in order to correct this market failure. This view prevailed for a long time, and in many ways remains the basis of Innovation Policy. In contrast, proponents of evolutionary economics seek to reintroduce real-world conditions, to which we should add that "while all perfect markets are alike, each imperfect market is imperfect in its own way" (Krugman 1996: 110). In addition, by extending the list of relevant factors to include, for instance

> incomplete information, rationality shaped and bounded by social institutions, the existence of cooperative connections between market players, and knowledge as a form of capital, the rationales for policy intervention to encourage innovation are significantly expanded over those suggested by neoclassical approaches.
>
> (Bryant and Wells 1998: v)

7.2.1.2 Failures at the level of the NSI construct

These may take place because a key element – a node in the system –is missing, or has limited competences or has low absorptive capabilities. This may apply to a multitude of such nodes in the system, i.e. firms, universities, research institutes, etc. The performance of each of the nodes contributes to the overall performance of innovation but the state of the system of finance (as to extent, vitality, versatility, flexibility, responsiveness, abundance and accessibility), which is linked to all the elements of the construct, is believed to be by far the most crucial factor as to success or failure with respect to innovation. This is the reason why a whole subsection of this chapter is devoted to finance.

Failures may also occur because connections among heterogeneous agents and complementary activities are not present, or do exist but are weak. The competence of innovating entities, be they firms, universities, research institutions or other knowledge-producing units, is important, but equally important is the competence of the remaining links, that is, suppliers, users, knowledge-producing institutions and policy makers (Malerba 2009: 33–45). Additionally, the strength of the links among the components of the system is also important. In chapter 2 innovative activity was found to be a function of how well integrated an economy is, how well the different sectors of the economy are connected. Finally, the density of such links is of paramount importance, as also seen in chapter 2, where innovation was found to be a linear function of the link density for low densities, rising to an exponential dependence for high densities.

AREAS OF SYSTEMIC FAILURE

Failures of infrastructural provision and investment More or less absent from the Neoclassical market-failure approach, infrastructure is very much emphasized in the systems approach in terms of its interaction with the rest of the nodes of the system, and refers to both physical and science-technology infrastructure. The former usually has to do with energy and communications; the latter with universities, technical institutes, regulatory agencies, libraries and data banks. Both types are associated with large-scale indivisibilities and long-term horizons, and are unlikely to produce adequate returns and lead to major difficulties in their financing. Policy intervention in support of infrastructure may take the form of regulation to set up incentives and controls for private provision, subsidies to private provision or direct public provision (Smith 1998).

Transition failures Firms may encounter technological problems outside their existing capabilities or face discontinuous shifts in technology, or shifts in technological regimes or paradigms. Here, policy may intervene in order to correct for mismatches or misalignments among actors within an established system as they may fail to emerge and develop – as in the case of standard-setting policies.

Lock-in failures A strong feature of systems theories is the notion of path dependence, or lock-in to existing technologies. One of the basic reasons for path dependence is the existence of system or network externalities, combined with the fact that technologies exist in close relation to their social and economic environment. This means that technological alternatives must compete not only with components of an existing technology, but also with the overall system in which it is embedded.

Industries, and in fact the whole socio-economic system, can be locked in to a particular paradigm. It is very unlikely that movement away from such a paradigm can be induced by, for example, tax policies on a particular input. The elements of a technological paradigm interlock with each other, and with the social and technical infrastructure built up over a long period.

Actions at the level of individual agents are unlikely to overcome lock-in. External agencies, with powers to generate incentives, to develop technological alternatives, and to nurture emerging technological systems are required.

Institutional failures A key aspect of the institutional context is the framework of regulation which involves technical standards, risk-management rules, health and safety regulations, and so on. The regulatory system also includes the general legal system relating to contracts, employment and the IPRs within which firms operate. Finally, there is the context of political culture and social values, which shapes public policy objectives and in particular the macroeconomic policy environment. Taken together, this integrated set of public and private institutions, regulatory systems and the policy system make up the overall context of economic and technical behavior, which shapes the technological opportunities and capabilities of firms. It thereby shapes firms' economic performance and the macroeconomic evolution of the economy as a whole.

POLICY RECOMMENDATIONS

Changing a policy with the concomitant changes in the NSI construct is easier said than done. National innovation systems are very persistent in their fundamental characteristics shaped by the given economic structure and research infrastructure. Minor policy changes and policy instrument fine-tuning abound. However, (infrequently) an NSI can also change as to goals, institutional arrangements and sources of dynamism – as in the case of Finland in the late 1980s and China during the last six decades and especially since 1985.

In contrast with the policy maker bent on addressing market failures, in the context of the NSI the policy maker is expected to foster the system's ability to adapt to changing conditions in order to maintain and enhance accumulated knowledge and capabilities. This means that the focus is on infrastructure provision, network problems, transition problems and technological lock-ins (Woolthius and Lankhuizen 2005).

The policy maker is expected to address the most salient characteristic of innovation, namely, uncertainty. The extent of uncertainty varies with the sector, the type of innovation and the different stages of the innovation process. It is more pronounced in new fields, in radical innovations and at the early stages of the innovation process: "private organizations perform least well in situations where uncertainty is largest" (Borras et al. 2008 : 13).

THE NSI CONSTRUCT

Before considering applying any policy, policy makers should first make sure of addressing the appropriate construct. Otherwise, even the best policy's results will range from less efficient than expected to outright harmful.

The NSI should be up to date as to the evolving nature of the innovation process so as to accommodate even its latest forms such as "open innovation", "user centered innovation" and "new combinations". Additionally, to be effective, policy should recognize the rising importance of non-technical innovation and encompass the wider range of activities in which innovation occurs beyond R&D, and should include, among others, organizational changes, testing, marketing, design and training (OECD 2010: 25).

We should remember that an NSI construct is based on assumptions that depart from perfect competition and attempts to embrace real-world conditions. In the laissez-faire version, market and competition are the main prerequisites, and essentially there is a single recommendation for institutional design for all. In the case of the systems approach, competition is still important but so is the need for close cooperation between producers and users to facilitate knowledge flows. In the case where generic technologies need to be developed, cooperation needs to be extended to competitors as well.

A crucial task of innovation policy is to maintain variety and competition, served by the creation of incentives and regulation – for which there should be a provision in an NSI construct and which are additional departures from the dictates of perfect competition. As "competition consumes its own fuel" (Metcalfe 1994: 938) (in the sense that the selection process gradually weakens), beyond setting up the rules of the game, there is a need for monitoring and assessment of regulatory performance and, if necessary, changes in regulatory systems.

At the design level of the NSI, it should be remembered that the procurement of information and knowledge, and their dissemination and processing, which are of no consequence in the Neoclassical view, are (from an evolutionary perspective) central policy recommendations. These measures to increase the adoptive skills of participants contribute to faster diffusion and, consequently, to a quicker conversion into economic outcomes. They include qualification and further education, management style and employee relations, easing of access to socially distributed knowledge, i.e. cooperation between firms as well as reverting to the publicly provided "technological infrastructure" (universities, research institutes, patent offices, science parks, etc.).

Globalization pressures have acted differently upon each of the components of the existing national system of innovation, resulting in a gradual weakening of the institutional matching between its various components. This calls for a constant reexamination of NSI and policies pursued, and reassessment of their effectiveness. Additionally, there is a need for readiness to bring about changes in the NSI, its policies and instruments so as to accommodate changing goals or serve existing ones better, as demonstrated in the cases of China and Finland.

And a word of caution: provided the data exist, some problems are rather straightforward to diagnose, as for instance investigating whether universities perform adequately in research or whether they have sufficient links with industry and the rest of the elements of the NSI. However, in other problems the answer is more elusive. For instance, if we were to diagnose and prescribe Innovation Policy so as to close the technology gap between Europe and the US, we would find that there are two conceptions in the literature that seem to give contradictory diagnoses – and hence differing prescriptions (Encaoua 2009). The first (Dosi et al. 2006) highlights insufficiency of public funds dedicated to research and private investment in R&D in Europe coupled with the weak position of European firms in global oligopolistic markets. According to this view, the implications for Economic Policy are that Europe should increase public investment in human capital formation and research, and embark on an Industrial Policy to reinforce leadership in large firms and to define priority areas for state intervention in innovation.

The other idea (Aghion 2006) emphasizes structural rigidities that prevent the process of "creative destruction" from stimulating growth of the knowledge-based economy as the source of European weakness. The Economic Policy implications thus consist of a set of complementary structural remedies to alleviate these rigidities (Sapir et al. 2004; Kok 2004). These measures are aimed at labor market rigidities and the maintenance of under-performing elements of the NSI such as universities, research labs and organizations, etc.

So which of the two policies is the correct one? The answer will remain elusive unless we can design a methodology that makes it possible to compare them.

Box 7.1 Finance

Ever since innovation has demanded more than rudimentary means for its realization, finance has been one the main prerequisites for its success (Lamoreaux and Sokoloff 2007; O'Sullivan 2006). As a consequence, a lasting issue pertaining to finance and innovation is whether there is adequate financing of R&D, and if not, to determine, and in terms of policy address, the causes for this inadequacy.

Some of the obstacles raised as to financing are attributed to the characteristics of the innovation process; others to issues deriving from the financial system itself. Indeed, the most salient characteristic of innovation, as we know already, is uncertainty both at the technical and the market levels. Innovation most likely involves R&D activities, both formal and informal, which require sunk costs before production can take place and which rise steeply as we proceed from the cheapest (but most uncertain as to outcome) initial stage to the subsequent development and diffusion stages, which involve less uncertainty but are much more costly, often ten to twenty times the R&D costs. So it is only natural that "the gap between the end of an R&D project and the development of a product prototype is sometimes called 'the valley of death' in the US" (Borras et al. 2008: 14).

Besides uncertainty, there is another reason why economists have long held the view that innovative activities are difficult to finance in a freely competitive marketplace. Innovation consists of non-rival knowledge, in the sense that its use by one firm does not preclude its use by another. To the extent that knowledge cannot be kept secret, the returns on the investment in it cannot be fully appropriated by the firm undertaking the investment, and therefore such firms will be reluctant to invest, leading to underinvestment in R&D.

Beyond uncertainty, and supposing that the firm is willing to overcome considerations of partial appropriability, the firm still has to consider whether it is within its possibilities and whether it is willing to withstand a financial drain which may be considerable in depth and stretch over some period of time, sometimes several years.

If this hurdle too is successfully reached, then the firm will have to consider the source of finance. This is the point where the roads of firms seeking and those not seeking external finance part. The large firm, eager not to divulge its future plans lest it may attract the interest of would-be imitators and competitors, will attempt to proceed with internal financing and will seek external financing only as a second choice when unable to do so. This does not mean that the large firm is immune to the need for external financing but it is in a much better position than the small firm to rely on internal funding. Besides, if it does have to rely on external finance, it will be met with a wholly different response – qualitatively and quantitatively – than the small firm would.

Innovation may occur at any of the stages of the life cycle of the firm. It may appear in large and also in small firms, but it may also coincide with the inception (and actually may be the cause) of a startup. For small firms – which incidentally are also not eager to divulge their future plans – the requirement for substantial investments in R&D precludes the use of internally generated resources (McNally 1995). It is only natural then that a large proportion of small firms cite access to finance as a considerable barrier to innovation (Hoffman et al. 1998). Financial constraints have been found to have more of an impact on not starting, delaying or postponing projects than other internal or external hampering factors (Canepa and Stoneman 2002), and most importantly the presence of financial constraints has been found to reinforce the hampering effect of other obstacles (Mohnen et al. 2008: 207). Financial constraints may also help explain why many innovations originating in small firms reach completion in the hands of large firms, as well as the reason for some of the acquisitions of small firms by larger firms.

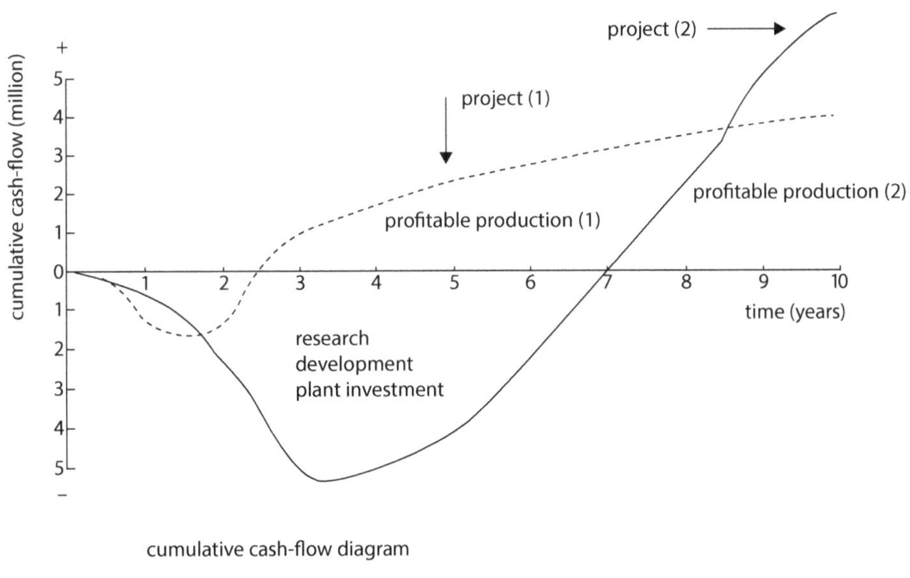

Figure 7.2 Innovation and its financial requirements over time

Now why is it so important to differentiate between seekers and non-seekers of external funds? Because in addition to the gap due to partial appropriability, there is a gap between the private rate of return and the cost of capital when the innovation investor and financier are different entities. Thus, even if problems associated with incomplete appropriability of returns to investment in R&D are solved through the use of intellectual property protection, subsidies or tax incentives, it may still be difficult to finance R&D using sources external to the firm.

> That is, there is often a wedge, sometimes large, between the rate of return required by an entrepreneur investing his own funds and that required by external investors. By this argument, unless an inventor is already wealthy, or firms already profitable, some innovations will fail to be provided purely because the cost of external capital is too high.
>
> (Hall 2009: 170)

This second gap is traced to three sources: information asymmetries, moral hazard and taxes. Innovative activities entail by necessity a fundamental information asymmetry between the inventor and external agents such as prospective lenders, i.e. funding bodies, banks. Presumably the inventor has intimate knowledge of the technology and the details of the planned innovation and his own estimate of the probability of its success. However, even if the information on those crucial matters is well documented, there will always be a significant gap between what the inventor knows and what an external agent can assess.

> In particular, there will be significant information asymmetries in this respect between the inventor and mainstream financial intermediaries like banks and institutional investors, who lack the capacity to verify the information and claims of the entrepreneur. Potential investors will therefore be skeptical of the likely returns on investments in developing new technologies, and therefore, entrepreneurs who could offer attractive returns may have no credible way of conveying such potential to risk-averse investors.
>
> (Trajtenberg 2009: 374)

This means that in the absence of cash flows or other collateral, a typical startup company or individual innovative entrepreneur will not have access to traditional sources of finance – this is the so-called "funding gap".

As for moral hazard problems, these are the second reason for the existence of the second funding gap in R&D investing; it arises from the separation of ownership and management that characterizes most firms. It leads to a principal-agent problem when the goals of the two conflict. In other words, this may lead either to managers spending on goals that are dear to them, thus forcing them to finance R&D through the use of higher cost external finance, or to managers becoming risk averse – more than shareholders – and avoiding investments in projects that increase the risk to which the firm is exposed.

Finally, taxation may be an additional contributor to the reason why external finance is more costly than internal funding. Indeed, this happens very often because in most of developed economies the tax treatment of R&D is very different from other kinds of investment. This is because R&D expenses are disbursed as they occur, implying an effective tax rate on the corresponding assets lower than in other types of investment; their economic depreciation is less than the depreciation allowed for tax purposes (Hall 2009: 178).

To sum up our findings so far, the presence of either asymmetric information or a principal-agent conflict implies that new debt or equity finance will be relatively more expensive for R&D than for ordinary investment, and that lack of collateral further reduces the possibility of debt finance. These arguments suggest, and several empirical studies confirm, the important role of retained earnings in the R&D investment decision, while debt is a disfavored source of finance (Hall and Lerner 2010). In addition, studies have shown that positive cash flow may be more important for R&D than for ordinary investment.

The problems associated with financing investment in R&D are further amplified and are most apparent for new entrants and startup firms. The information asymmetries are particularly stringent at the very early stages of the innovative process (so-called "early-stage technological development"), and it is at these stages that the "funding gap" is most acute and where the market may be particularly prone to failure (Traijtenberg 2009: 372–374). In addition, such firms are not well suited to providing the type of collateral or assets required by banks and other financial intermediaries, since their value is embodied in human and intellectual rather than physical capital. As a result of all these handicaps in relation to the financial system, the financial establishment tends not to be particularly interested in financing such firms (Bank of England 2001: 27–36, 98). Indeed, a study by Branscomb and Auerswald (2002) for the US shows that mainstream intermediaries like banks, private equity and other institutional investors are entirely absent in the funding of early-stage technical development.

Considering overall the issue of financing investment in R&D, it would be fair to say that with the exception of the US, and to a lesser extent the UK, which could both benefit from marginal improvements, there seems to be substantial room for improvement if financial markets are to serve investment in R&D effectively and adequately.

In recognition of this assessment many governments, including those in the US and the UK, currently have programs consisting of tax incentives to companies in order to reduce the cost of capital they face for R&D investment, especially for small and new firms. Such a policy approach hinges on the observation that the cost of capital is relatively high for R&D, and tries to close the gap through a tax subsidy. However, over the years venture capital, an alternative approach relying on the private sector, has developed. It attempts to close the financing gap by simultaneously reducing the degree of asymmetric information and moral hazard, rather than by simply subsidizing the investment.

"Venture capital" can be thought of as a general title covering a number of specialized mechanisms that have materialized to meet the different needs of financing not met by traditional sources of finance. As a general definition, we may say that it provides support to a company in the form of participation in its equity or an option to convert a loan to equity. Figure 7.3 links each phase of development of an innovation to particular financial needs as they evolve over the duration of the innovation process, and to the corresponding financing mechanism.

Proceeding from left to right, that is, from high- to low-risk investment in Figure 7.3, the main investment actors are:

The 3Fs (family, friends, fools) refer to the initial stage where the innovator-to-be scrapes together his savings, exhausts his possibility for overdrafts and gets all the help he can from his entourage.

Business Angels are wealthy individuals prepared to use their own financial resources to make risk investments based on their own experience and interests. They are often retired senior executives or people who have sold their companies. They tend to invest locally in sectors in which they have some experience, and they participate actively –"hands on" – not just by giving advice. Their investments are quite small but when they act in syndication they may raise substantial sums.

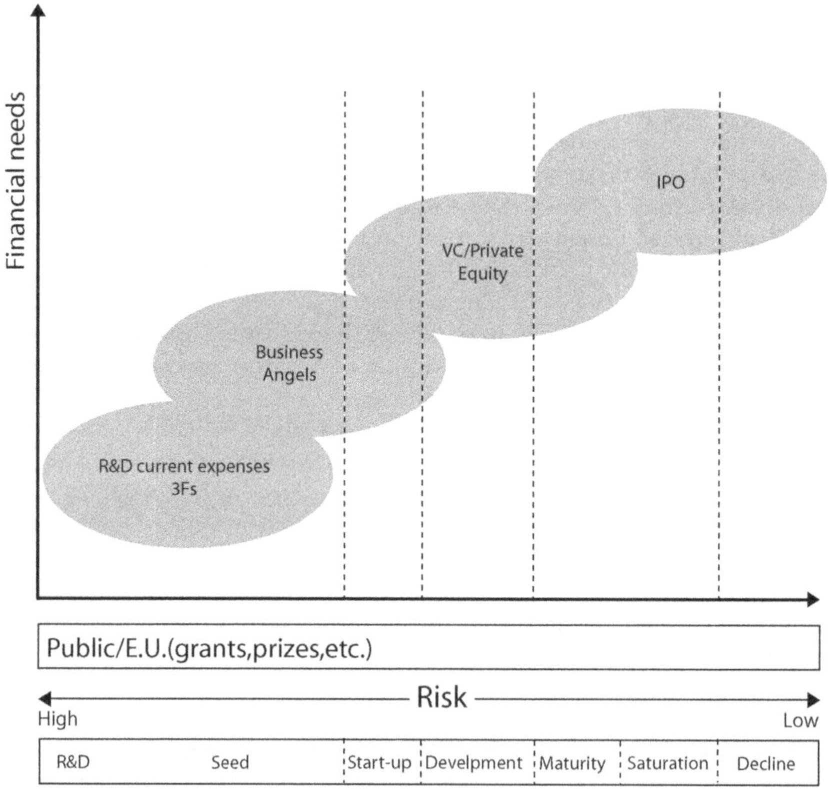

Figure 7.3 Financial needs at different stages of an innovative product

Fund management companies constitute the conventional venture capital sector. They manage the funds of their shareholders – wealthy individuals and institutional investors – and return them their profits after deduction of expenses and fees. They have a fixed term, usually ten years after which term the fund is closed and the profits are distributed. They operate mostly as limited partnerships, which in the US have the advantage of not paying capital gains taxes – only the taxable investors in the fund are liable to such taxes. In order to maintain their status of limited liability, investors must not become involved in the management of the fund.

They invest in young firms, usually in exchange for preferred stock with various special privileges. They are unique in performing some key functions, which help raise the value of the firm they invest in, including meticulous screening, active monitoring and advice, and setting up the proper exiting procedure and above all the timing for exiting. In the end, they have the choice either to sell these firms to corporate acquirers or liquidate their holdings after taking the firms public. Either way, the possibility of exiting is very important to them as the proceeds from exiting are reinvested in the fund to serve other new projects. They manage small sums but larger than those the Business Angels put up – a few million dollars – though the largest handle several billions. They have a clear focus for their investments and concentrate on management buyouts and development capital or seed startup.

Large companies are another source of venture capital. "Corporate venturing" means the taking of a minority equity stake by a large company in a small unquoted firm. It can take either of two forms. The first consists of the company investing directly through a venture capital fund. The second involves the large firm investing in a small unquoted firm for strategic purposes so as either to obtain a window for monitoring emerging technologies or to use a more economical form of R&D (small firms are more efficient in the use of R&D resources). The small firm or startup may be either a spin-off or a prospective acquisition of the larger firm.

Public sector funds are provided from regional, national or supra-national (EU) sources, sometimes supplemented with private funds, and their purpose is often not for profit as their goal is job creation at the regional and national levels. Any surplus is reinvested. Such funds have been active throughout the entire scope of the development of the innovation as shown in the graph in Figure 7.3, though some specialize in funding specific levels of development as in the case of early-stage state or incubator funds.

As for the main types of investment, again proceeding from left to right in the graph:

- Pre-startup or seed finance refers to researching, assessing and developing an initial concept before the startup phase.
- Startup is the finance provided for product development and initial marketing. The company being financed may be in the process of being set up or may have been in business for a short time but with no sales yet.
- Other early-stage finance refers to financing companies that have completed the product development stage and require further funds to initiate commercial manufacturing and sales.
- Development or expansion financing refers to financing the growth and expansion of a company, which is already breaking even or trading profitably. Capital may be used to finance increased production capacity, market or product development, and to provide additional working capital.

- Mezzanine or Bridge finance refers to finance made available to a company during the period of transition from being privately to becoming publicly owned.
- Management buyout refers to finance provided to enable current operating management and other investors to acquire an existing product line or business. Management buy-in is finance to enable a manager or group of managers from outside the company to buy into the company with the support of venture capital investors.

The US has by far the largest venture capital market, followed at some considerable distance by the UK. Europe is far behind, with its countries in order of size of market, Germany, France, Italy and the Netherlands. Apple, Google, Yahoo, Cisco Systems, Genentech, Netscape, Microsoft and many more by now famous companies have been supported by venture capital firms ever since the first venture firm, American Research and Development, which was formed in 1946 and invested in companies commercializing technology developed during the Second World War. Following the clarification of the US Department of Labor's Employee Retirement Income Security Act's "prudent man" rule, the venture industry grew dramatically in the 1980s. The rule clarification allowed managers of pension funds to invest in venture capital and other high-risk assets up to 5 percent of their assets.

The venture capital industry is cyclical both in terms of funds invested and also in terms of investment activity. In the US there is a close association between the venture capital industry and the stock market, with rapid growth in venture capital activity occurring in bull markets. Venture capital funds can more easily raise finance in "hot" initial public offering (IPO) markets when there are financial gains to be reinvested in venture capital funds.

The US venture capital sector is not only much larger than the European one but enjoys the comparative advantage of relying on a much larger, thicker and more transparent IPO market. (In comparing the US and the European venture capital industries one should bear in mind that in Europe the term refers to all private equity, including buyout, late stage and mezzanine financing, while in the US these are separate classes; Hall and Lerner 2010). A significant difference between the two industries is that in Europe the venture capital industry plays a less significant role than in the US in the provision of seed and startup investment as it shows a preference for expansion capital and management buyouts. Another significant difference between the US and the European venture capital industries is that the European invests primarily in mainstream industrial sectors while technology-based sectors are much more significant to the US venture capital industry. Finally, banks are the main investors in European venture capital funds, while in the US the main investors are pension funds, endowments, private individuals and insurance companies. (It is only in the UK amongst European countries that pension funds represent a significant source of funds.) The implication is that the shorter investment horizon of banks as compared to pension funds and insurance companies influences the nature of investments made; and there is concern about whether the dominance of banks may be an impediment to the development of the venture capital industry (OECD 1996: 7).

RATIONALES FOR POLICY ACTION

Reasoning in terms of the NSI construct, one may realize the paramount importance of finance, as one may say that the better the financial subsystem performs and the better its links with the rest of the components of the NSI, the better it allows the whole NSI to perform.

Beyond the rationale for action aimed at a well functioning NSI there is the rationale for growth. An obvious link between finance and growth is entrepreneurship or, to put it differently, the financing of R&D – the critical input to innovation – and hence to productivity increases and growth. The empirical literature (Levine 2005) establishes a strong connection between the level of financial development and economic growth – the more developed the financial system, the more it allows an economy to realize its potential for growth. There appears to be considerable room for policy as the findings from a number of studies "suggest that government policies toward financial systems may have an important causal effect on long-run growth" (King and Levine 1993: 540).

Growth by itself is a worthy goal, but as observed in many instances in Europe, it either occurs at the expense of jobs or it is "jobless growth", that is, it does not add to existing jobs. The goal would be worthy indeed if growth were to result in job creation. This leads us to focus mainly on new technology based firms (NTBFs) which not only achieve on average higher growth rates than non-innovative firms, but also achieve positive net employment (Almus and Nerlinger 1999).

POLICY RECOMMENDATIONS

Given the state of financing of R&D and the higher financing costs for this type of investment relative to others, policy makers justify, and rightly so, such interventions as the intellectual property system, government support of innovative activities, R&D tax incentives and the encouragement of research partnerships of various kinds. Further arguments in favor of such policy is that government-funded or -supported R&D (in several European countries) has been not completely to displace but rather to add to private expenditure on R&D, and to be productive, in the sense of resulting in patenting by the firm. Finally, firms receiving SBIR grants grow significantly faster than non-recipients of such grants (Lerner 1999), an outcome mostly attributed to the role of "quality certification" that receipt of the grant confers upon the recipient, who is thus enabled to raise further funds from private sources.

In recognition of the difficulties faced by small firms and startups in particular, many governments provide some kind of assistance to such firms. In the US, the Small Business Investment Company (SBIC) and Small Business Innovation Research (SBIR) provide substantial funds though less than venture capital funds. In Sweden, a series of investment companies has been set up since the early 1980s with a considerable government share of venture capital funding. Such initiatives in public support of R&D seem most productive when focused upon these firms as they are in Germany and the UK (Storey and Tether 1998: 944). In Germany several hundred federal and state government financing programs have been established for new firms. In the UK, the Enterprise Fund, which includes a series of government programs, allocates funds and guarantees loans to small and medium-sized firms in high technology in some regions.

Another way to ameliorate finance of R&D would be to improve financing from the banking system. We already know that banks are reluctant to finance R&D, and they are not keen to finance new, small, innovative firms. It is usually said that they do not have the expert personnel to evaluate R&D projects properly. A policy step in the right direction, then, would be to set up technology-rating organizations, as has been done in the Netherlands (OECD 1996: 52).

Some writers have suggested that it is not so much the technology that worries the banks but rather that their "concerns relate to the successful commercialization of innovation rather than the ability of the technology to function as proposed" (Freel 1999: 715). A policy suggestion, then, would be to consider that it is not technology specialist loan officers who are desperately required but rather industry specialist loan officers.

Finally, we should realize that there are limits to bank financing due to serious drawbacks such as the fact that banks' ability to hold shares is limited by regulation, and that they are in no position to monitor firms.

A further way to improve the financing of R&D is by increased recourse to an expanded venture capital industry. This may be a more efficient way than depending on bank financing, as venture capital has proven to be superior to other forms of financing due to its strong positive impact on innovation. It was found to be three to four times more potent than other forms of financing in stimulating patenting, and while it finances only about 3 percent of R&D, it may have caused as much as 10 percent of US industrial innovation in the last decade (Mollica and Zingales 2007).

Government policy can dramatically affect the health of the venture capital sector (Jeng and Wells 2009). Lower capital gains taxes have particularly strong effects on venture capital supplied by tax-exempt investors, and thus help the venture capital sector expand through increased supply of funds. An additional way to enhance the expansion of the venture capital sector is to allow the participation of pension funds in such investments in countries that have not done so already – practically all countries with the exception of the US and the UK. Although these are steps in the right direction, such measures do not guarantee an even share to all concerned. The venture capital industry has been observed to expand repeatedly while smaller firms do not seem to have benefited proportionately (Freel 1999: 718)

Another way to support the venture capital sector is through the enhancement of an IPO market (Black and Gilson 1998: 243–277) since the existence of a thick and robust IPO market is a prerequisite for the functioning of the venture capital sector by assuring exiting from investments undertaken. As the European experience demonstrates, the setting up of durable dedicated trading platforms for quotation of TBSFs is no easy feat. In fact, all newly created European markets since the 1996 inauguration of EASDAQ – a NASDAQ-like pan-European exchange for new high-tech firms created and promoted by the European Commission and the European Venture Capital Association (EVCA) – collapsed after the Internet bubble (Revest and Sapio 2010). EASDAQ itself collapsed in 2003, the same year as the German Neur Markt. The French Nouveau Marché and the Italian Nuovo Mercato were replaced by successor schemes in January and September 2005 respectively. A policy prescription for the countries mentioned above, plus Belgium, the Netherlands, Switzerland, Austria, Spain and the Nordic countries which have set up their own markets, is to pool their resources into a single market so as to achieve critical mass. If insistence on national markets prevails, then they should at least abandon the NASDAQ architecture and apply the entry requirements and information standards of the British AIM (Alternative Investment Market) – a non-NASDAQ-type market set up by the London Stock Exchange in anticipation of EASDAQ, and the sole survivor of the Internet bubble.

To be realistic, even if the formation of a durable, thick and efficient IPO market were within reach, as in cases mentioned above, Jeng and Wells' finding that the IPO market does not influence commitments to early-stage funds as much as those to later-stage ones would still hold. Thus, once again, while the later-stage investments in venture capital could benefit from the measures proposed so far, the earliest and most fragile parts of venture capital and those initiated by small undertakings, especially TBSFs, would stand to gain much less. It follows that an appropriate policy aimed at this portion of the market would be to encourage Business Angels, or to promote their syndication.

The policies suggested above are appropriate for the most advanced economies, that is, the ones mentioned above, plus a few more. But for the rest, that is, the vast majority of countries, they are far off the critical mass of candidates for a market formation and they cannot really depend on Business Angels, whose existence presupposes a tradition in entrepreneurship lacking in varying degrees in developing economies.

The only solution left, and the only one so far that offers some chances of success and has proved its mettle in practice, is a public policy aimed directly at the formation of venture capital funds accompanied by links to a strong and transparent market abroad. This is precisely what Israeli policy planners achieved through the Yozma program aimed at the creation of funds and links to NASDAQ for exits. Moreover, by allowing for the co-evolution of the venture capital industry with the booming ICT sector it has provided a self-reinforcing mechanism and the vehicle for a catching-up process (Avnimelech and Teubal 2006a, 2008).

This policy consists of an intricate procedure involving several consecutive stages, each including a number of preconditions for the next stage. More specifically there are a background-conditions, a pre-emergence, an emergence, a restructuring and a consolidation stage (Avnimelech 2008: 85). This is a delicate overall procedure requiring sequencing at some stages and simultaneity at some others (Avnimelech and Teubal 2006b: 21). It triggered the creation of a venture capital industry and transformed Israel's high-tech industry from a military-dominated one to a startup-intensive high-technology cluster. This is a difficult proposition but one that other policy planners should at least try to emulate.

7.2.2 Achieving a hospitable ecosystem

A hospitable ecosystem is one that welcomes innovators, provides for a friendly environment for them to operate and encourages them to innovate. Most developed countries have a more or less friendly environment for innovation and entrepreneurship consisting of not necessarily low but steady taxation, an efficient civil service with minimal bureaucracy and a low level of corruption, an Immigration Policy that accepts innovators and promising entrepreneurs, and IPRs to encourage innovation.

7.2.2.1 Immigration Policy

However, few countries welcome immigrants and those that do show a strong preference for skilled ones. The fact that so few countries welcome immigrants, even skilled immigrants, is intriguing, as the host country stands to reap multiple benefits. The US experience is telling. "Skilled immigrants have achieved great success in founding US engineering and technology startups, which have in turn contributed significantly to the country's economic growth over time" (Wadhwa et al. 2008: 6). In education in STEM (science, technology, engineering and mathematics) fields, which tends to be correlated with high rates of entrepreneurship and innovation, immigrants tend to hold more advanced degrees and start their companies much sooner after graduation than US-born entrepreneurs. Saxenian (1999) found that Indian and Chinese engineers ran a growing share of Silicon Valley companies started during the 1980s and 1990s and were at the helm of 24 percent of the technology businesses started from 1980 to 1998. Nationwide, 25.3 percent of the companies interviewed reported that at least one of their key founders was an immigrant. According to Partnership for a New American Economy, some 40 percent of the Fortune 500 firms were founded by immigrants or their children and so were the firms behind seven of the ten most valuable brands in the world (*Economist*, April 13, 2013: 55).

It is interesting to note that even those immigrants who returned to their home countries to take advantage of opportunities there were building links to the US and spurring technological innovation and economic expansion for the US.

Skilled immigrants are a blessing for the developed countries offering a hospitable ecosystem, and a loss for their country of origin. Nevertheless, in most cases the reason for their migration is that their home country offers them an ecosystem, which is if not inimical at least not very hospitable to innovation. If they had remained at home, it is very likely that most of their innovations would not have materialized.

Box 7.2 Intellectual Property Rights

IPRs are incentives to innovate. Some of their older forms have been with us since antiquity, as rulers would confer a "right" to an innovator, producer or seller. Over the centuries, the meaning of the right evolved (Rosen 2007), while the state took an increasingly active interest in using IPRs as an incentive for innovation. The first patent law dates back to Venice in 1474 and the English Parliament passed the Statutes on Monopolies in 1623. In the New World, the need for such a policy tool was felt from the start to be so pressing that it was included in the US Constitution (Article 1, Section 8; Greenhalgh and Rogers 2010: 35).

Innovators, fearing that their findings might be imitated, need some guarantee of private ownership so as to invest in innovative activity. The IPR, in effect, gives the innovator a monopoly right for the duration of that right; in other words, the right to raise the price and limit output in contrast to what would have been the case under conditions of perfect competition.

Consumers lose out since not enough of the innovative good is being sold.

> The knife-edge on which the intellectual property law tries to balance is that of defining enough property rights to preserve adequate incentives for innovation while avoiding the gift of excessive monopoly power, which will lead to socially inefficient exploitation of that creation.
>
> (Greenhalgh and Rogers 2010: 27)

The policy maker is torn between rewarding the innovator sufficiently and allowing the innovation to enhance productivity and growth at the macro level, which would allow the innovation to diffuse throughout society. As it turns out, the balance may not be easily gauged. The IPR may simply prove not to be large enough, i.e. a sub-optimal incentive. This means that the mere presence of an IPR does not by itself preclude a market failure in investment in innovative activity.

While interest in IPRs has risen, the types of right have proliferated to meet the specialized expanding needs arising from the ever-increasing complexity of economic, cultural, literary, technological, scientific and business activity. More recently, R&D and innovation traditionally dealing with the production of tangible goods expanded their horizons to intangible products and services (e.g., the Internet and its swarm of related services). All these changes have prompted the emergence of an impressive array of IPRs. At present, the main forms protected by law are patents, trademarks, designs and copyright. Forms of IPR less often encountered include trade secrets, database rights, plant variety rights and performers' rights.

Patents are the most common, well known, analyzed and possibly oldest type of IPR. They are conferred by specialized government agencies (the US Patent Office, the British Patent Office, the European Patent Office) provided applications meet three conditions: novelty, in the sense that it is a worldwide first; non-obviousness, even to experts in the field; and capability of industrial application. Although the US allows patents for computer software and for business methods, these are excluded in most other countries, together with scientific theories, mathematical methods, aesthetic creations, methods of treatment and diagnosis and plant and animal varieties. Some of these categories may be covered by other IPRs such as design or copyright.

In most countries patents last up to 20 years, are renewable each year for a gradually rising fee, are limited geographically and are awarded to the applicant applying first for a patent – except for the US which rewards the first to invent.

Possibly the most significant aspect of a patent, and the subject of most of the litigation arising with respect to filing one, is its breadth, i.e. how close it can get to an already existing invention without being judged by the examining authority to have infringed the right of the existing patent holder.

How effective are patents as an appropriation mechanism for innovators?

Applying for a patent requires information about the innovation to be disclosed and in most countries – the US is an exception – this information is published 18 months after the application, thus releasing it into the public domain.

This allows other interested parties to challenge the application, and restricts the possibility of duplication of research effort, but also gives the opportunity for learning and invention to other researchers, who, with modest additional research, may reach a breakthrough – referred to as "standing on the shoulders of giants". Thus patents are at best an imperfect appropriation mechanism compensating innovators for some fraction of the benefits they create.

Yet despite their limited effectiveness as an appropriation mechanism, the fact that most innovations are in any case imitated within four years on average, despite the cost of obtaining one and especially despite the cost of defending it in court should the need arise, applicants do not seem to have been deterred. The cost of defending a patent has always been an expensive proposition and, judging from the proliferation of firms whose purpose is to capitalize on the shortcomings of the patent system, this cost must be on the rise. Such firms, referred to as "patent trolls" and "patent warehouses", specialize in the collection of patents, which they acquire not for their technological prowess but in order to enforce them in relation to actual or potential infringers and to take advantage of licensing opportunities (*Economist*, April 12, 2012).

Given all these shortcomings, how can one explain the fact not only that patents are used so extensively but also that the number of patents awarded each year has been rising, especially in the last two decades? Is it possible, for instance, that their extensive use is due to their being overrated? Consecutive surveys of US manufacturing firms have found that patents are thought to be a relatively unimportant way to exploit innovations in several industries and that " 'the patent premium' was highest in pharmaceuticals, biotech and medical equipment, followed by machines, computers and chemicals" (Arora et al. 2008). Asked to evaluate the effectiveness of methods for maintaining and increasing the competitiveness of product and process innovations, respondents rated lead-time advantages first, followed by secrecy, then complexity, then patents followed by design registration.

To put in perspective the extensive use of patents, one has to realize the fact that patents are one of a set of opportunities within which firms have to construct their strategy for appropriation of returns, and that they also constitute an integral and necessary ingredient – a necessary first step – to be used in conjunction with other factors in the application of the firm's overall strategy. Consequently, the value of a patent to the patent holder may differ significantly from its estimated value both on the basis of importance, that is, in terms of frequency of subsequent citations, and on the basis of patent renewals (or values reported in PatVal-EU). To put it simply, with respect to the patent holder, there may be more to a patent than the direct appropriation of returns; alternatively, we may think of what the patent holder would lose if he did not own the patent (Scotchmer 2006: 275). This is most obvious in cases where successful defense of an incumbent innovator against a newcomer depends additionally on complementary assets – capacity to produce using the relevant technology, distribution facilities, marketing, after sales services, advertising and other factors that are specific to the particular industry (Teece 1986, 2006; Nelson 2006; Winter 2006).

For most patents their value is reportedly very small but there are exceptions such as the patent for Lipitor, the best-selling drug in history, which expired in November 2011. Lipitor earned Pfizer $11 billion in revenues for 2010. Another instance indicative of the value that patents can have is the $1 billion penalty imposed on Samsung in August 2012 for infringement of Apple's patents.

As to the relatively recent rise of the number of patents conferred, this can in part be attributed to the rise of R&D but far more importantly to the legal and policy changes made to the US patent system. A number of facts support this statement. One is the Supreme Court's gradual extension of patent protection to biotechnology, software – so far covered by copyright and which now accounts for about 15 percent of the yearly crop of patents – business methods, and scientific research methods (Hall 2007). Another is the Bayh–Dole Act – which allows patenting by universities of publicly funded innovations – and its positive impact on patenting by universities. Above all there has been the strengthening of patent holders' rights through the creation of a Court of Appeals to handle cases of patent infringement and validity.

These facts leave little doubt that the observed rise in patents has been mostly policy led (Bessen and Meurer 2008).

Are patents effective as an incentive to innovation and growth? More than six decades ago Penrose summed up the contemporaneous situation by saying that "if the patent system did not exist it would be very difficult to justify creating it" (Hall 2007). Nowadays this statement is still appropriate as the subject is as controversial as ever. If anything, opponents of the existing system have scored some points by drawing attention to negative aspects of patents such as the reduction of competition through the strategic use of patents; the hindrance to sequential innovation; the adverse effect on small firms and startups; and the inefficiency of patent races. In sum, policy makers have been wise occasionally to reconsider the patent system (e.g. US Patent Reform Act 2009) and above all to have developed a number of alternative incentive tools since Penrose's statement.

The 1994 Agreement on Trade-Related Aspects of Intellectual Property Rights (TRIPS), the brainchild of globalization, signing up to which is a prerequisite for membership of the World Trade Organization (WTO), is at least as controversial as the patent system, only for different reasons. TRIPS affects economies in different ways, with a strong predilection for the technologically advanced that stand to gain the most, either through FDI or trade or technology transfer (Lall and Albaladeho 2002). For the rest, adherence to a uniform minimum standard of IPRs is seen to be unfair as these restrictions did not exist at the time the present technological leaders started developing. Besides, enforcing TRIPS, that is, erecting and maintaining an IPR system, is an expensive proposition. For most of the countries of the developing world adherence to TRIPS is not only costly, but would be unlikely to turn them into hotspots of nanotechnology anytime soon.

A hotly debated issue within TRIPS was compulsory licensing, whereby a patent holder is ordered to license a product against his will. Long accepted in international treaties in cases involving emergencies (the US threatened to use it in the anthrax scare of 2001), it is controversial in less pressing situations.

Trademarks attest to a certain consistent quality of a product or service and are mostly used in differentiation from other products, especially imitations. So it is only natural that consumer product firms, pharmaceutical firms, cosmetics and luxury goods firms, and retail firms are the heaviest users of trademarks. While there is no need to register a trademark, which like patents involves application, examination and the conferring of the trademark, it is wise to do so in case it proves useful in case of litigation; besides, it is cheap. The trademark itself refers to a logo, picture, word, shape, or color, even sound, that is distinctive, not in common use and not deceptive.

The yellow and red Shell emblem, Microsoft's fluttering flag and Apple's bitten apple are trademarks. In contrast with patents, trademarks last indefinitely, provided the producer continues to produce and trade the item(s) covered – and, again unlike patents, there is no such thing as thinking of a trademark, registering it and then selling it.

Design rights refer to the visible features of a product and items whose design is decisive to the product's performance, as in the case of semiconductor chips. Design rights can be registered – upon provision of drawings, and photographs – but they need not be. They offer a varying period of protection depending on the country, and are conferred mostly on textile firms and firms producing furniture and toys.

Copyrights refer to literary and creative work. They include books, plays, published editions, performances, dance performances, music, sound recordings, films, videos, broadcasts, paintings, sculpture and computer software. Registering is not required and for literary and art works the copyright is valid during the author's lifetime plus 70 more years. For music and recordings the copyright is valid for 50 years. What is protected by copyright is not an idea but the way it is expressed, whether in writing or through artistic creation. Royalties for copyright holders are either paid directly (for instance to authors from their publishers) or through collecting organizations (as in the case of music). Of course, there is a need for the "updating" of legislation on IPRs in general in order to keep up with modern technological developments, as these can easily allow a breach of copyright as in the by now famous Netscape case of the late 1990s where a file-sharing service allowed the copying of music from peer to peer without payment of royalties.

7.2.3 Balance between variety and selection

Every industry (sector) is subjected to two forces pulling in opposite directions, variety and selection (Metcalfe 1995). The generation of variety through innovation, firms' entry into the industry, new knowledge and behavior each introduces novelty in an industry. Firms may be characterized by a drive for too much exploitation (that is, small modifications in existing technologies and a focus only on incremental innovations) at the expense of some exploration. Alternatively, in the opposite case (which is, however, more rare) they can be driven by too much exploration at the expense of full exploitation of what has been explored.

Market selection restricts the range of variety through the elimination of unsuccessful firms, technologies, products and behavior, and changes the relative weights of established entities through the growth of some and the decline of others. Thus, we may encounter cases of tough selection, where competition is intense, and low barriers to entry, as well as situations of weak selection with mild competition and substantial barriers to entry.

In sum, we may end up with some industries that may be characterized either by a lot of generation of variety with weak selection processes, or by tough selection with little generation of variety (Malerba 1996). Figure 7.4, may help visualize the situation.

The rationale for policy intervention is that situations of very tough selection combined with very low variety eliminate trial and error processes, and may, at the extreme, produce lack of progress and innovation because the industry may become rapidly dominated by a few firms or even by just – a monopolist. Moreover, in these situations the working of path dependency, global or local positive feedback and network externalities may lead an industry to be locked into an inferior technology. Potentially superior technologies may not take off and the generation of diversity may be reduced or blocked. On the other hand, situations with weak selection combined with very high variety may lead to too much experimentation and exploration, with the presence of slack behavior, wasted resources, the survival of too many ineffective firms and limited industry growth (Malerba 1996).

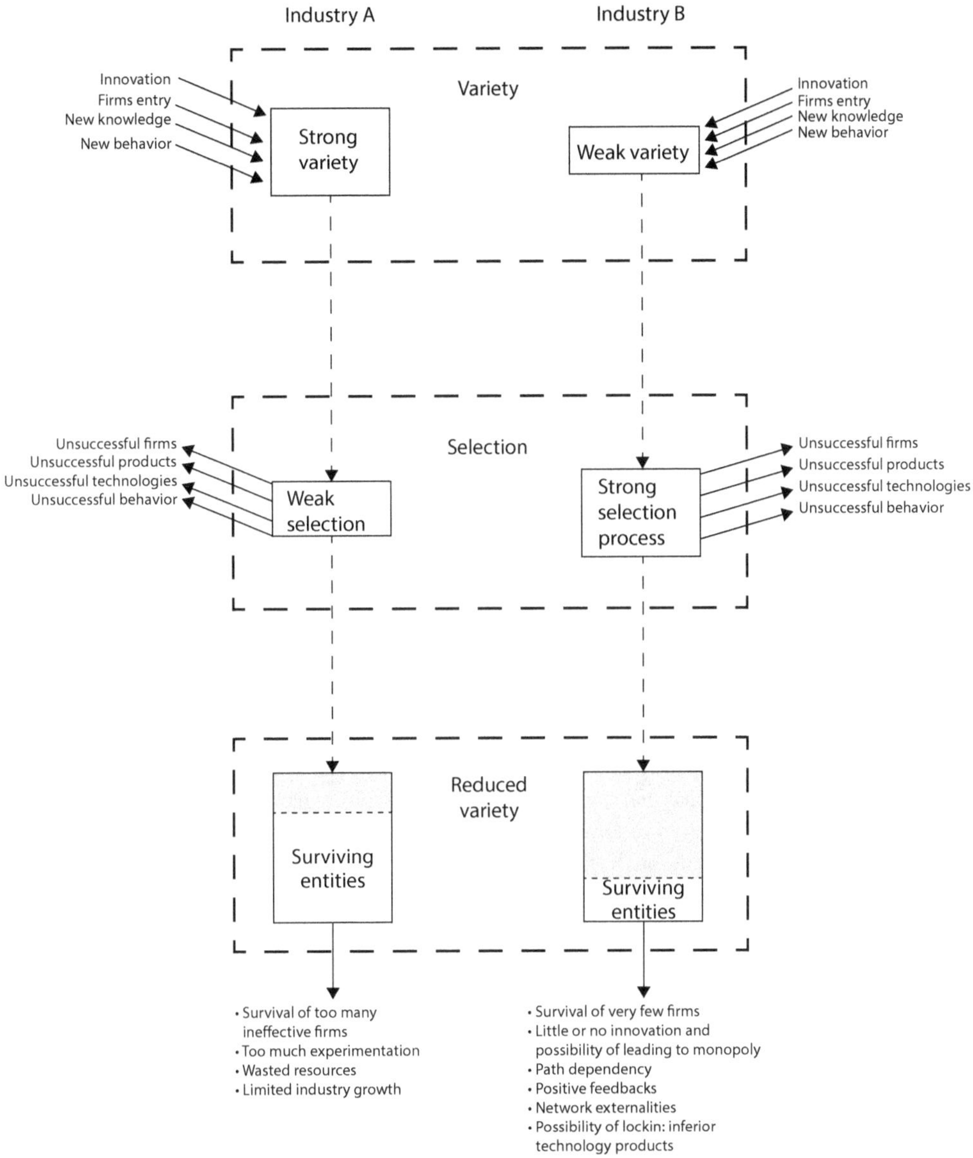

Figure 7.4 Innovation Policy and industrial dynamics

7.2.4 A productive structure based on innovation: able to meet external threats such as globalization, and able to take advantage of the opportunities it offers

7.2.4.1 Globalization

Globalization presents a formidable challenge to the very concept of a national Innovation Policy, questions the need for it, challenges its effectiveness and may ultimately call for a change in its role.

7.2.4.2 Rationales for policy action

There are certainly aspects of policy – such as the creation of international institutional arrangements or the pursuit of industrial policies or of competitive advantage or of policies directed towards competence – where policy at the national level, in the presence of globalization, has become more important and desirable. This is particularly true in cases where, with increased globalization, any loss of competitive advantage is translated all the more rapidly into declining market share, output, employment and living standards.

Globalization generates changes that have a beneficial impact on the economic performance of some countries and an adverse impact on that of others – there are "winners and losers" as the benefits of trade integration may not be evenly spread (Sawyer 1994: 20), "leaving some countries with catering and retail sales as the primary employment opportunities for the next generation".

However, seeing through the necessary policy will require more far-reaching and radical intervention on the part of national governments than would have been the case in previous, easier circumstances. An increase in globalization certainly renders less effective various national Economic Policy instruments such as interest rate or exchange rate changes.

More closely related to Innovation Policy proper, there is a gradual weakening of the institutional matching between the various components of countries' national systems of innovation because globalization pressures have acted differently upon each of the components of the existing national system of innovation. This calls for a constant reassessment of their effectiveness.

The process of globalization exacerbates competition and makes more evident the relevant advantages associated with "structural" competitiveness and the disadvantages associated with traditional industries (Pianta et al. 1996). This is especially true with respect to the productivity–employment tradeoff, as the resulting growth from innovation and productivity increases may in several instances do very little to help or may even exacerbate the problem of unemployment. In fact, in most countries in Europe, the rise in productivity is earned at the expense of employment. This is not true in the case of the US where over the past 20 years manufacturing employment has barely changed nor was it true in the case of Japan where it rose in the same period by 2 percent. It is the EU which has had the big manufacturing job losses, amounting to 20 percent over the past 20 years.

The need for action is obvious but so is the need for an explanation for the productivity–employment tradeoff and its worsening in the face of rising globalization. It turns out that economies, both developing and developed, rely on either one of two "engines of growth" which differ as to the embodied or disembodied characteristics of their innovative activities. In the first instance, they rely on a strong research system with a large generation of innovations, display a distinct productive structure with a predominance of science-related activities, and pursue mostly strategies of technological competitiveness, such as innovation in products and markets. Such economies fare much better in terms of new employment creation than economies pursuing the second alternative – as the former create new jobs while the latter aim at reducing costs of production. More specifically, the latter rely on the rapid diffusion of innovations produced in other countries and sectors. They display a productive structure characterized mostly by supplier-dominated activities, pursue strategies of cost competitiveness such as innovation in processes, and experience rises in productivity at the expense of employment increases (Bogliacino and Pianta 2010, 2011). (Of course, at the product and firm levels these strategies – product and process – coexist and are complementary, as many firms that produce product innovations also produce process innovations (Reichstein and Salter 2006). However, at the industry level, there emerges a relative prevalence of one of the two strategies.)

Globalization emphasizes these structural differences and reinforces their impact as to net impact on employment. Thus, in the case of the "disembodied" NSI, the job gains due to the employment-creation effect of technological change (mainly product innovations) are supplemented by the additional jobs created due to expanding sectors and supported by demand abroad. Similarly, in the case of the "embodied" NSI, the job losses due to the direct labor-displacing effect of innovations (mainly in processes) are augmented by job losses due to the decline of particular sectors associated with the shift abroad of production and jobs and to weak demand patterns. The message here is clear, as there is a definite rationale for action in the direction of the structure of production.

Fortunately, there is another side to the coin, which mitigates, but only partially and only selectively, the productivity–employment tradeoff. Contrary to what is sometimes alleged, globalization and national specialization are complementary aspects of a common process, and not conflicting trends (Archibugi and Michie 1997). The globalization that has occurred has not necessarily meant a reduction in national differentiation; on the contrary, it has been accompanied by an increase in technological specialization (increased sectoral differentiation; Guereri 1999), and by global winners and losers in terms of economic growth (Kitson and Michie 1995). The process of international integration is leading to an increased division of production and employment and this implies that each country is focusing on selected industries and relying on trade for others. Again, there is a clear case for action to move towards specialization, which is another way of saying that comparative advantage should be fostered.

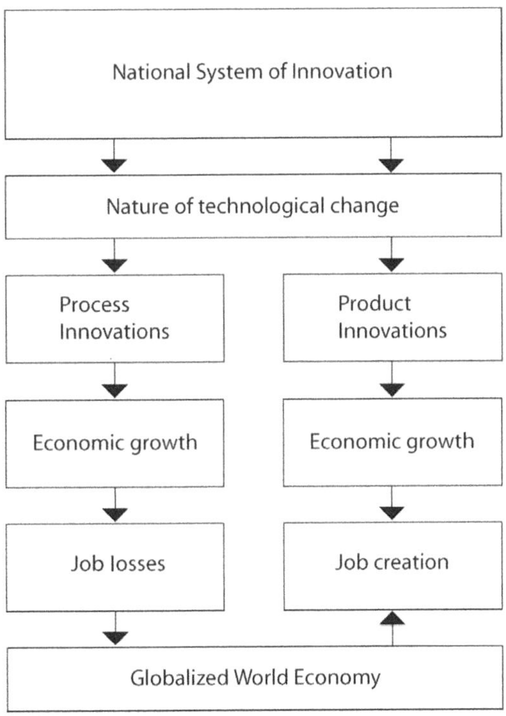

Figure 7.5 Globalization, NSI and employment change

There is still one more aspect (referred to in chapter 1) to consider: if globalization is here to stay and cannot be fought back into retreating, then, how could it best be accommodated and if possible how could it be taken advantage of.

MNCs are rather reluctant to locate technological activities in host countries as the geographical concentration of innovative activities in the developed world and its stability over the years suggest (Castellacci 2006). Core competences, including R&D and innovation centers, are still concentrated in the companies' home countries. (Pavitt and Patel 1999) and their foreign production is in general much less innovation intensive than their domestic production, as the vast majority of foreign investments perform no innovative activity in the host country. Nevertheless, the globalization of innovation is changing (Narula and Zanfei 2006) and a rising (still a limited phenomenon) proportion of MNCs' innovative activities is performed in host countries; within this minority, two types of investments can be distinguished. The first refers to foreign direct investments, which do perform innovative activities that are limited largely to the adoption and diffusion of technology developed in the home base, such as the modification of products and processes to make them competitive in the local market.

The second category of foreign investment (asset-augmenting activity also referred to as home-base-augmenting activity) concerns MNCs that perform innovative activities in the host country in order to improve their existing assets, to acquire or to create new technological assets, through access to complementary location-specific advantages. Local contexts are considered to be sources of competencies and technological opportunities. Comparative advantage is thought of as no longer residing in any one country but in many, and may be exploited on a global scale.

The reason why MNCs engage in asset-augmenting activities in areas where the knowledge for innovative activities is clustered is to benefit from external economies and spillovers generated in the relevant clusters already formed locally to exploit mostly endogenous capabilities.

More explicitly, the localization of innovative activities, that is, the clustering and the co-location of research and productive activities that is of particular importance in high-technology industries (Saxenian 1994) is explained firstly through the fact that knowledge does not travel as freely as once thought. The tacit nature of knowledge associated with these production and innovation activities in these sectors means that proximity is important in accessing and absorbing it. The second reason why knowledge-related activities appear to localize is because of district features (external economies), the localized accumulation of specialized inputs (including skilled labor), and the related creation of collective knowledge.

Localization appears to lead to a new kind of unbalanced economic development in which activities within the host countries are becoming centralized at the same time as industries develop more extensive international links and interdependencies (Cantwell and Iammarino 2003; Steinmueller 2009: 21). Through the process by which MNCs organize affiliate specialization in order to tap into the locally specific and differentiated stream of innovation in each center, in accordance with local strengths, the latter are reinforced. Thus, this process tends to reinforce and not to dismantle nationally distinctive patterns of development and national systems of innovation (Cantwell 1995).

In an extensive empirical study, the R&D investment strategies of the 345 MNCs with the greatest patenting activity in Europe during the period 1988–1996 were analyzed (LeBas and Sierra 2002). The most frequent strategy is the one in which not only the MNC but also the host country has a relative technological advantage. This indicates the formation of "centers of excellence" in which strong domestic research environments function as global attractors. Achieving that status would be a worthy goal of policy in support of innovation by national states, a way to benefit from accommodating globalization and by the same token minimizing the productivity–employment trade off, putting to good use locally developed assets and rendering the country a better magnet for further future investments, and thus enhanced growth.

7.2.4.3 Policy recommendations

Policy (in general) is needed, in the sense of national action having greater payoffs and inaction greater costs, although due to the rise of globalization, this is becoming more difficult to implement because policy instruments are rendered less effective. In addition, precisely in the context of globalization, innovation policy itself becomes an important part of the response of the system, as it aims at shaping the conditions for learning, increasing absorptive capacity, attracting foreign knowledge and innovating (Borras et al. 2008).

There are signs that globalization has stalled ever since the financial crisis of 2007–2008, but past experience tells us that this is a temporary pause. Whether that turns out to be true or not, globalization has already attained such levels that a realistic assessment of the present situation and the foreseeable future suggests that globalization is a force to reckon with and may require accommodation or attenuation. It cannot be fought into retreating, and it may be worth joining it. If done correctly, some advantages may accrue.

It is also realistic to expect that the state will continue to provide infrastructure, incentives and institutions that strongly influence the rate and direction of innovative activities in locally based corporations. The efficiency gains from the geographical concentration of innovative activities will remain, even if firms seek out unique skills in foreign countries. Core competences, including research, will continue to be concentrated in the companies' home countries.

This is due to the fact that the intensity of linkages between science and technology, which still tends to decline with geographical distance (Jaffe et al. 1993; Feldman 1994), will continue to do so because these linkages have remained essentially localized due to the importance of face-to-face contact in communicating the results of complex learning processes embodying tacit elements. Despite advances in information technology, they are unlikely to eliminate the need for geographical proximity in the launching of major innovations.

Nevertheless, a rising proportion of MNC innovative activities mostly intended to address local needs will be performed in host countries. A much smaller but rising minority of MNCs will wish to perform truly innovative activities in their host countries (not necessarily confined to local needs) by capitalizing on existing local technological assets in order to improve their existing ones.

On the basis of the above assessment of things to come, this writer suggests that with globalization, national policies should be increasingly oriented towards what we have referred to as "disembodied" types of strategy for growth rather than the "embodied" alternatives. This would ameliorate the situation of the productivity–employment tradeoff and might even reverse it. Of course, shifting from the second to the first strategy is not only a very difficult undertaking, but may also require several years to achieve.

For this purpose two broad categories of government action could be envisaged (Metcalfe 1995): firstly, financial incentives to companies to attract their innovative activities and, secondly, public supply of infrastructure to make a country attractive for the deployment of such activities. The latter approach, compatible with what we have called "disembodied" strategy, includes investment in education, communications and university–industry links, which are likely to be more effective than the financial incentive approach. On the other hand, the effectiveness of financial incentives may be greater in attracting lower-tech activities, which are more cost sensitive and more internationally mobile.

Adopting such a strategy is more likely to appeal to MNCs and in this instance an industrial policy role will still exist, even in a global economy, and consists essentially of facilitating the most appropriate pattern of national specialization. Put differently, it concerns questions of how best to build upon established local strengths in innovation, and how to encourage greater international coordination of productive activity in such a way as to improve the ability to learn locally from what is being done elsewhere.

More specifically, a policy implication is that action is needed to support and encourage specialization; this means supporting the creation of competitive advantage, which is less and less God-given. In fact, beside through the reallocation of resources, it is created through learning the activities of firms, national governments and national institutions.

Furthermore, to maximize benefits, this specialization, should aim for the achievement of the status of "center of excellence", a state that confers the benefits of localization – attracting foreign investments and the accompanying innovation-orientated research. Beyond placating the employment–productivity issue, this aim allows for the use of locally developed technological assets, thus rendering them even better magnets for future foreign investments and improved assets for local production.

In order to achieve the goals set out above even partially, it is still worth governments backing the local establishment of basic research facilities, even if some of the knowledge generated has spillover benefits abroad, since the greatest and most immediate impact will be concentrated locally, owing to the tendency for the linkages between science and technology to be geographically localized (Frost 1996; Almeida and Kogut 1997). We should make a note here of the fact that less developed economies are at a disadvantage in this respect, since in their case the leakages are substantial due to their inarticulate and non-integrated character.

In addition it is crucial that firms and public organizations do benefit from the technological opportunities offered by globalization to upgrade the capabilities of national innovation systems (Archibugi and Michie 1997). However, the ability to utilize increasingly globalized technology will depend on national systems' "social capability" (Abramowitz, 1986). Thus, policies directed towards competence (such as those favoring training, education, the acquisition of managerial skills and encouraging technological change) become crucial instruments to allow national communities to face the processes of globalization.

7.3 Some further prescriptions for the design and implementation of Innovation Policy

Specificity, both with respect to the NSI construct as well as the policies pursued, emerges as one of the strongest lessons to be drawn. The need for specificity – "one size does not fit all" – at the design level has already been discussed at some length. At the policy level, it translates into the fact that best practice cannot be transplanted. Institutions are situated within the entire historical and cultural framework of particular countries and therefore cannot usually be reliably transferred, adapted or translated into other contexts. It also follows that benchmarking is not recommended.

Even a well conceived and designed policy prescription runs the risk of missing its target if its articulation and the implications of its implementation have not been foreseen down to the last detail; this may happen if insufficient attention has been paid to the peculiarities of the situation at hand.

In an excellent article Trajtenberg (2009) depicts Israel's astonishing success in innovation (actually ranking fourth internationally in terms of patents in the US in 2004) but elusive overall growth. This success was due to the prowess of its high-technology sector, which grew at more than 10 percent per year, earning $15 billion from exports annually. In contrast, the rest of the economy, poorly linked to the high-technology sector, grew only 2 percent annually during the corresponding period. Given what we learned in chapter 2 about links between sectors and the level of integration of the productive system, asymmetry was to be expected. What begs for an explanation is its extent. The author argues that it was mainly the previous application of policy that is to be held responsible (though locally performed R&D by multinationals and the mode of financing should also receive some blame). Innovation Policy, which was intended by the policy makers to be neutral with respect to R&D, in practice turned out to be heavily weighted in favor of the ICT sector because of a bias towards export orientation and products rather than processes. Thus, a policy that was originally intended to enhance innovation throughout the economy was apparently confined to a sector, albeit a very promising one: general-purpose technology (GPT).

Finally, the need for specificity with respect to both the NSI construct and the policies and instruments used is widely acknowledged by economies exhibiting many similarities (Boyer 2005), even when they share the same type of innovation (Whitley 2002), such as members of the same so-called "variety of capitalism".

Flexibility is also needed both in the interpretation for rationales for intervention and in policies. In our by now familiar Israeli case study, it was hoped that the successful development of the ICT industry, joining forces with ICT MNCs and otherwise encouraging the ICT-producing sectors, could have played an important role in the process of development. This expectation rested on the assumption that the spillovers from this success would then have prompted the rest of the economy to follow suit. However, the point to be made is that this latter stage may not happen by itself (as indeed it has still not happened), or may take too long, and may therefore require government intervention. With a bit of flexibility, we may classify this case as a systemic failure and, more specifically, a transition failure.

Let us now consider flexibility in policies. The diffusion of innovations is significantly slower in countries in earlier stages of development (Comin and Hobbijn 2004). Thus, in those countries it may be justified to try to support early adopters who may positively impact the decisions of later adopters, particularly those who can potentially enhance productivity in a wide range of sectors (Trajtenberg 2009: 387).

The nature of policy: we know from recent endogenous growth literature (Acemoglu et al. 2006) and from empirical evidence (surveyed in Aghion and Griffith 2005, chapter 4) that the nature of innovation policies depends crucially on the countries' distance from the technological frontier. While investment in research and higher education is an essential determinant of growth in countries that are close to the technological frontier, capital accumulation and technology adoption are more beneficial policies in countries that are still far from this frontier. By the same token, the same is true regarding the choice of activities, as some activities are more suited than others to specific country cases. The EU's "fascination" with high technologies does not serve all of its members equally well.

The degree of intervention: both Switzerland and Israel are success stories. Yet Switzerland has possibly the least interventionist policy in the world, with innovation purely driven by market forces, while Israel's policy is an example of an interventionist approach. Switzerland, a developed economy, with high motivation towards innovation on the part of its business community and already well placed in terms of specialization in the globalized division of production and employment, fits well with the non-interventionist approach. Israel, on the other hand, with a developing and not well integrated economy, with motivation towards innovation limited to the ICT sector and facing a maze of rigidities and deficiencies common to economies at a similar level of development, fits well with an interventionist approach.

The level of public funding and government initiative: Finland, which encourages clusters, displays one of the highest levels of public support. What is interesting is that it also has one of the highest levels of privately funded R&D – thus raising the question of whether there is a complementarity between the two. On the other hand, Japan, which more than any other market economy uses an explicit national policy to promote specific sectors and industries, does not rely on massive subsidies. In fact, its public sector is much less involved in R&D funding in the private sector, but reserves for itself the role of coordinator, adviser, consultant and planner – a much more diversified role than in most other countries.

Beside the potential role of government to meet uncertainty, an issue to which we have already referred, the combination of government-funded R&D with public procurement of technology has had a very significant role in the development of new emerging technologies such as hardware, software, semiconductors, commercial aircraft, machine tools, etc. In fact, government intervention has not always been restricted to the role of "midwife" and has extended to the role of "initiator", as in the case of the Internet (Day et al. 2000) and nuclear power.

Governance: in implementing Innovation Policy it is essential for it to be considered as a central component of government policy. As such it cannot be limited to a specific ministry or agency such as the Ministry of Technology or Education or Economics, as used to be the case in most instances and often still is. An Innovation Policy may involve almost every aspect of the economy, from R&D to education, taxation, foreign investment, even immigration policies. However, the instruments of Innovation Policy must remain in tune with each other so as to ensure coherence, which implies that efforts must be coordinated. As for its implementation, responsibility should be clearly defined and should rest with a strong leadership. Such coordination can take different forms, but in most countries it has taken the shape of Science and Innovation Policy Councils. To guarantee a strong leadership, a few countries such as Hungary and China have gone so far as to put the prime minister at the helm of the relevant authority. An often-cited example where such a step met with success is Finland (OECD 2010: 186–210).

Innovation Policy requires concerted action by many entities acting in unison.

7.4 Salient features of Innovation Policy across countries

Practically all countries are to some degree in favor of an Innovation Policy. For some, the extent of its use is limited by lack of knowledge, realism, vision or commitment. Others, due to ideology have a predilection for "small government" and use Innovation

Policy reluctantly and as sparingly as possible in order to satisfy a rising level of development and the demands of an evolving society. In general, conservative administrations are in favor of income tax reductions – especially on higher income – resulting in squeezes on the innovation effort. This situation is further aggravated in times of budgetary restraint. There are also those that have an aversion to "picking winners".

Britain and Germany provide a good example of opposite poles in terms of ideology as to Innovation Policy in the early stages of its application. In Britain, during and after the Industrial Revolution, innovation was not considered to be a matter of direct government intervention. It was the job of government policy to maintain a "free market" environment within which market forces would shape industrial and technical change. Abroad, "Britain's position as a laissez-faire industrial nation was developed by imposing trading relations upon other countries through the use of military force" (Hodgson 1984). "The British government spent about ten per cent of GNP in maintaining 'free trade' and an environment in which industry was able to operate as it saw fit" (Coombs et al. 1989: 201).

It is only natural then that British policies for science and technology centered mainly on commercial expansion and military needs. Of course, there was the rewarding of inventors and the financing of technical developments in power sources, dyestuffs, fire proofing and land reclamation (Inkster 2002). Additionally, specific acts were passed to encourage innovation in new products such as flax, or in order to exploit colonial ventures (Lopez-Martinez 2006: 66).

The laissez-faire ideology at the time was so prevalent that even the relatively indirect action of changing the patent laws to encourage British innovators was resisted for many years. Nevertheless, by the 1870s it was established that the control of some of the worst side effects of technology was a task for the government. It was much longer, however, before it was accepted that it was within the realm of government activity to try and promote or stimulate technology by any means more direct than the opening up of markets and sources of raw materials.

Very little was done in Britain until the First World War brought home the fact that British industry had been so thoroughly overtaken by German industry in some sectors that "Britain was dependent on Germany for imports from dyes for uniforms, drugs, explosives, rubber and other chemicals, to … range-finders for guns and other optical and precision engineering equipment" (Rose and Rose 1970). It was not until the end of the nineteenth century that the British government's own research effort became organized, British patent law was not amended until 1907 and the British government was not convinced into sponsoring research, technical and scientific education and the rationalization of industry until the First World War. It was only during the interwar years that sentiments in Britain started changing towards more active state involvement in Science Policy (King 1974; Elzinga and Jamison 1995).

It is usually stressed that this attitude contrasts with those held in other countries, "notably Germany after 1871 and subsequently the United States, which industrialized later than Britain and on the basis of a great deal more direct government assistance than was considered proper in Britain" (Coombs et al. 1989: 201–202). However, it should be remembered that at the time of the Industrial Revolution, which was an entirely British affair, and even later, innovations were simple and based on ingenuity and experience. They did not rely on science, the process of innovation was rudimentary, and the need for supporting organizations, research institutes, links and intricate policies was still not strongly warranted.

In Germany, which industrialized later, the circumstances were different. By now Science was starting to influence innovation to some degree, especially in industries arising from the new scientific disciplines, and the need for supporting organizations started being felt. In parallel with the need to support organizations there was a need for policies to orchestrate them.

The German government took a leading role in stimulating industrial production, and their policies played a major role in the success of national industry. Policy instruments included the state-backed development of a banking system that provided the necessary capital for industrialization, transport concessions, preferential duty on certain raw materials, favorable amendment to the patent law, and state support for scientific and technical education (Coombs et al. 1989: 217).

What follows is not intended to describe contemporary Innovation Policy as applied generally, but to draw attention to some of the salient features exhibited by a few actors, namely Switzerland, Israel, China, EU, USA.

One may be temped to consider Switzerland a paradise for the lazy policy maker. Switzerland has the least interventionist Innovation Policy in Europe, yet it is amongst the world's innovation leaders with R&D spending at 2.9 percent of GDP (of which 70 percent is spent by two MNCs, Novartis and Hoffmann-LaRoche; Arvanitis and Sydow 2009).

This "non-activist" policy, which provides primarily for the improvement of the framework conditions for private innovation activities, is supplemented by the activities of the CTI (Commission for Technology and Innovation) whose use as the main instrument of policy promotion is unique in Europe. Under this scheme, private companies get no public subsidies or funding for their R&D. Instead, the CTI supports mainly R&D cooperation projects from all scientific fields by funding the public partners of such cooperation – a university or public research institution – the private partner being an enterprise that agrees to match its contribution to the project to at least the amount of funds offered by the CTI (a contribution of at least 50 percent). CTI promotion was found to significantly improve the innovation performance of supported firms. In addition, the CTI runs a federal program to give coaching support for startup projects: CTI Startup and CTI Invest. There is a kind of private–public partnership: the public authorities help to build up the project and to improve its quality; the private side is the investors who allocate seed and venture capital (Hotz-Hart 2009: 272–279).

Innovation Policy in *Israel* has been an unqualified success, with the country rising to fourth place internationally in terms of patents, after the US, Japan and Taiwan in 2004. That year R&D/GDP ratio reached a high of 4.6 percent, the world's highest; the number of high-tech companies is estimated at 4000.

Innovation policies in Israel consist mainly of a matching grant program in which firms submit proposals for R&D projects. Those that qualify receive a grant of up to 50 percent of the R&D costs; if the project succeeds, the recipient pays back the grant in installments. Another policy instrument used is the Magnet industry–academy consortia program, which supports the formation of consortia in order to encourage generic pre-competitive technologies; there are grants for up to two-thirds of the total approved budget, and no repayment is required. A further instrument is the incubators program, intended to provide fledgling entrepreneurs with the basic means required in the very early stages, in order to develop their innovative ideas and set up new businesses, including financial support, physical installations and advisory services.

Last, and most importantly, there is the Yozma program which was intended to jump start the venture capital market in Israel. It managed to establish a number of venture capital funds that were initially funded by the government but also included local and foreign private investors. It relied on the issue of options to buy Yozma's shares in these funds in five years' time at a predetermined price.

The Israeli policy planners realized (in contrast with most other planners) that for Yozma not only to be successful but also to have a lasting effect, that is, to create a local venture capital market, it would have to be linked to a strong financial market: in this case the US NASDAQ.

The program was a great success. It managed to attract prominent international investors with financial resources and expertise; it set up 10 venture capital funds and helped raise $200 million. In initiating the creation of a venture capital industry, a high-technology startup-intensive cluster has injected new life into Israel's high-tech industry, which had previously been dominated by the military. In addition, by teaming the evolving venture capital industry with the booming ICT sector it provided a self-reinforcing mechanism that could usher in a catching-up process (Avnimelech and Teubal 2006a, 2008).

Since the end of the Yozma program, the venture capital market in Israel has boomed with 80 funds in operation. It raised close to $10 billion during the period 1993–2000, with actual venture-capital backed investments reaching a high 2.7 percent of GDP in 2000 – a world record (Avnimelech and Teubal 2006a). Since then, capital markets have also expanded and international access to funds has improved spectacularly. Israel is the foreign country with the largest number of initial public offerings (IPOs) in NASDAQ. "In 2010 high-tech firms in Israel attracted $1.3 billion in venture capital, nearly twice as much as in Britain, whose economy is ten times as big" (*Economist*, August 6, 2011: 12).

Possibly the most important characteristics in the case of *China* are the readiness to recognize the need for change in its NSI construct, as well as the policies and the willingness along with commitment to do so.

In tune with the phenomenal political changes that have marked China in the last six decades, its NSI has been in constant flux ever since its inception in the early 1950s. At first the NSI was designed in accordance with the centrally planned regime, and characterized by its huge size and the separation of R&D centers from productive enterprise users.

The crucial event for R&D system reform came in 1985 following the agricultural and industrial reforms, at which point the intent was to rearrange the relationship between knowledge producers and users and their relationships with the government (Gu and Lundvall 2006). The process of sales and training (S&T) system reform proceeded through trial and error with continuous adjustments in policies. The changes brought about were successful in safeguarding and recombining technological capabilities in the context of market reform and opening up to the global economy. However, these reforms did not address the system's major weakness, which has a negative impact both on the absorption of foreign technology and on domestic innovation: the fact that the economic structure does not support learning by interaction. This dimension was fully recognized in the most recent policy shift in 2005 calling for "endogenous innovation" and "harmonious development". Gu and Lundvall suggest that "endogenous innovation" should be translated as innovation-driven growth and learning-based economic development.

In the *EU*, responsibility, incentives and inducement mechanisms are being transferred from the national to the European level in areas such as competition, trade, regulatory aspects of energy and the environment, and a variety of R&D programs related to firms. But others, such as education and training, corporate governance, natural resources, and defense procurement, are likely to remain stubbornly national.

There are only a few signs of the emergence of a European system of innovation. The closest thing to it is likely to be the German system of innovation, just in terms of sheer weight as it accounts for 40 percent of all of Europe's technological activities.

Referring to the EU, one should bear in mind that the Union is composed of 28 countries whose single common characteristic is the desire to belong to that Union, and even that to varying degrees. Otherwise, they differ significantly in terms of level of development, structure of the economy, financial markets, laws, institutions, history, language, skills, resources, ways of meeting social problems, and of course NSIs.

There are no mechanisms to ensure that national systems of innovation converge in either their characteristics or their performance. Empirical evidence as to technological and economic convergence within the EU is disappointing (Clarysse and Muldur 2001). One obvious reason for this is that the total of the EU budget is tiny, given that the EU's share of total European R&D budget remains very small. For instance the US NIH budget represents by itself more than one-half of the global FP7 budget for seven years.

R&D networks still tend to be mainly national and, with the exception of the Netherlands, business R&D is still performed in the home country. But while larger member countries depend only marginally on the EU budget, smaller and peripheral economies are strongly affected by it because "the subordinate position of the latter compels them to import more or less irrelevant patterns of thought and action and this without any pressure on the part of the leading regions" (Liagouras 2010). Additionally, the EU's emphasis on R&D-based technical change to the detriment of other factors and its insistence on high-tech and best-practice policies become problematic in the case of the laggard economies.

In general, the EU does not have an environment as favorable to SMEs as the one in the US nor has it reached a similar level of achievement in the internal market, in both products and services. European SMEs seem to benefit much less from the size of the internal market in Europe than US firms do from their own market. Beside higher barriers to entry in the European markets, there are also higher barriers to growth for SMEs in Europe. In addition, Europe does not possess the equivalent of the US Small Business Act.

The direct R&D funding instruments available from the EU budget include the research Framework Program (FP) and three smaller programs: the Competitiveness and Innovation Framework Programm, Structural Funds for regional development, and European agencies funds dedicated to specific programs such as Galileo.

Private R&D investment in Europe is still too low: 55 percent in contrast to 65 percent in the US. Further, most of it is realized by the largest firms, which are the primary benficiariesof public R&D subsidies. Even if member countries' R&D expenses represent more than ten times the Community level R&D effort, the global R&D investment in Europe still lags behind that of the US – it amounts to only two-thirds of the US effort.

With respect to the overall target for R&D spending, some European countries are already above target while others remain far below. The different starting positions are such that defining the same target for every country would be completely inappropriate. In addition, their requirements for growth differ vastly as technological leaders need investment in research and higher education while technological laggards need capital accumulation and technology adoption. Finally, the structural heterogeneity between the EU country members requires some flexibility, allowing different policy reforms with specific agenda determined by the peculiarities of each country – turning away from the "one size fits all" principle.

The *US* exhibits what is usually referred to as a "public mission Technology Policy" organized in parallel industrial complexes with public and private user interests driving policy, and defense and health as the main recipients for public support in R&D. A strong element is diversity, supported by the fact that both different agencies and big private foundations are willing to fund good research.

The US has the undeniable advantage of being the center for most efficient reach to market for innovations. This means creating, building, extending and maintaining the best, most hospitable ecosystem, balanced between variety and selection, with an innovation-based productive sector to transform ideas into products and services in the market; the best support for small and starting businesses; as well as legislation and regulation to support the edifice. If we realize what this involves, then we must acknowledge that this is hardly a coincidence; these things simply could not have materialized on their own. The US possesses the best university system, is still by far the largest producer of technology, the most admired and the most spied upon, and boasts the largest and most specialized financial markets and institutions that may support innovation and the largest and most efficient venture capital market in support of startups. All of these were not created by themselves as each carries the state's indelible mark, some from the moment of their inception, others through the state's continued support, while others are heavily influenced by the state's actions, decisions, legislation and regulation. The state's Innovation Policy renders it the leading agent in achieving innovations that benefit the business community and allow the economy to grow. The state proactively creates new high-growth areas before the business community has begun to suspect their potential existence, much less their potential benefits. The Internet, biotechnology and nanotechnology are examples. The state does not just fix markets, it also creates markets.

To see the state in the role of ex post support may not be surprising, but it is a surprise to see the state playing an entrepreneurial role – "the entrepreneurial state" as Mazzucato calls it, "funding the most uncertain phase of the research that the private sector is too risk averse to engage with, seeing and commissioning for further development and often even overseeing the commercialization process" (Mazzucato 2011: 19). The Defense Production Act, passed in the 1950s, allows the Defense Department to back nascent industries important to national security. In the past, the state has played this role, investing in and helping to develop technologies such as microchips, radar and GPS, up to the point where price came down enough for these products to take off in the private sector. In the 1960s, the Department of Defense invested in the then embryonic semiconductor industry until prices fell sufficiently for them to become affordable. At present, and for the next few years, the navy is likely to keep investing in private-sector companies that are building advanced biofuel refineries, receiving equity in return. This is a coordinated effort with the Department of Energy and the Department of Agriculture, which are also contributing. There are more than a hundred companies in the advanced biofuels industry. How long it will take them to drive the price down is a matter of sufficient demand since scale of production is the main constraint. This amounts in part to whether the navy will keep supporting this nascent industry. For the moment it is doing its best, and has even provided something of a marketing stunt, a massive demonstration in July 2012 called the Great Green Fleet, in which the aircraft carrier *Nimitz*, accompanied by a task force of destroyers running on biofuels, and more than 70 F-18s similarly powered, showed off the navy's vision of the future: a military force no longer dependent on imported oil.

The level of sophistication of US policy intervention and a lightness in handling it, as witnessed in detail, is equally impressive. In 2009 the Defense Advanced Research Projects

Agency (DARPA) launched Blue Angel, a program that challenged three tobacco "pharming" facilities to produce 10 million doses of flu vaccine in one month! The reward was tens of millions in funding. There are many, many other incidents like this one.

Israel could be considered to be one of the most successful countries in the design and implementation of Innovation Policy given the difficulty of prevailing circumstances. On the other hand, the US has exhibited by far the most sophisticated and subtle Innovation Policy: it is ready to intervene, proactive and committed, with a vision, and while not alone in addressing risk aversion, it is close to being the only state actively engaged in the perpetuation of the system.

7.5 Current problems and future challenges for Innovation Policy

There are many challenges to innovation policy but nowadays the main issues are: firstly, a lukewarm stance towards "real" innovation expressed either as lack of interest or risk aversion or lack of true entrepreneurship; secondly, a lack of interest of policy planners at the world level (with one notable exception) in the regeneration and perpetuation of the system.

Countries call for made-to-order, individual innovation policies as they face differentiated problems depending on their level of development, type of production, degree of integration, economic structure, history, culture and institutions. Above all, however, it is a question of whether a true entrepreneurial spirit exists and if so to what extent. For most countries around the world, entrepreneurship is synonymous with peddling or at best shopkeeping. These are the less developed countries, which are also bent on rent-seeking activities. The more a society values rent-seeking activities, the less innovative it is. This may be taken to be lack of entrepreneurship or risk aversion.

In the least developed countries there is no local innovation, and consumption of (foreign) technology takes place in the form of "embedded" knowledge mostly in the form of consumer goods. In fact, in most developing economies one of the basic premises of NSI, namely, that innovation is an interactive process between producers of innovation and users, cannot be met, as there are neither "producers" nor "users" (Unger 1988). There is no entrepreneurial spirit to speak of and entrepreneurship is synonymous with peddling.

A few steps above these most abject economies, there is a large group of countries, latecomers to the development process where there is (1), to a varying degree, insufficient awareness of the need for innovation, or pure lack of interest in such an activity, and (2) a lack of understanding of what innovation implies, for example cooperation and interaction. Here, entrepreneurship means mostly shopkeeping, there is strong risk aversion and a penchant for rent-seeking economic activities.

It may just be that in these situations the difficulties to be overcome by innovation are simply too formidable (importers, traders and a banking system long accustomed to favoring trade) and/or the risks attached to innovation in those environments are much too high to overcome higher expected returns in other long-established traditional activities.

So what is it that can transform a society from a rent-seeking, risk averse and non-entrepreneurial society into an entrepreneurial innovative one? Mokyr's answer is the exposure to what he calls the Industrial Enlightenment: "the Enlightenment had another impact that is hard to quantify: It affected the institutional structure of society ... it redirected creativity and energy away from rent-seeking and towards activities that increased national prosperity" (Mokyr 2010: 63). While other variables may also have a role, we already have an important ingredient at hand in explaining the change in a society. Another policy action that emerges, then, is the emphasis on knowledge – education, higher education, technical education – and making sure that those able to make best use of knowledge are able to access it. This also calls for open science, for the placement of useful knowledge in the public realm and for its codification to be made easily accessible.

Yet there are countries that have been exposed to some degree of industrial enlightenment and to a reasonable level of educational achievement which still cling onto rent-seeking values. It may turn out that for some of these cases incentives to combat risk aversion, similar to those suggested earlier, may work: venture capital firms, technology-rating agencies, low capital gains taxes, limited participation by pension funds and links to a strong foreign market.

For most of the less developed world Innovation Policy, with its emphasis on innovation proper, may sound like a luxury, and we shall argue that a policy with a different emphasis, like diffusion, in the form of technology transfer, may turn out to be more appropriate. Besides, within this range of development, capital accumulation takes precedence over knowledge accumulation, as the distance to the technological frontier is enormous.

In most of the developing world and even in parts of the developed one, due to lack of knowledge there is no real appreciation of what it takes to innovate, and hype tends to overshadow real substance. A simple smartphone application produced locally may end up in the media as the object of national pride. Realism and sophistication, which are needed to recognize the country's strengths, weaknesses and peculiarities and to determine what is achievable and what is not, are also absent. It is not unusual to hear a minister in a developing country promise to turn it into a Silicon Valley.

For some countries Innovation Policy is for all practical purposes in the hands of MNCs who carry out most of the "national" innovation effort. Two groups can be distinguished. The first consists of countries such as Ireland, Hungary, the Czech Republic, the Slovak Republic and Poland, where innovational effort is at an average level and carried out by foreign MNCs. The second group consists of the strongest innovation effort achievers, mainly small developed countries such as Finland, Sweden, Switzerland and to a lesser extent Denmark, whose R&D effort is mostly carried out by indigenous MNCs. The blessings of this second group are certainly substantial but the tyranny of small numbers, as exemplified by the demise of Nokia and its huge impact on the Finnish economy (of which it used to represent up to 20 percent), should also be heeded.

At the level of more developed countries risk aversion finds several expressions. One such expression, and a problem in the making, is the fact that large corporations are less and less willing to conduct basic research, and that large research institutions and organizations are no longer what they used to be in terms of research. Pharmaceutical firms are reexamining their dustbins for previously discarded projects, in case they look more promising now, or produce "me too drugs". Innovations shelved for years reappear, or rather reach the market, some with great success. In 2006, a certain Steve Jobs visited Corning Glass and asked for an especially strong thin glass for his iPhone, saying he wanted it in production in six months. This is how "Gorilla glass" started; it began with the iPhone and soon became the standard for all touch screens, not only for smartphones but also for all sorts of gadgets including iPads, tablets, and so on. "Gorilla glass" had been created 50 years earlier by Corning Glass for the automotive industry, which had turned it down at the time, and was now dusted off and modified. A similar phenomenon happens with venture capitalists, who make money by recycling old ideas – bringing old ideas to new markets (*Economist*, June 2, 2012: 72).

Another expression of the risk aversion tendency is that innovation is becoming more and more marginal in nature and as close to market as possible. We are told that innovational effort is on the rise as the number of patents rises. They do rise in numbers but in the era of the smartphone we tend to confuse innovations with "apps" (applications, for the few uninitiated). You can even lock or unlock your front door using an "app" on your smartphone and grant access to friends with in-app invitations; customized virtual keys during set hours or keys that expire after a specific date can also be made available. You can even download an app to your iPhone and shout commands at it to locate shows on television. In fact, apps on smartphones are so much in vogue that course CMSC498I on iPhone programming at the University of Maryland has been oversubscribed ever since 2009; by now similar courses are available at many US universities.

Large firms shy away from research, take advantage of research results obtained by state laboratories or state-funded research in universities, sit on the pile of huge profits they make, and proceed with share buy-backs or perform reverse takeovers, in which a smaller foreign firm takes over a much larger firm which thus avoids paying tax in its home country.

With some notable exceptions the world over, the firms have been mostly left to decide on the type, direction and volume of innovation. But firms do not seem particularly engaged, their R&D is a pale shadow of its past, and they are less and less preoccupied with basic research. In addition, conglomerates and hedge funds with substantial stakes in the largest firms are already in a position to influence decisions on their R&D. Some, like Black Rock, have very wisely apparently refrained from doing so, at least until now. An interesting question is what would have happened if it had meddled with decision-making in the R&D of Apple or Microsoft or GE, just some of the firms in which Black Rock is the largest shareholder. It may turn out that hedge funds like Third Point may be more in tune with the future; Third Point has not been so discrete vis-à-vis Sony. Many innovations to be, no matter how promising they may seem in the long run, are sacrificed by firms if they fail to deliver profits fast. It is only natural for this outcome to be reinforced if these firms belong to conglomerates or, even worse, to hedge funds.

In addition, the use of patents is problematic. Microsoft has claimed for 245 infringements of its patents by Linux. There are patent trolls, as we have said, and industrial espionage is widespread. Small innovators are increasingly put in the difficult position of either facing exhausting litigation battles or the prospect of being acquired by large firms seeking an oligopolistic throne on what they consider their "turf". Lately, the contours of these turfs seem to have become rather easily penetrable (robotics and Google) or encompass newly discovered areas (cloud computing, drones and Amazon), allowing these large players to move into territories beyond their original reach. Besides, the situation in acquisitions is not particularly promising, as Google's acquisition of Android, and its successful subsequent use, is not the norm if we consider the industry as a whole or even if we limit ourselves to the fate of some of Google's other acquisitions.

7.5.1 Emerging challenges for Innovation Policy

Over the last two and a half centuries economists have been pondering whether jobs destroyed by technological change outnumber the jobs it creates. For the frontrunners in technology in the West, which are also the most developed, there have been periods when jobs created were not in balance with jobs destroyed, but in the long run they seem to have stayed in balance. (For the laggards this has not been the case, as unlike the frontrunners they produced no innovation that could generate jobs, but this issue has not been of much concern until more recently.)

Historically, when productivity in agriculture rose substantially, surplus labor passed to industry. When productivity rose in industry, surplus labor passed to services, the last stand. Nowadays, the labor-saving innovation bias seems more threatening as it besieges services. Many of today's innovations actually eliminate jobs – mostly menial ones and a few high-tech jobs in services. Uber displaces taxi drivers, Airbnb reduces hotel personnel, Locals affects tourist guides, Robot garage reduces the need for parking attendants. There are driverless buses, robot nannies and robot hospital care personnel, robots in transportation in warehouses, hospitals and stations.

In essence, software of the digital age at the heart of present-day innovations is eating away at jobs created in the analogue age. Jobs destroyed by the "app innovation" seem set to become a multiple of the "roughly 466,000" overall jobs so far created since the introduction of the iPhone in 2007 (Mandel 2012). Globalization became a threat as it exported jobs abroad, but today's innovations send jobs to oblivion.

Innovation together with globalization reduce employment and income in the West. Additionally, as a result of the fact that productivity improvements are not totally translated into rising incomes, the shares of labor and capital are no longer stable, with capital gaining over time. During the period 1980–2008, the share of labor in GDP declined in the West by 14.6 percent in Ireland, 10.5 percent in Austria, 6.7 percent in Greece, 6.6 percent in Japan, 6 percent in Italy, 5.9 percent in Spain, 5.5 percent in France, 3.8 percent in Germany, 3.2 percent in the Netherlands, 2.5 percent in Sweden, 2.4 percent in the US, 2.1 percent in the UK and 1.5 percent in Belgium, with the only exception, Portugal, experiencing a rise of 5.4 percent (Milanovic, 2013).

Inequality is further generated as wages are affected by technologies like IT that favor skilled labor by raising its relative productivity and therefore wages. While "knowledge" workers in the US have experienced a tripling of their income in the last 30 years, unskilled labor income remains at the level of 1970 or below. Thus, there is a working elite, which falls within the top 5 percent of wealth holders, a new arrival on the list which was previously dominated by old family wealth (Piketty 2014). But the 0.1 percent of the wealthiest includes those individuals who have found new ways of making riches, such as through financial markets, mergers and acquisitions, golden parachutes, hedge funds, but also increasingly innovators-entrepreneurs. It used to be said that innovation makes millionaires. Now we say it makes billionaires.

7.5.2 *Innovation Policy for the future: additional, and concluding remarks*

First, a hopeful note on a policy measure with universal applicability. The emergence of the Internet added one more impediment to the development of most countries, especially the least developed ones, and a "digital divide" was erected between the developed and the less so. Fortunately, thanks to cheap hardware and telecommunications the divide has shrunk considerably over the last decade.

A very positive development especially for the less developed countries has been the realization that the benefits to be derived from the Internet come through its use and not through mastering the art or achieving the capability of producing ICT goods and equipment. This is a welcome change and a much more accessible goal – and an exception to the rule so far – than having to be able to produce the means of production, i.e. capital goods, to achieve developed status. So an Innovation Policy action point that emerges is for the spread of the use of the Internet to all countries, especially the less developed ones. They all stand to gain in terms both of innovation and growth.

The most pressing problem is the waning of benefits from innovation in terms of jobs and income. Most writers reason that creating more startups will create jobs and generate innovation (Kane 2010). The rise of dynamism in the US economy in the 1990s and up to the financial crisis is usually associated with the increase in the numbers of new startups. And the decline in the number of new startups since then coincides with what is currently described as "the loss of dynamism" in the US economy (Haltiwanger 2012). It is only natural that there should be claims for the support of the generation of more startups. But the typical startup is very small, operates in retail or personal services (Hurst and Lusardi 2004), is not innovative and creates few jobs. It also has low productivity (Haltiwanger et al. 1999), for which it does not make up as it gets older because the typical startup is dead within five years of its creation. In fact, startups mostly amount to "wage-substitution businesses that have more in common with self-employment than with the creation of high growth companies" (Shane 2009: 3).

Most observers seem not to notice the fact that "a tiny sliver of companies accounts for the vast majority of the contribution to job creation and economic growth ... These gazelles more than make up for the lack of job and wealth creation of the typical start-up" (Henrekson and Johansson 2009). According to the National Venture Capital Association, out of the more than two million startups every year since 1970, venture capital has supported 820 – which amounts to less than 0.05 percent of the total population of firms. This miniscule number of companies has a tremendous impact on the economy both in terms of employment and output. According to Venture Impact, in 2003 companies that were backed by venture capitalists employed 10 million people or 9.4 percent of the labor force in the US, and generated 9.6 percent of business sales.

What is needed is support for the creation not of any startups but of those startups that will eventually turn out to be gazelles. How can they be recognized? There are many criteria to work with. For instance, "the typical start-up founded between 1998 and 2003 required $11,400 in capital. So, even at the time that SAP, or Google or EasyJet were founded they weren't anything like the typical new business" (Shane 2009: 146). But there are even better criteria for spotting gazelles as their source of financing, venture capital, uses well-known criteria for their identification: the human capital of the founder and his motivations, the industries in which they are founded, their business plans and strategies, their legal forms and capital structure, etc.

Since encouraging more people to become entrepreneurs or encouraging all startups is bad policy (Shane 2009), a selection system should be set up to identify those startups that should be supported, the prospective gazelles, those that have the characteristics appreciated by venture capitalists. Selection could be carried out by the agency administering the program along the lines of a policy of shifting resources, from programs supporting generic entrepreneurship to those supporting high potential businesses such as the Small Business Innovation Research Programs (SBIR).

Yet another way of responding to the waning of the benefits on income and employment from innovation is to turn to the "entrepreneur of last resort". Mazzucato, in her persuasive book (Mazzucato 2011), like several writers in the past, including this writer (Vernardakis 1988; Branscomb 1993), treats the role of government in a similar way. Like some of her predecessors, she calls for a proactive government role that involves investment in areas where the private sector would not take the risk, beyond the concept of market failure, even beyond research that is not basic. She calls for a more entrepreneurial economy that does not require the state to withdraw but to lead (Mazzucato 2011: 15).

An avid admirer of US Innovation Policy – and who wouldn't be? – she notes that the US government does not just fund innovation and create the right conditions for it, but that it also envisions the opportunity space, engages the most risky and uncertain early research, oversees the commercialization process, does not just fix markets but creates them. Quoting the findings of a recent study on the development of the biotech industry, "the knowledge economy was not born but made" (Vallas et al. 2011).

Impressive as the role of the state in the US may be, it seems to be rather precarious as it is also self-effacing – possibly so as not to attract the criticism of opponents clamouring for tax reductions, especially in times of budget tightness. Many programs, in retrospect very successful, would have been stopped in their tracks if they had not been branded as critical by the Department of Defense (DoD). Mazzucato ponders whether an innovation system based on government support can be sustainable with such a system of rewards. "The lack of knowledge in the public domain about the central entrepreneurial role that government plays in the growth of economies worldwide, beyond Keynesian demand management, and creating conditions for growth, is currently putting the successful model in major danger" (Mazzucato 2011: 109).

The question of whether the system is sustainable does not refer to the US alone but to the rest of the world as well. So far, world leadership in innovation in the sense of demonstrating the ability to show the way, has been well served by the technological hegemony of the US. The US government thinks of and creates new activities and opportunity spaces, with strange-sounding names like biotechnology or nanotechnology or space travel, a decade or two before businesses hear about them; then, once the path has been created it steps aside, making room for its business community to follow. The rest of the world, whose businesses are further away from the frontier of knowledge also stand to benefit from these new vistas with some greater delay. If the US government stops or falters in advancing in these uncharted territories, then innovation will be seriously adversely affected, unless some other country is willing and able, even after a long delay, to perform the task the US government has been performing so far. For the moment, the US government is the only one interested in the perpetration and constant regeneration of the system.

For the moment the US is still at the forefront of science, technology and innovation. However, indicators on business innovation and on the growth of multi-factor productivity suggest that the US lead is narrowing. R&D and patenting by businesses have also grown less rapidly than in the past and the quality of patents has been declining, a fact that may be of some relevance to the jobless growth and the waning of the benefits of innovation experienced at present. China is expected to become the top spender on R&D in less than a decade, but it will still be a long way from technological leadership in terms of accumulated knowledge and willingness to show the way and take risks. Under these circumstances, would the US still be playing its role of pathfinder?

While innovation has been deified, diffusion has been neglected in Innovation Policy. Models of diffusion, especially the epidemic model, lead us to believe that diffusion is a very slow process, which is not always the case. As a result, policy is usually geared in specific directions and towards particular actions (Geroski 2000). Typically, the aim is to hasten the rate of diffusion, as Innovation Policy is generally interested in encouraging the adoption of particular new technologies for welfare-enhancing reasons, either because it serves particular public policy goals or because certain technologies are viewed as conferring externalities on society as a whole (adoption of vaccination). What schemes there are largely counter problems of imperfect information and are directed at SMEs (Stoneman and Diederen 1994: 927). In several countries, such as the UK, France and Germany, regional innovation centers have been set up offering consultancy support; the same goes for the European Union as a whole, e.g. SPRINT. There have been attempts to create clusters to encourage the interchange of knowledge, and attempts to stimulate the formation of networks to link science and technology more closely. Judging from the funds involved in supporting these programs, the effort spent on diffusion is token in comparison with the effort spent on innovation.

It has been generally assumed that diffusion is a straightforward affair, providing poor countries with plenty of opportunity to catch up with the frontrunners, as long as they copy Western political and legal systems, do not interfere with markets, promote free trade and have a liberal stance towards FDI. This package amounts to the policies proposed by the World Bank and the IMF to developing countries and came to be known in the 1990s as the "Washington Consensus". Such policies are by now highly questionable, as several countries have in the last few decades succeeded in joining the group of technological champions by ignoring the gospel of the Washington consensus. "FDI, for example, played virtually no role in the spectacular catch-up by Japan and Korea, while government interference in the economy was very frequent" (Fagerberg 2010:8). In addition, some countries, like China, whose efforts to catch up are being met with success, have political systems that are far removed from Western democracies.

Within the deification of innovation there is the deification of high technology. But there is plenty of room for achieving significant overall improvements through innovations in medium- or low-technology activities. For most of the less developed world Innovation Policy may sound like a "provocation" (Johnson et al. 2003: 14). Yet, a rudimentary, very specific and narrow Innovation Policy in the form of diffusion of foreign technology, such as water purification, is a possibility.

An alternative policy for latecomers to development (other than the highly interventionist road followed by Korea and the Asian Tigers) which have managed to acquire a semblance of an NSI and a well-educated scientific community, but lack domestic demand and a substantial entrepreneurial class, is to follow Israel's interventionist road and reliance on venture capital. A word of caution: venture capital can fund only a portion of total innovation in a country as it expects high rates of return, has a penchant for specific fields (at present ICT and biomedicine), has an affinity with specific regions and needs access to a thick stock market (Kenney 2011).

A topic which has attracted very little interest till now in Innovation Policy but promises to do so in the future is the latter's social impact. In the past, its impacts could be devastating but for some reason attracted hardly any attention. They may have been too far in the past or are not well known, or sporadic and localized, as in the case of the mechanized tomato harvester introduced in order to replace the flow of immigrant seasonal land workers for the collection of tomatoes in California in the late 1960s. The mechanical harvester turned out to be too harsh and crushed the tomatoes. So another innovation appeared: the genetically engineered tomato. It is hard and can be picked by the mechanical harvester; it looks like a tomato, but it neither tastes nor smells like one. It stopped the flow of immigrant workers and 90 percent of the tomato farmers in California who could not afford the harvester lost their farms (Hightower 1972).

The rise of inequality as a consequence of innovation is another issue largely ignored so far, but which will have to be addressed sooner or later. However, Innovation Policy is not a panacea and as such it should not be expected to cover or pre-empt Development Policy in its entirety. Inequality is a challenge that could not be well met through Innovation Policy alone, or primarily through it. Of course, it could offer some limited help. For instance, Mazzucato (2011:109) asks "Is it right that the NSF did not reap any financial return from funding the grant that produced the algorithm that led to Google's search engine?" The state could collect billions from all those companies that benefit from government-funded research and then use those funds to pay for further research, thus lightening the tax burden on citizens and performing a sort of redistribution. This is not an easy exercise and may be limited to those benefiting directly from government-funded research but will be difficult to collect from indirect beneficiaries and spillovers. This would not reduce inequality but it would at least tend to lessen contributing to it through innovation.

Increasingly companies innovate through methods that may deliver value for the company but less so for the consumer. A new marketing strategy may amount to pestering consumers. A new marketing approach aimed at 19-year-old girls buying all sorts of candy by It's Sugar specifically plays on their desire to feel free by enjoying something that is admittedly unhealthy – and emphasizing the illicit allure of the product (Bradley 2013: 12). This is an issue that will call for some attention.

In this book, we have already encountered the fact that innovation may lead to situations where the "winner takes all" or most, where a lesser innovation or one with fast results may be adopted at the expense of better innovations, to situations in high-technology industries governed by increasing returns where the market is unable to decide which is the better technology, and to quasi-monopolistic or monopolistic situations. Should not Innovation Policy have something to say about the problematic use of patents, or about "fair play" among innovation producers, or whether the "best" innovations should win, for the sake of both innovators and the public at large – not to mention the waste of resources? Our present line of defense rests on the regulatory system. Regulation is based on price or market share considerations in a unique and relatively stable equilibrium point framework. But competition depends less and less on price and more and more on innovation, and market share is determined in a not necessarily unique or stable framework of equilibrium. Should not the regulatory system, which is already under the strain of present problems, especially in cases involving high technologies, be alerted and made fit to deal with anticipated future developments? The tendencies are already understood – and some writers claim that the methodological tools to foresee the outcomes of such developments are also available (Arthur 1989).

It is worth keeping in mind that competition is not uniformly a stimulus to innovation and hence each case should be considered on its merits. Standards can play an important role in Innovation Policy. Open standards can enhance competition in open markets, whereas closed standards can act as a barrier to entry. They can also help to reduce transaction costs and so develop specialization and the division of labor, which plays an important role in increasing innovation and productivity. Open standards play a still more important role in innovation, as they make it easier for small-scale entrants to sell innovative add-on products; and they make it easier for producers and users to exploit network effects.

From a policy point of view, it is better that no proprietary product should win a standards race, but that a public and open standard should emerge to be used and produced by all firms. In any case, if a standards race should take place, there should be a policy already in existence to see to it that it does not last and to regulate the field afterwards.

In the same spirit, there is a growing need to limit waste of innovative effort. Measures are needed to deal with the problematic use of patents, and the use of exhaustive litigation and sheer market power by large firms resulting in the smothering of innovation originating in small firms. Only a small share of what is acquired by large firms is actually put to use.

Finally, we need to design new antitrust laws in the future (Sidak and Teece 2010), and we need to better understand a kind of competition different to the one we have been used to, specifically "dynamic competition". In Schumpeter's words " The first thing to go is the traditional concept of ... competition ..." (Hall and Rosenberg 2010: 7).

Note

1 Science, Technology and Innovation Policy: the evolving discourse in a nutshell. Ever since the 1960s there has been an evolving discourse around Science, Technology and Innovation Policy. A convenient way to trace the changes in aims, direction and emphasis as well as new ideas, concepts and methodology brought forward throughout this discourse is to consult the reports, policy analyses and memoranda of the Organization for Economic Cooperation and Development (OECD) and the contents of the conferences which have taken place under its auspices. Its consecutive contributions read like a diary of this discourse, and report on the state of the art at the time.

The OECD 1963 report, *Science, Economic Growth and Government Policy*, stresses the importance of Science Policy by linking it to economic growth. Another report, almost a decade later (OECD 1971) introduced a broader concern for human and social considerations to Science and Technology Policy. A later report (OECD 1980) treated Innovation Policy as an aspect of Economic Policy, a legitimate policy field. As in previous reports, increases in investments in science and R&D are still recommended but there is an additional focus, society's capacity to absorb new technology. A further report (OECD 1992) recognizes the emergence of a theoretical divide within Innovation Policy – to which we shall refer shortly – between the Neoclassical and the systems approach and makes a clear choice in favor of the latter. In fact, it gives the systemic version of Innovation Policy an analytical foundation by resting on the position that innovation is an interactive process. There is added reliance on network formation and new forms of organization as well as absorptive capacity of firms and feedbacks from users. Prompted by the succession of such reports, some scholars have gone so far as to claim that almost from the start the OECD promoted a systemic model of Science and Innovation Policies whose evolution is a clear antecedent to the NSI approach (Lopez-Martinez 2006, p. 121). The next report (OECD 2001) is a further assertion of the importance of innovation as a factor that deserves analysis and further understanding. There is still a further report, actually the most recent position of OECD on Innovation Policy (OECD 2010). Once more, not only does it side with the innovation systems' approach, which it describes in detail as it has evolved up to that time, and to which it makes important additions. Areas to which previous reports had alluded are now explicitly dealt with: such as the role of services, of users and open innovation in innovation and the paramount need to coordinate all types of policies with the aim of designing and implementing Innovation Policy. In this sense, this report presents the state of the art as to the discourse but also the state of the art on research on innovation in most of its aspects. Additionally, it prescribes the use of this approach as a best practice example.

Of course, the extent of the application of this state of the art by each individual country, that is, how far each country stands from best practice, is another story, as each country exhibits a different level of awareness on these issues, and allots them different levels of importance. And of course, each country's efforts are met with varying degrees of success.

Bibliography

Abramowitz, M. (1986) "Catching up, forging ahead and falling behind", *Journal of Economic History*, 46, 299–313.

Acemoglu, D., Aghion, P. and Zilibotti, S. (2006) "Distance to Frontier, Selection and Economic Growth", *Journal of the European Economic Association*, 4, 37–74.

Aghion, P. (2006) "A Primer on Innovation and Growth", *Bruegel Policy Brief*, 2006/66, October, Brussels, <http://www.bruegel.org/publications/publication-detail/publication/233-a-primer-on-innovation-and-growth/>.

Aghion, P. and Griffith, R. (2005) *Competition and Growth, Reconciling Theory and Evidence*, Cambridge, MA: MIT Press.

Almeida, P. and Kogut, B. (1997) "The Exploration of Technological Diversity and the Geographic Localization of Innovation", *Small Business Economics*, 9, 21–31.

Almus, M. and Nerlinger, E.A. (1999) "Growth of New Technology-based Firms: Which Factors Matter ?", *Small Business Economics*, 13, 141–54.

Archibugi, D. and Michie, J. (1997) "The Globalization of Technology: A New Taxonomy", in D. Archibugi and J. Michie (eds.), *Technology, Globalization and Economic Performance*, Cambridge: Cambridge University Press.

Arora, A., Ceccagnoli, M. and Cohen, W.M. (2008) "R&D and the Patent Premium", *International Journal of Industrial Organization*, 26(1), 153–179.

Arrow, K. (1962) "Economic Welfare and the Allocation of Resources for Invention", republished in Nelson R. (ed.), *The Rate and Direction of Inventive Activity*, Princeton: Princeton University Press, 609–625; republished in N. Rosenberg (ed.), *The Economics of Technological Change* (1974), 164–181.

Arthur, W.B. (1989) "Competing Technologies, Increasing Returns, and Lock-in by Historical Events", *Economic Journal*, 99, 116–131.

Arvanitis, S. and Sydow, N. (2009) "How Effective are the R&D Promoting Activities of the Swiss Innovation Agency CTI ? An Evaluation Based on Matched-pairs Analysis", in D. Foray (ed.), *The New Economics of Technology Policy*, Cheltenham, UK: Edward Elgar.

Avnimelech, G. (2008)"A Five-phase Entrepreneurial Oriented Innovation and Technology Policy Profile: The Israeli Experience", *European Planning Studies*, 16(1), 81–98.

Avnimelech, G. and Teubal, M. (2005) "Evolutionary Innovation and High Tech Policy: What can we learn from Israel's targeting of venture capital?", *Samuel Neaman Institute STE Working Paper Series*, 25.

Avnimelech, G. and Teubal, M. (2006a) "Creating Venture Capital Industries that Co-evolve with High Tech: Insights from an Extended Industry Life Cycle Perspective of the Israeli Experience", *Research Policy*, 35(10), 1477–1498.

Avnimelech, G. and Teubal, M. (2006b) "Evolutionary Innovation and Technology Policy: A Four-Phase High Tech Policy Model", paper presented at the DRUID Summer Conference on Knowledge, Innovation and Competitiveness: Dynamics of Firms, Networks, Regions and Institutions, Copenhagen, Denmark, June 18–20, 1–23.

Avnimelech, G. and Teubal, M. (2008) "From Direct Support of Business Sector R&D/Innovation to Targeting Venture Capital/Private Equity: A Catching-up Innovation and Technology Policy Life Cycle Perspective", *Economics of Innovation and New Technology*, 17(1–2), 153–172.

Bank of England (2001) *Financing of Technology-Based Small Firms*, Domestic Finance Division, February, 27–36, 98.

Bessen, J. and Meurer, M. (2008) *Patent Failure*, Princeton, NJ: Princeton University Press.

Black, B. and Gilson, R. (1998) "Venture Capital and the Structure of Capital Markets: Banks versus Stock Markets", *Journal of Financial Economics*, 47, 243–277.

Bogliacino, F. and Pianta, M. (2010) "Innovation and Employment: A Reinvestigation Using Revised Pavitt Classes", *Research Policy*, 39, 799–809.

Bogliacino, F. and Pianta, M. (2011) "Engines of Growth: Innovation and Productivity in industry groups", *Structural Change and Economic Dynamics*, 22, 41–53.

Boorstin, D.J. (1985) *The Discoverers*, New York: Vintage Books, 46–53, 67–72.

Borras, S., Chaminade, C. and Edquist, C. (2008) "The Challenges of Globalization: Strategic Choices for Innovation Policy", in J. Gustafsson (ed.), *The Innovation Imperative: Globalization and National Competitiveness*, Cheltenham, UK: Edward Elgar.

Boyer, R. (2005) "How and Why Capitalisms Differ", *Economy and Society*, 34, 4, 509–557.

Bradley, R. (2013) "A Candy Chain's Sweet Success", *Fortune*, July 1, 12.

Branscomb, L.M. (ed) (1993) *Empowering Technology, Implementing a US Strategy*, Cambridge, MA: MIT Press.

Branscomb, L.M. and Auerswald, P.E. (2002) "Between Invention and Innovation, an Analysis of Funding for Early-stage Technology Development", Advanced Technology Program, *National Institute of Standards and Technology*, November.

Bryant, K. and Wells, A. (ed.) (1998) *A New Economic Paradigm? Innovation-based Evolutionary Systems*, Discussions in Science and Innovation 4, Department of Industry, Science and Resources, Science and Technology Policy Branch, Canberra, Australia, v.

Bush, V. (1945) *Science: The Endless Frontier*, A Report to the President by Vannevar Bush, Director of the Office of Scientific Research and Development, Washington, DC: US Government Printing Office.

Canepa, A. and Stoneman, P. (2002) "Financial Constraints on Innovations: A European Cross Country Study", Kiel Institute of World Economics, Working Paper no. 02–11.

Cantwell, J.A. (1995) "The Globalization of Technology: What Remains of the Product Cycle Model?" *Cambridge Journal of Economics*, 19, 155–174.

Cantwell, J. and Iammarino, S. (2003) *Multinational Corporations and European Regional Systems of Innovation*, London: Routledge.

Castellacci, F. (2006) "Convergence and Divergence among Technology Clubs", paper presented at the DRUID Conference, Copenhagen, June 2006. Available at <http://www3.druid.dk/wp/20060021.pdf>.

Clarysse, B. and Muldur, U. (2001) "Regional Cohesion in Europe? An Analysis of How Public RTD Support Influences the Techno-economic Regional Landscape", *Research Policy*, 30(2), 275–295.

Comin, D. and Hobbijn, B. (2004) "Cross-country Technological Adoption: Making the Theories Face the Facts", *Journal of Monetary Economics*, 51(1), 39–83.

Coombs, R., Saviotti, P. and Walsh, V. (1989) *Economics and Technical Change*, London: Macmillan, 201–209.

Courvisanos, J. (2009a) "Optimize versus Satisfice: Two Approaches to an Investment Policy in Sustainable Development", in R. Holt, S. Pressman and C.L. Spash (eds.), *Post-Keynesian and Ecological Economics: Confronting Environmental Issues*, Cheltenham, UK: Edward Elgar, 279–300.

Courvisanos, J. (2009b) "Political Aspects of Innovation", *Research Policy*, 38, 1117–1124.

Day, G.S., Shoemaker, P.J. and Gunther, R.E. (eds.) (2000) *Wharton on Managing Emerging Technologies*, New York: John Wiley & Sons.

Dertouzos, M.L, Lester, R.K., Solow, R.M. and the MIT Commission on Industrial Productivity (1988) *Made in America, Regaining the Productive Edge*, Cambridge, MA: MIT Press.

Dodgson, M. (ed.) (1989) *Technology Strategy and the Firm: Management and Public Policy*, London: Longman.

Dosi, G., Llerena, P. and Sylos-Labini, M. (2006) "The Relationship between Science, Technologies and their Exploitation: An Illustration through the Myths and Realities of the so-called 'European Paradox'", *Research Policy*, 35, 1450–1464.

Economist, The (2013) "The Jobs Machine: Immigration and America's High-tech Industry," April 13.

Economist, The (2012) "When Patents Attack", April 12.

Economist, The (2011) "Where's Britain's Bill Gates?" August 6.

Economist, The (2012) June 2, 72.

Edquist, C. (1997) "Systems of Innovation Approaches: Their Emergence and Characteristics", in C. Edquist (ed.), *Systems of Innovation, Technologies, Institutions and Organizations*, London and Washington: Pinter, 1–35.

Edquist, C. (2006) "Systems of Innovation, Perspectives and Challenges" in J. Fagerberg, D. Mowery and R.R. Nelson (eds.), *The Oxford Handbook of Innovation*, Oxford: Oxford University Press, 181–208.

Elzinga, A. and Jamison, A. (1995) "Changing Policy Agendas in Science and Technology", in S. Jasanof, G.E. Markle, J.C. Petersen and T. Pinch (eds.), *Handbook of Science and Technology Studies*, Thousand Oaks, CA: Sage, 572–597.

Encaoua, D. (2009) "Nature of the European Technology Gap: Creative Destruction or Industrial Policy?" in D. Foray (ed.), *The New Economics of Technology Policy*, Cheltenham, UK: Edward Elgar, 281–314.

Fagerberg, J. (2010) "The Changing Global Economic Landscape: The Factors that Matter", in R.M. Solow and J-P. Touffut (eds.), *The Shape of the Division of Labour, Nations, Industries and Households*, Cheltenham, UK: Edward Elgar, 6–31.

Feldman, M.P. (1994) *The Geography of Innovation*, Dordrecht: Kluwer Academic Publishers.

Freel, M.S. (1999) "The Financing of Small Firm Product Innovation within the UK", *Technovation*, 19, 707–719.

Freeman, C. (1979) "The Determinants of Innovation", *Futures*, II, 206.

Freeman, C. (1988) "Japan: A New National System of Innovation", in G. Dosi, C. Freeman, R. Nelson, G. Silverberg and L. Soete (eds.), *Technical Change and Economic Theory*, London: Pinter, 330–348.

Frost, T. (1996) "From Exploitation to Exploration: The Geographic Sources of Subsidiary Innovations and the Evolutionary Theory of the Multinational Enterprise", Stockholm, paper presented at the EIBA Annual Conference, December.

Geroski, P.A. (2000) "Models of Technology Diffusion", *Research Policy*, 29, 603–625.

Greenhalgh, C. and Rogers, M. (2010) *Innovation, Intellectual Property and Economic Growth*, Princeton, NJ: Princeton University Press.

Gu, S. and Lundvall, B.A. (2006) "China's Innovation System and the Move Toward Harmonious Growth and Endogenous Innovation", Danish Research Unit for Industrial Dynamics, DRUID Working Paper no. 06-7.

Guereri, P. (1999) "Patterns of National Specialization in the Global Competitive Environment", in D. Archibugi, J. Howells and J. Michie (eds.), *Innovation Policy in a Global Economy*, Cambridge: Cambridge University Press, 139–161.

Guidici, G. and Paleari, S. (2000) "The Provision of Finance to Innovation: A Survey Conducted among Italian Technology-based Small Firms", *Small Business Economics*, 14, 37–53.

Hall, B.H. (2007) "Patents and Patent Policy", *Oxford Review of Economic Policy*, 23(4), 568–587.

Hall, B.H. (2009) "'The Funding Gap': Financial Markets and Investment in Innovation", in D. Foray (ed.), *The New Economics of Technology Policy*, Cheltenham, UK: Edward Elgar, 169–192.

Hall, B.H. and Lerner, J. (2010) "The Financing of R&D and Innovation", in B.H. Hall and N. Rosenberg (eds.), *Handbook of the Economics of Innovation*, Amsterdam: Elsevier-North Holland.

Hall, B.H. and Rosenberg, N. (2010) "Introduction to the Handbook", in B.H. Hall and N. Rosenberg (eds.), *Handbook of the Economics of Innovation*, Amsterdam: Elsevier-North Holland, 7.

Haltiwanger, J. (2012) "Job Creation and Firm Dynamics in the US", *Innovation Policy and the Economy*, 12, 17–38.

Haltiwanger, J., Lane, J. and Speltzer, J. (1999) "Productivity Differences across Employers: The Roles of Employer Size, Age, and Human Capital", *American Economic Papers and Proceedings*, 89(2), 94–98.

Henrekson, M. and Johansson, D. (2009) "Gazelles as Job Creators: A Survey and Interpretation of the Evidence", *Small Business Economics*, 35, 227–244.

Hightower, J. (1972) *Hard Tomatoes, Hard Times: The Failure of America's Land Grant Complex*, Cambridge, MA: Schenkman.

Hodgson, G. (1984) *The Democratic Economy*, Harmondsworth: Penguin.

Hoffman, K., Milady, P., Bessant, J. and Perren, L. (1998) "Small Firms, R&D, Technology and Innovation in the UK: a Literature Review", *Technovation*, 18(1), 39–55.

Hotz-Hart, B. (2009) "Small and Medium Sized Enterprises: The Promotion of R&D and Innovation Behavior in Switzerland", in D. Foray (ed.), *The New Economics of Technology Policy*, Cheltenham, UK: Edward Elgar.

Hurst, E. and Lusardi, A. (2004) "Liquidity Constraints, Household Wealth, and Entrepreneurship", *Journal of Political Economy*, 112(2), 319–347.

Inkster, I. (2002) "The National Imperative: The State, Science and Technology Policy Evolution circa 1400–2000", in *Science and Technology Policy*, from Encyclopedia of Life Support Systems (EOLSS), Oxford: UNESCO, EOLSS, available at <http://www.eolss.net>.

Jaffe, A.B., Trajtenberg, M. and Henderson, R. (1993) "Geographical Localization of Knowledge Spillovers, as Evidenced by Patent Citations", *Quarterly Journal of Economics*, 58, 577–598.

Jeng, L. and Wells, P. (2000) "The Determinants of Venture Funding: Evidence Across Countries", *Journal of Corporate Finance*, 6, 241–289.

Jeremy, D.J. (1973) "British Textile Technology Transmission to the United States: the Philadelphia Region Experience, 1770–1820", *Business History Review*, 47, 26.

Johnson, B., Edquist, C. and Lundvall, B.A. (2003) "Economic Development and the NSI Approach", First Gobelics Conference, Rio de Janeiro, November 3–6, 14.

Kane, T. (2010) "The Importance of Startups in Job Creation and Job Destruction", *Firm Formation and Economic Growth*, Kauffman Foundation Research Series, July.

Kenney, M. (2011) "How Venture Capital Became a Component of the US NSI", *Industrial and Corporate Change*, 20(6), 1677–1723.

King, A. (1974) *Science and Policy: The International Stimulus*, New York: Oxford University Press.

King, R.G. and Levine, R. (1993) "Finance, Entrepreneurship and Growth", *Journal of Monetary Economics*, 32, 513–540.

Kitson, M. and Michie, J. (1998) "The Political Economy of Globalization", in D. Archbugi, J. Howells and J. Michie (eds.), *Innovation Policy in a Global Economy*, Cambridge: Cambrige University Press, 163–184.

Kok, W. (2004) "Facing the Challenge: The Lisbon Strategy for Growth and Employment", Report from the High-Level Group, <http://ec.europa.eu/growthandjobs/pdf/kok_report_en.pdf>.

Krugman, P. (1996) *Pop Internationalism*, Cambridge, MA: MIT Press.

Lall, S. and Albaladeho, M. (2002) "Indicators of the Relative Importance of IPRs in Developing Countries", Working Paper QEHWPS85, University of Oxford, Queen Elizabeth House.

Lamoreaux, N.R. and Sokoloff, K.L. (2007) "Introduction: The Organization and Finance of Innovation in American History", in N.R. Lamoreaux and K.L. Sokoloff (eds.), *Financing Innovation in the United States, 1870 to Present*, vol. 1, Cambridge, MA: MIT Press, 1–37.

Le Bas, C. and Sierra, C. (2002) "Location versus Home Country Advantages in R&D Activities: Some Further Results on Multinationals' Locational Strategies", *Research Policy*, 31(4), 589–609.

Lerner, J. (1999) "The Government as Venture Capitalist: The Long-run Effects of the SBIR Program", *Journal of Business*, 72, 285–318.

Levine, R. (2005) "Finance and Growth: Theory and Evidence", in P. Aghion and S. Durlauf (eds.), *Handbook of Economic Growth*, Amsterdam: Elsevier–North-Holland.

Liagouras, G. (2010) "What Can we Learn from the Failures of Technology and Innovation Policies in the European Periphery?", *European Urban and Regional Studies*, 17(3), 331–349.

Lopez-Martinez, R.E. (2006) "A Systems Approach to Innovation Policy", unpublished PhD thesis, Manchester Business School, University of Manchester.

Lundvall, B.A. (1988) "Innovation as an Interactive Process: From User-Producer Interaction to the NSI", in G. Dosi, C. Freeman, R. Nelson, G. Silverberg and L. Soete (eds.), *Technical Change and Economic Theory*, London: Pinter, 349–369.

Lundvall, B.A. (1992) *National Systems of Innovation: Towards a Theory of Innovation and Interactive Learning*, London: Pinter.

Lundvall, B.A. and Borras, S. (2006) "Science, Technology and Innovation Policy", in J. Faberger, D.C. Mowery and R.R. Nelson (eds.), *The Oxford Handbook of Innovation*, Oxford: Oxford University Press, 599–631.

Malerba, F. (1996) "Public Policy and Industrial Dynamics: An Evolutionary Perspective", ISE report, TSER/4thFP, DGXII/EC.

Malerba, F. (2009) "Increase Learning, Break Knowledge Lock-ins and Foster Dynamic Complementarities: Evolutionary and System Perspectives on Technology Policy in Industrial Dynamics", in D. Foray, *The New Economics of Technology Policy*, Cheltenham, UK: Edward Elgar, 33–45.

Mandel, M. (2012) "Where the Jobs Are: The App Economy", *TechNet*, February 7, 3.

Mazzucato, M. (2011) *The Entrepreneurial State*, London: Demos.

McKelvey, M. (1991) "How do National Systems of Innovation Differ? A Critical Analysis of Porter, Freeman, Lundvall and Nelson", in G.M. Hodgson and E. Screpanti (eds.), *Rethinking Economics: Markets, Technology and Economic Evolution*, Cheltenham, UK: Edward Elgar, 117–137.

McNally, K. (1995) "Corporate Venture Capital: The Financing of Technology Business", *International Journal of Entrepreneurial Behaviour and Research*, 1(3), 9–43.

Metcalfe, J.S. (1994) "Evolutionary Economics and Public Policy", *Economic Journal*, 104(425), 931–944.

Metcalfe, S. (1995) "The Economic Foundations of Technology Policy Equilibrium and Evolutionary Perspectives", in P. Stoneman (ed), *Handbook of the Economics of Innovation and Technological Change*, Oxford: Basil Blackwell, 513–557.

Milanovic, B. (2013) Branco Milanovic on Twitter: "The Universally Declining Share of Labor", <https://twitter.com/BrancoMilan/status/> accessed 12/1/2015.

Mohnen, P., Palm, F.C., Van Der Loeff, S. and Tiwari, A. (2008) "Financial Constraints and Other Obstacles: Are They a Threat to Innovation Activity?" *De Economist*, 156(2), 201–214.

Mokyr, J. (2010) *The Enlightened Economy: An Economic History of Britain 1700–1850*, New Haven: Yale University Press.

Mollica, M. and Zingales, L. (2007) "The Impact of Venture Capital on Innovation and the Creation of New Business", Unpublished Working Paper, University of Chicago.

Mowery, D. and Rosenberg, N. (1979) "The Influence of Market Demand upon Innovation: A Critical Review of Some Recent Empirical Studies", *Research Policy*, 8, 102–153.

Narula, R. and Zanfei, A. (2006) "Globalization of Innovation: The Role of Multinational Enterprises", in J. Fagerberg, D. Mowery and R.R. Nelson (eds.), *The Oxford Handbook of Innovation*, Oxford: Oxford University Press, 318–346.

Nelson, R. (2006) "Reflections on David Teece's 'Profiting from Technological Innovation'", *Research Policy*, 35(8), 107–109.

Nelson, R. (1959) "The Simple Economics of Basic Scientific Research", *Journal of Political Economy*, 67, 297–306, republished in N. Rosenberg (ed.), *The Economics of Technological Change* (1974).

O'Sullivan, M. (2006) "Finance and Innovation", in J. Fagerberg, D. Mowery and R.R. Nelson (eds.), *The Oxford Handbook of Innovation*, Oxford: Oxford University Press, 240–265.

OECD (1963) *Science, Economic Growth and Government Policy*, Paris: OECD.

OECD (1971) *Science, Growth and Society*, Paris: OECD.

OECD (1980) *Technical Change and Economic Policy*, Paris: OECD.

OECD (1992) *Technology and the Economy: The Key Relationships*, Paris: OECD.

OECD (1996) *Venture Capital and Innovation*, OCDE/GD(96)168, Paris.

OECD (2001) *The New Economy : Beyond the Hype*, Paris: OECD

OECD (2010) *The OECD Innovation Strategy: Innovation to Strengthen Growth and Address Global and Social Challenges*, SG/INNOV(2010)2, Paris, February 26, 2010, 186–208.

Pavitt, K. and Patel, P. (1999) "Global Corporations and National Systems of Innovation: Who Dominates Whom?", in D. Archbugi, J. Howells and J. Michie (eds.), *Innovation Policy in a Global Economy*, Cambridge: Cambridge University Press, 94–138.

Pianta, M., Evangelista, R. and Perani, G. (1996) "The Dynamics of Innovation and Employment: An International Comparison", *Science, Technology Industry Review*, 18, 67–94.

Piketty, T. (2014) *Capitalism in the 21st Century*, Cambridge, MA: Belknap Press.

Reichstein, T. and Salter, A.J. (2006) "Investigating the Sources of Process Innovation among UK Manufacturing Firms", *Industrial and Corporate Change*, 15(4), 653–682.

Revest, V. and Sapio, A. (2010) "Financing Technology-based Small Firms in Europe: What Do We Know?", *Small Business Economics*, published online July 11.

Rose, H. and Rose, S. (1970) *Science and Society*, Harmondsworth: Penguin.

Rosen, W. (2007) *The Most Powerful Idea in the World*, London: Jonathan Cape.

Rosenberg, N. (1974) *The Economics of Technological Change*, Harmondsworth: Pelican.

Ruttan, V.W. (2001) *Technology, Growth, and Development: An Induced Perspective*, Oxford: Oxford University Press.

Sapir, A., Aghion, P., Bertola, G.,Hellwig, M., Pisani-Ferry, J., Rosatti, D. and Wallace, H. (2004) *An Agenda for a Growing Europe, The Sapir Report*, Oxford: Oxford University Press.

Sawyer, M. (1994) "Obstacles to the Achievement of Full Employment in Capitalist Economies", in P. Arestis and M. Marshall (eds.), *The Political Economy of Full Employment*, Aldershot: Edward Elgar.

Saxenian, A. (1994) *Regional Advantage: Culture and Competition in Silicon Valley and Route 128*, Cambridge, MA: Harvard University Press.

Saxenian, A. (1999) *Silicon Valley's New Immigrant Entrepreneurs*, San Francisco: Public Policy Institute of California.

Scotchmer, S. (2006) *Innovation and Incentives*, Cambridge, MA: MIT Press.

Shane, S. (2009) "Why Encouraging More People to Become Entrepreneurs is Bad Public Policy," *Small Business Economics*, 33, 141–149.

Sidak, J.G. and Teece, D.J. (2010) "Dynamic Competition in Antitrust Law", *Journal of Competition Law and Economics*, 5(4), 581–631.

Smith, K. (1998) "Innovation as a Systemic Phenomenon: Rethinking the Role of Policy", in K. Bryant and A. Wells (eds.), *A New Economic Paradigm? Innovation-based Evolutionary Systems*, Discussions in Science and Innovation 4, Department of Industry, Science and Resources, Science and Technology Policy Branch, Canberra, Australia, 19–48.

Steinmueller, W.E. (2009) "Technology Policy: The Role of Industrial Analysis and Innovation Studies", in D. Foray (ed.), *The New Economics of Technology Policy*, Cheltenham, UK: Edward Elgar.

Stoneman, P. and Diederen, P. (1994) "Technology Diffusion and Public Policy", *Economic Journal*, 104, 918–930.

Storey, D.J. and Tether, B.S. (1998) "New Technology-based Firms in the European Union: An Introduction.", *Research Policy*, 26, 933–946.

Teece, D. (1986) "Profiting from Technological Innovation", *Research Policy*, 15(6), 285–305.

Teece, D. (2006) "Reflections on 'Profiting from Innovation'", *Research Policy*, 358(1), 131–146.

Trajtenberg, M. (2009) "Innovation Policy for Development: An Overview", in D. Foray (ed.), *The New Economics of Technology Policy*, Edward Elgar, UK: Cheltenham, 367–395.

Unger, K., (1988) "Industrial Structure, Technical Change and Microeconomic Behaviour in LDCs", in G. Dosi, C. Freeman, R. Nelson, G. Silverberg and L. Soete (eds.), *Technical Change and Economic Theory*, London: Pinter, 480–495.

Vallas, S.P., Kleinman, D.L. and Biscotti, D. (2011)"Political Structures and the Making of US Biotechnology", in F.L. Block and M.R. Keller (eds.), *State of Innovation: The US Government's Role in Technology Development*, London: Paradigm, 57–76.

Vernardakis, N. (1988) *The Train of Development and Greece in the Fifth Kondratiev Cycle*, Athens: Papazisis (in Greek), trans. into English by the services of the European Commission, Doc.Int. 132/88.

Wadhwa, V., Saxenian, A., Rissing, B. and Gereffi, G. (2008) "Skilled Immigration and Economic Growth", *Applied Research in Economic Development*, 5(1), 6–8.

Whitley, R. (2002) "Developing Innovative Competences: The Role of Institutional Frameworks", *Industrial and Corporate Change*, 11(3), 497–528.

Winter, S. (2006) "The Logic of Appropriability: From Schumpeter to Arrow to Teece", *Research Policy*, 35(8), 100–101, 106.

Woolthius, R.K. and Lankhuizen, M. (2005) "A System Failure Framework for Innovation Policy Design", *Technovation*, 25, 609–619.

8 Epilogue

Innovation has become a household word, a panacea for all ailments, even for those we don't know much about. Nowadays, many advertisements, whether they refer to financial "products" or to painkillers, travel packages or car rentals, use the word "innovation"- or "innovative" – preferably accompanied by the word "disruptive."

Today, when we refer to a rise in productivity or to income and job creation we think of innovation. Less than two decades ago we would have thought of investment. During the same time span innovation has changed so much as a concept that we can no longer use the terms "innovation" and "technological change" interchangeably. Other changes abound: the topics of location of innovation and size of firms have become irrelevant and fallen out of fashion because innovation is pervasive and can now be produced by anybody and anywhere. Networks, standards and regulation have gained importance. Economic Theory has taken a few steps in the right direction but needs to do a lot more, starting from the realization that knowledge is much more than information. The firm is learning to live with uncertainty, ephemeral goals, an oligopolistic environment, must stand ready to face dynamic competition and is condemned to keep innovating. Innovation Policy has developed into a potent tool to deal with productivity, growth, income and jobs, especially when implemented systematically, i.e. via a national system of innovation. With the rise in importance of Innovation Policy and the realization that there is nothing automatic about innovation, the role of the state has been enhanced: it must act not only as pathfinder and midwife but also as coordinator, builder and repairer of innovation's fragile ecosystem, and as investor of last resort.

Although innovation has almost been deified, we have been late in trying to understand it and it is still not well understood. Innovation and the innovation process itself, which are constantly evolving, have been changing the firm, sectors, economies, the world and its economy – as well as Economic Theory, which attempts ex ante to explain what has been happening and why. Innovation is literally knocking on our door and is making its weight felt in our everyday lives. In the meantime, every aspect on the stage of innovation has changed, including the actors, even the process itself. As a concept, it has broadened, has become more pervasive, more open, more accessible.

Innovation has to reach the market in order to qualify for the title. Hopefully it will be met by economic success but not necessarily so. Profit is the usually accepted motive for innovation but there are exceptions to this rule. The single most important characteristic of innovation is uncertainty, which may arise for reasons related either to technological or to market success. Indeed a very large proportion of attempted innovations fail.

From the mid eighteenth century, when science has started playing a role in innovation, the dependence of important new technological opportunities on advances in scientific knowledge has been rising. There is a tendency for more and more R&D, which is becoming more and more expensive. Historically, the time lag between a scientific discovery and its application has diminished, in some fields dramatically.

It is difficult to forecast the use of an innovation through time. It is equally or even more difficult to forecast the importance of an innovation. Not all activities are equally important, as there are some activities the technological progress of which significantly affects the technological progress of others, while others have absolutely no effect on other developments. It may also be the case that increases in productivity are realized in activities not directly related to the innovation. This kind of innovation is not particularly visible because it mainly concerns intermediate goods.

Innovations very seldom function in isolation. Sometimes, the realization of a particular innovation has to wait for the development of a particular input. This is why some important innovations initially bring only marginal productivity increases.

Timing is of utmost importance in innovation, as an innovation has to reach the market at exactly the right moment, not before nor after. At this point society decides as to its need and is also in a position to afford it. The environment – social, religious, ethical, geographical, etc. – impacts innovation. It is the society in which an innovation develops that imparts upon it its value. Societies bent on preserving tradition are not particularly keen on innovation.

At first innovation referred to new methods of production; later, it included new products as well. Innovation gradually expanded as a concept to include technological change, that is, new methods of production and new products, as well as non-technological change such as to organization, marketing or distribution. Loosely speaking, innovation is invoked whenever a "new combination" leads to a new business model.

Additional sources of innovation include users, especially the group referred to as "lead users." There is a new type of innovation, "open innovation" – as in open code in informatics. There is also serendipitous innovation of two kinds, both arising in the lab. The first kind refers to an innovation originally meant for a specific use but later found to be more effective in other uses (i.e. Viagra). The second kind refers to an accidental event, as in the case of the melting of a chocolate bar in the pocket of a researcher at Raytheon when passing in front of cathodic rays, leading to the accidental innovation of the microwave oven.

Almost all radical innovations such as electricity, the telegraph, the iron ship, and the Internet belong to the category of innovation by mandate. "Technology-push," a broad term, refers to innovation in general, innovation produced for its own sake, innovation which may not have profit as its target, or produced with no specific potential use in mind, like the photocopying machine. "Demand-pull" refers to all innovations that attempt to respond to demand. Within this category there is a large group of innovations known as induced innovation – innovation that attempts to cover perceived or foreseen needs or provide solutions to problems. Throughout the lifetime of an innovation it is subject to reinvention, which means diversification, change, evolution of an innovation from its initial state during the process of its adoption and application.

"Disruptive innovation" is not yet another type of innovation in the same way as product or process innovation. It does not characterize an innovation itself but rather its impact on markets in terms of dynamism, structure and market dominance: it either creates new markets or reshapes existing ones.

Marginal innovations emerge almost constantly and concern mostly improvements in products, methods of production or services. The impact of each of them on productivity may be imperceptible but cumulatively their impact may turn out to be significant.

Radical innovations are much less frequent than marginal ones. "Technological revolutions" are what Schumpeter calls the "gales of creative destruction." Such changes entail many groups of radical and marginal innovations, and have diffuse consequences for the entire economic system.

A "technological paradigm," in the sense of an era, defines the needs that have to be addressed, the scientific principles to be used in finding solutions and the technology that will be used. There are strong paradigms, with many technological opportunities, which generate fast and protracted rates of technological change. Other paradigms are weak and may offer few opportunities and hints on how to exploit them. Innovation is the result of the matching of opportunities and appropriation possibilities.

Innovation does not precisely follow market signals or needs. (1) The direction of innovation depends on the "state of the art" of technologies being used. (2) Very often the nature of technologies defines the products and production methods that "fit" changing economic conditions. (3) Innovation is cumulative by nature because it is based on technological capabilities, that is, accumulated knowledge, both tacit and codified.

Research over the years has come up with a number of observations: Successful innovations originating from the "market-pull" tend to be marginal with smaller long-term value than innovations originating from "supply-push." Innovation is a very complex procedure, which cannot be explained by a single factor, as success requires the simultaneous presence of both "supply-push" and "demand-pull." Rather than being competitive, the two forces are complementary. The majority of commercially successful innovations may originate from "demand-pull" but nevertheless rest on the existing knowledge base. There is further complementarity between supply and demand over time, as one precedes the other. During the first stages of development of an industry, "technology-push" tends to be more important. "Demand-pull" gains importance as higher levels of maturity are reached.

Until fairly recently we used to think that innovation is generated almost exclusively in industry but gradually we have come to realize that innovation is generated practically in any type of private or public activity including services, and irrespective of the level of technology intensity exhibited by the activity. In addition, it has been observed that there is an ordinal ranking of innovativeness among sectors that is very similar across countries – in other words, that innovativeness derives more from the nature of sectors than from a country's level of development. It has also emerged that the more backward and forward linkages there are in an economy, which means the more developed and integrated its production system, the higher is its innovative activity.

We also used to think that innovation is the domain of large firms. Now we know that large firms still account for the majority of innovations but the share of small firms is rising steadily and accounts for a significantly larger share than used to be thought. Innovations by smaller firms differ from the ones created by larger firms in many ways. They tend to produce marginal innovations – product innovations – close to the market, and their sources of innovation tend to be their customers and suppliers and information gathered from exhibitions. In other words, their sources have little in common with R&D, very much in contrast with the larger firms that are more focused on official research. It has also been found that large, incumbent firms tend to pursue relatively more incremental and relatively more process innovation than smaller firms. The application of scientific discoveries seems to belong to large firms as a result of their privileged relationship with official research. However, small firms often spend more in relative terms than large firms on innovative investments in production so as to hasten their marginal product innovations to market.

A salient difference between small and large firms relates to efficiency in innovation. Several studies confirm, firstly, that smaller firms tend to account for a disproportionately large share of innovative output relative to their size, and, second, that R&D productivity tends to decline with firm size. A similar result was found with respect to patents, as the number of patents per R&D dollar for smaller firms is considerably higher than for larger firms.

As for the relationship between market structure and innovation, there is hardly any evidence of a positive relationship between innovative output and market structure. There is no one-way causal relationship running from market structure to innovation as market structure is itself affected by innovative activity. Instead, there is a two-way causal relationship as both variables are endogenously determined within a complex system of interactions with numerous other variables.

Innovation and diffusion are of complementary importance. In the absence of diffusion, innovation would have little economic or social impact. It is diffusion rather than innovation that determines the rate of economic growth and the rate of the rise in productivity. Diffusion should attract substantial research in the future. The importance of network effects has increased in recent decades with the rise of ICT and the proliferation of network products. Besides, standards will keep playing an increasing role, and so will technology transfer. We are still short of a theory on diffusion across sectors.

The characteristics of the innovative process, namely, non-specificities, time delays, uncertainty and cost, have serious implications for the innovation strategy of the firm. They work against the first firms to innovate, against small firms and the conduct of basic research.

Because innovative activity is eclectic, specific, cumulative, not accessible to all, the research impetus of the firm takes place in specific areas of technology not far from its existing activities, and the technology the firm is likely to develop over time is not at all random. The technological search process is cumulative, so what the firm can hope to achieve technologically in the future is narrowly defined by what it has achieved in the past.

The innovations of firms are based to a great extent on their own research but also on some contributions from other firms and from public knowledge. Firms produce products through methods that are firm-specific, that is, technologically differentiated from the methods and products of other firms.

It is important to realize that innovation is not a reaction to relative prices and demand. On the contrary, it is a process to a great extent endogenously defined by competition, where firms constantly attempt to improve their technologies, routines, dexterities and artifacts independently of whether market signals change or not. In every paradigm the opportunities and boundaries within which the market can exert the incentive mechanism are delineated.

The constant migration of all activities, albeit with different speeds, through innovation towards higher technology intensity levels has heightened competition and has encouraged oligopolistic practices and behaviors, especially in high-technology industries which are increasingly subjected to non-traditional economic principles. A breed apart from the rest of firms, they differ in behaviour and culture and call for new management techniques and strategies. The firm is constantly being transformed by innovation, which provides it with new challenges and opportunities and also imposes upon it a number of imperatives: for flexibility, cooperation, competitiveness, enlistment to the international environment, recognition of the value of human capital, and above all, to innovate and to keep on innovating. These imperatives proliferate and become more demanding as the firm moves into high-technology territory.

The firm used to attempt to optimize mass production of specific goods, and the challenge for its management then was to forecast demand in slowly rising markets. Internally, respect for hierarchy and loyalty were valued most. Externally, the firm's relations with its environment could be described as "business is war." Gradually, over a timespan of five or six decades, the firm has moved from being a producing entity to a learning one. It has learned to face more ephemeral opportunities in fast-growing markets while large profits await whoever dares make the first move (innovation). Internally, the firm has learned to value its human capital more and has proved willing to trade respect for hierarchy with innovativeness – a parallel development to the firm's flattening structure. As to the firm's external environment, companies tend to congregate in ephemeral groupings and in "galaxies" of collaboration and "strategic alliances – constellations of bilateral agreements among firms – which are increasingly necessary to support innovative activities. As a result, discussions on the link between firm size and innovation are rather outmoded, as the boundaries of the firm have become much more fuzzy in recent decades. Competition is accompanied by cooperation; we have a state of "co-opetition."

Innovation has immensely transformed the real economy at world, sectoral and firm level. It has been instrumental in bringing about a new international division of production and employment, based on a hierarchy of products in terms of technology intensity corresponding to a hierarchy of technological frontrunners followed by technological laggards, at a widening distance. The high-technology-intensity products produced by the frontrunners command a high value added and compete through innovation in contrast with the low-technology-intensity products that are produced by the laggards, command a low value added and compete on price.

Innovation has helped us to realize the change in the substance of comparative advantage from being based on natural resources to knowledge, and it has enhanced globalization. It has compelled specialization to move up to value chain production leading to unprecedented levels of interdependence among countries, as witnessed by the speed of the spread and severity of the international recession of 2008. In a process of "creative destruction" it is rejuvenating the capitalist system by replacing the old with the new, creating, destroying, merging and separating sectors and firms.

At first, Economic Theory hardly took notice of innovation and by the time it finally did decide to deal with it, it had already adopted hypotheses that are both rigid and hard to replace without destroying a major part of the theoretical edifice as a whole. Because of its rigid and simplistic assumptions, it cannot explain developments at the firm level, the sector or the national and international levels. Developments in convergence do not support theory and the same is true for comparative advantage, growth and trade. The only type of innovation that Microeconomic Theory can address is methods of production. Growth cannot be accounted for and neither can trade be explained. The way the firm is portrayed ignores the concept of the firm as the main generator of innovation. Time is ignored while information and knowledge are one and the same.

Innovation is forcing Economic Theory to revise some of its beliefs. Innovation should be considered to be at the heart of the socio-economic system from which it emerges, and the impact goes in both directions. The concept of technology as information with general application, easily produced and reproduced, and easily accessible to firms, which simply absorb data at no cost from a reservoir of knowledge, is highly irrelevant. Innovation that is knowledge is much more than information; it is properly processed information. Firms develop their own technologies using knowledge in the public domain, as well as in collaboration with other firms. A system that is constantly evolving, with bursts of innovation that act as shocks to the system, is at odds with the notion of "equilibrium" in a world heading towards increasing returns and "winner takes all" (or most). In practice, innovation is an element of non-perfect competition. Economic Theory should realize that we are moving towards a world of increasing returns in which "winner takes all," a world not compatible with comparative advantage the way we know it. The world we live in is an imperfect world.

There is nothing automatic about the process of innovation. Some of the steps along the chain from the conception of an idea to its realization have been gradually simplified, broken down into smaller steps, and have been assisted in their execution, thus rendering the whole process more accessible. There is no need to repeat the heroic deeds of the romanticized intrepid innovator-turned-entrepreneur character of the past. However, innovation remains misunderstood. It is often confused with the hype and its fragility is not always recognized, nor is it seen as a part and a result of a ecosystem as a whole.

For innovation to keep alive, strong, efficient and meaningful to society, an Innovation Policy is needed whose aims of generating growth, raising income and creating employment are best served through the provision of a well-functioning ecosystem that extends beyond meeting individual "market failures" (failures to provide adequate investment). It must be hospitable to innovators through IPRs and skilled immigration laws, with a productive structure biased towards innovation. It must be able to withstand external threats like globalization, balanced between variety and selection, and supplemented by standards, and updated regulation and antitrust laws.

This means facing market failures through a systemic, well-planned strategy construct: a National System of Innovation (NSI) built on sharing knowledge, consisting of universities, institutions, organizations, research units, laboratories, firms, finance, and the state. All these must be supplemented by a maze of laws – ranging from skilled immigration policy to IPR – regulations and public programs, linked together, kept in tune and orchestrated by the state so as to serve the goals of generating growth, raising income and creating employment.

In concrete terms, this amounts to setting up the NSI construct, maintaining and keeping it in running condition with all its paraphernalia, and incorporating policy recommendations: support to university-performed R&D; application of programs for small businesses; SBIR-type programs; encouragement for the formation of venture capital firms; links to thick markets; technology rating agencies; low capital gains taxes; limited participation by pension funds; measures to identify and support gazelles. Other recommendations are more difficult to implement and may require time, such as the steps needed to raise the level of education and acquire dexterities on the part of the working population in order to take advantage of globalization, i.e. to attract foreign investment on localized production and research. Still other recommendations are not only very difficult to implement but may also require a very long time scale. An example is the shift from one engine of growth to another, meaning the change of productive structure in a country from (very broadly speaking) process innovation to product innovation. Other recommendations sound straightforward but call for extreme care in their application. For instance, attempting to match variety with selection requires careful case-by-case consideration.

The presence of the state is necessary not only to enforce the above but also in order to orchestrate the whole system by coordinating, organizing, directing and above all supporting it on a constant long-term basis. For the role to be potent, it must be proactive, and it must not limit itself to creating the right conditions for innovation and then waiting for it to happen, but must instead take the right measures to *make* it happen. Among the technological frontrunner(s) the state is also expected to play the role of midwife and initiator by extending old markets and creating new ones. Countries with a vibrant entrepreneurial community, especially where this is small, already developed and specialized, with R&D covered by large multinationals, have less room for state intervention. On the other hand, technological laggards lacking a vibrant entrepreneurial community and latecomers to development can attempt to emulate the Israeli model.

These countries may also benefit from innovation policies relying on diffusion in the form of technology transfer coupled with domestic efforts in R&D and innovation, not necessarily in the high-tech domain. Prescriptions for the poor countries such as the ones offered by the "Washington Consensus," consisting of copying Western political and legal systems, non-interference with markets, promotion of free trade and a liberal stance towards foreign direct investment, are by now highly questionable.

Innovation has provided a longer, healthier and more prosperous life to part of the world population, is currently offering some of the best deals ever to consumers, and has offered mankind unprecedented gifts in terms of rises in productivity and improved standard of living. But innovation is facing definite threats and challenges. First, there is the lukewarm stance towards "real" innovation both at the level of entrepreneurs and policy makers, which is expressed as risk aversion. As a result, basic research is avoided, and innovation has come to be associated more with the hype (the apps innovation), has become more marginal, and has less substance, while, as patents show, quality has suffered. At the same time we have both too little and too much innovation.

Innovation has also become more laborsaving than ever. Software of the digital age is destroying jobs created in the analog age. Innovation is replacing old businesses with new ones as bricks and mortar firms are replaced: the SMS replaced the mail, Netflix replaced Blockbuster, Amazon and iTunes respectively displaced Borders and CDs. Many of today's innovations eliminate jobs in the service sector, even menial ones: taxi drivers are displaced by Uber, Airbnb reduces hotel personnel, Locals affects tourist guides, robot parking reduces the need for parking attendants. Globalization became a threat as it exported jobs abroad, but today's innovations send jobs to oblivion.

Together with globalization, which it enhances, innovation diminishes employment and labor income (hence income share too) in the West and contributes to the rise of inequality. It is only natural to consider that as of now, the most pressing problem is the waning of the benefits of innovation in terms of jobs and income.

Another serious threat to innovation is the problematic use of patents. Excessive litigation and trolls stifle innovation, especially innovation by small firms. Conglomerates and hedge funds have been building substantial stakes in the largest firms and are already in a position to influence their research efforts, a practice that should be avoided at all costs. Finally, industrial espionage is an issue that will have to be addressed.

Looking at the future, the worldwide spread of the Internet (itself the result of innovation), which at first threatened to erect a digital divide between the developing world and the rich countries, is turning out to be a very positive development for all, but especially so for the less developed countries. This is because the benefits to be derived from the Internet come through its use and not through the much more demanding chore of mastering the art or achieving the capability of producing ICT goods and equipment.

The question of whether the system is sustainable is relevant not just to the US but to the rest of the world as well. The US government originates and creates previously unheard-of activities with strange-sounding names like biotechnology or nanotechnology a decade or two before businesses hear about them; then, once the path has been created, it steps aside, making room for its business community to follow. The rest of the world, whose businesses are further away from the frontier of knowledge also stand to benefit from these new vistas, though with greater delay. If the US government stops or falters in advancing into these uncharted territories, then innovation will be seriously adversely affected unless another country is willing and able, even after a long delay, to reach the present technological frontier and to carry out the role the US government has been performing until now. For the moment, the US seems to be the only government interested in the perpetuation and regeneration of the system, as it realizes that this domain is possibly its last comparative advantage. However, indicators of business innovation and of growth of multi-factor productivity suggest that the US lead is narrowing. R&D and patenting by businesses have grown less rapidly than in the past and the quality of patents has been declining, a fact that may be of some relevance to the growth in unemployment and the waning of the benefits of innovation currently being experienced. China, the imminent contender for the title of the greatest R&D spender, will be in no position to play the role of pathfinder for the foreseeable future.

A few final remarks: we are moving towards a world of heightened competition, where "winner takes all," a world passing from comparative advantage to absolute advantage. We are moving away from the concept of "equilibrium" and price competition towards competition through innovation and dynamic competition. We are moving towards a world of knowledge and increasing returns. In this world, regulation, which is already under stress from current problems, will need to be upgraded so as to carry out its future tasks. If Economic Theory is to be at all relevant as an analytical tool in relation to innovation, then it is high time for it to assimilate some of the concepts, ideas and propositions discussed in this book. It does not need to be rewritten from the beginning, but parts of it need revision – urgently. We still require a unifying theory of the origins of innovation and a convincing theory of diffusion across sectors. As for business and management theorists, they should recognize the uniqueness of high tech firms and their need for special management techniques and strategies. We should always remember that there is nothing automatic about innovation and its fragile ecosystem. Innovation Policy, when it is well designed and properly used, is a potent tool to generate growth and income and to create employment, but it is not a panacea for all ills.

Index

For Product Safety Concerns and Information please contact our EU representative GPSR@taylorandfrancis.com
Taylor & Francis Verlag GmbH, Kaufingerstraße 24, 80331 München, Germany

www.ingramcontent.com/pod-product-compliance
Lightning Source LLC
LaVergne TN
LVHW082342011026
849160LV00025B/1571
* 9 7 8 0 4 1 5 6 7 6 8 0 9 *